STRUCTURAL
GEOLOGY

Folds. Shales, sandstone, and marly limestone of Cretaceous–Eocene age. Looking southeast Gulf of Genoa, southwest of Camogli, Italy. Photo: Kurt Lowe.

Third Edition

STRUCTURAL GEOLOGY

Marland P. Billings

Professor of Geology
Harvard University

Prentice-Hall, Inc.,
Englewood Cliffs, New Jersey

18   17   16   15   14

ISBN: 0-13-853846-8

Library of Congress Catalog Card Number 73-167628

Printed in the United States of America

PRENTICE-HALL INTERNATIONAL, INC., *London*
PRENTICE-HALL OF AUSTRALIA, PTY. LTD., *Sydney*
PRENTICE-HALL OF CANADA, LTD., *Toronto*
PRENTICE-HALL OF INDIA PRIVATE LIMITED, *New Delhi*
PRENTICE-HALL OF JAPAN, INC., *Tokyo*

PREFACE

Much new information on structural geology has been published since the last edition of this book. Obviously the more significant new data should be introduced into an elementary text. But in order to keep the length within reasonable bounds it would be necessary to eliminate some of the older material. This is not easy to do. Some might be eliminated completely and some might be shortened, but presumably the original text was as concise as possible.

As in the past, the principal purpose has been to present the basic principles of structural geology. There has never been a pretense to cover global tectonics, such as the evolution of mountain ranges, continents, and ocean basins. In the second edition I said: "Oceanography . . . is discovering many exciting facts about the topography and composition of the floors of the oceans, facts that are already revolutionizing our thinking about the crust of the earth." But little did I anticipate the full magnitude and impact of these discoveries. However, a synthesis of the tectonics of the ocean basins is beyond the scope of this book.

The laboratory exercises and problems at the end of the book are much as they were in the last edition, except that the section on the use of the equal-area net has been expanded from one to three exercises. Serious consideration was given to reducing the coverage and length of the first eleven exercises. In fact, a completely new and shorter draft was prepared. But I became convinced that little of the material could be omitted and that the text was as concise as possible. Also, consideration was given to the preparation of several new exercises, involving the focal mechanisms of earthquakes, radiogenic data, Mohr's circles, and geophysical methods. Problems in the first two of these subjects would necessitate the presentation of much more background material than was possible. Inclusion of exercises on the use of Mohr's circles and geophysical methods would have lengthened the book still more.

A complete discussion of geophysical techniques is obviously beyond the scope of this book. But a discussion of the use of geophysical methods in structural geology is essential. The techniques of both geophysics and structural geology have become more sophisticated in the last two decades. Very few structural geologists are trained to do geophysical field work. Conversely, few geophysicists are able to conduct structural investigations of the type discussed in this book. To understand this relation of geophysics to structural geology, an analogy might be drawn with paleontology. Most structural geologists are not competent to do their own paleontology. But they should understand the basic principles and be prepared to challenge what appear to be incorrect conclusions by the paleontologist.

Although many references are listed at the ends of the chapters, they are of necessity incomplete. In a minor way these references should help emphasize the large number of publications available. In addition, an occasional ambitious student will be encouraged to undertake further reading. The critical reader will also notice that for some of the chapters the references are rather ancient. But very little new can be written on the geometry of faults, methods of measuring the thickness of sediments, and the character of unconformities.

Perhaps the author should have used only the metric system of measurement. But in many English-speaking countries, a dual system will be used for decades.

I want to thank the numerous geologists who have helped develop this book. Foremost among them are my former teaching assistants, who helped especially in the evolution of the laboratory problems: Randolph W. Chapman, Jarvis B. Hadley, Robert P. Sharp, Walter S. White, George E. Moore, Clyde Wahrhaftig, Laurence Nobles, Dallas Peck, and Bruce Reed. Exercises 12 through 14 in the present text are based on material prepared by Dr. James Stout. Mr. Claude Dean prepared the equations used in a discussion of the size of thrust blocks.

I am greatly indebted to Prof. John Haller and the United States Geological Survey for most of the photographs. Prof. Haller supplied many photographs from his own collection, as well as from those in his custody from the Lauge Koch expeditions to East Greenland. The staff of the Photographic Library of the U. S. Geological Survey in Denver, Irvil Shultz, Librarian, was exceedingly courteous and helpful. Other photos were supplied by Kurt Lowe, Bruce Reed, and Charles Doll.

As always, conscientious and efficient secretaries are essential to the preparation of any book. I am especially indebted to Mrs. Mary Maher in the preparation of the manuscript and to Mrs. Susan Williams in the later stages of getting permission to use figures from copyrighted articles and checking the galley proofs.

Marland P. Billings

CONTENTS

7. Joints, 140

8. Description and Classification of Faults, 174

9. Criteria For Faulting, 199

10. Reverse Faults, Thrust Faults, and Overthrusts, 214

1

STRUCTURAL GEOLOGY

Relation of Structural Geology to Geology

Geology has recently been defined as approximately synonymous with solid-earth sciences. Earth sciences are those sciences that deal with the earth; they include geology, geophysics, meteorology, and parts of oceanography. The geologist is concerned primarily with the solid part of the earth.

> The solid earth sciences are those that investigate the physical and chemical characteristics and processes of the earth and astrobodies; the origin, distribution, development, and utilization of earth materials and the land as a whole; and the interaction between the solid earth and the hydrosphere and atmosphere.[1]

Geology is such a large subject that it has inevitably been subdivided into many categories. But these subdivisions are arbitrary and many geolo-

1

gists specialize in several of these fields. There are as many combinations as there are geologists. Moreover, some geologists overlap into other fields of science, such as physics, chemistry, and biology. Nevertheless, the many aspects of geology must be classified to the best of our ability.

Structural geology is the study of the architecture of rocks insofar as it has resulted from deformation. *Tectonics* and *tectonic geology* are terms that many consider to be synonymous with structural geology. To some, however, structural geology is concerned primarily with the geometry of the rocks, whereas tectonics deals with the forces and movements that produced the structure.[2] The movements that affect solid rocks result from forces within the earth, causing folds, joints, faults, and foliation. The movements of magma, because it is often intimately associated with the displacement of solid rocks, is also a subject that lies within the domain of structural geology. The deformation of the rocks of extraterrestrial bodies is also the concern of structural geology, as well as the effect of collisions between bodies in the solar system. The aim of structural geology is to determine and explain the architecture of rocks as observed in the field; laboratory investigations are supplementary means to attain this primary objective.

In field work the solution of structural problems is only one phase of broader investigations. It is futile to try to study the structure of folded and faulted sedimentary formations without a knowledge of *stratigraphy*, that phase of geology treating of the sequence in which formations have been deposited. *Sedimentation*, which deals with the deposition of stratified rocks, may offer much evidence on the tectonic events in areas adjacent to the basins in which the stratified rocks accumulate. *Paleontology*, which is the study of fossils, is indispensable to the structural geologist who works in rocks containing organic remains. *Petrology*, a subject that includes the systematic description of rocks and the study of their origin, sheds much light on the structural history of igneous, metamorphic, and sedimentary rocks. *Mineralogy*, which deals with the classification, atomic structure, and genesis of minerals is an integral part of geology, inasmuch as most rocks are composed of minerals. *Volcanology* is the study of volcanoes and volcanic rocks. A knowledge of *geomorphology*, the study of land forms, is particularly important to the structural geologist who investigates regions of recent tectonic activity, where the topography is a rather direct expression of the structure. Even in those areas where the tectonic evolution ceased long ago, geomorphology may furnish important clues to the structural geologist.

Some subjects employ methods and instruments that associate them more closely with other sciences, notably physics and chemistry. *Geophysics*, which involves the application of physics to geological problems, has been successfully employed in helping to solve many problems of direct concern to the structural geologist. It includes studies in seismology, gravity, electricity, and magnetism. *Seismology*, which deals with the propagation of elastic waves through rock, whether caused by earthquakes or artificial means, aids in the solution of many structural problems. Moreover, it is the principal source of information on the nature of the interior of the earth, the source of tectonic energy. *Rock mechanics* deals with the physical properties of rocks and the significance of these properties in rock deformation and engineering projects. *Paleomagnetism* is the study of the ancient magnetic field of the earth and its tectonic significance. *Geochemistry* is concerned with the application of the principles of chemistry to geological problems. It overlaps many fields in geology, such as mineralogy, petrology, and weathering. *Geochronology* deals with the dating of geological events. For a century and a half the relative ages of geological formations was determined from an analysis of the fossils. In more recent years other means have been employed, such as tree rings and varved clay. But even more significant is *radiogenic dating*, which theoretically can determine in years the age of geological events.

Some terms refer to the application of scientific principles to specific domains. The science of geology evolved from observations made on the continents. Although the surface of the sea has been well known to man for many millennia, it is only recently that he has become concerned with the subsurface parts of the ocean. *Oceanography* is a large subject that involves many disciplines, notably biology, physics, chemistry, and geology. The structural geologist is especially concerned with the structure of the sea floor, as well as that of the underlying crust and mantle. The evolution of the sea floor is one of the most exciting subjects in modern geology. In recent years geologists have become involved in *lunar geology*, which involves especially mineralogy, petrology, structure, and geomorphology.

Objectives of Structural Geology

The structural geologist is concerned with three major problems: (1) What is the structure? (2) When did it develop? (3) Under what physical conditions did it form?[3]

In general, the first question must be answered first. It is essential to determine the shape and size of the rock bodies. Are they great flat-lying tabular masses covering scores of square miles? Or are they tabular masses that have been thrown into folds with a wavelength of several miles and an amplitude of thousands of feet? Or are they great cylindrical bodies thousands of feet in diameter and a mile or two deep?

Geological field work is indispensable to many such investigations, and it is this fact that distinguishes most phases of geology from many of the other sciences. Because the correct location of outcrops is of the utmost importance, accurate maps are essential. For many regions topographic maps are available, and by means of topography, drainage, and culture, a precise location is possible. Vertical aerial photographs are very useful in geological field work. These photographs, made from directly above, are essentially maps. In some respects they are superior to topographic maps because they not only portray all natural and artificial features with great accuracy, but they reveal, too, many features such as trees, forests, open fields, and fences that are not generally indicated on topographic maps. However, they lack contours; moreover, in mountainous regions, the scale is not constant. In regions for which suitable maps or aerial photographs are not available, it may be necessary for the geologist to prepare his own base map, usually by plane-table methods. A discussion of the technique of field methods is beyond the scope of this book, but is adequately discussed in several excellent texts.[4,5]

Successful geological field work consists of the accumulation of significant facts. At each outcrop the geologist records whatever data are pertinent to his problem, and, ideally, he should never have to visit an outcrop a second time. This is especially true in areas that are difficult of access, but even in accessible regions the work should be so planned that a second visit to an outcrop is unnecessary.

Geologic mapping, when properly done, demands skill and judgment. Such mapping requires keen observation and a knowledge of what data are significant. As the field work progresses and the larger geological picture begins to unfold, experience and judgment are essential if the geologist is to evaluate properly the vast number of facts gathered from thousands of outcrops. Above all, the field geologist must use the method of "working multiple hypotheses"[6] to deduce the geological structure. While the field work progresses, he should conceive as many interpretations as are consistent with the known facts. He should then formulate tests for these interpretations,[7] checking them by data already obtained, or checking them in the future by new data. Many of these interpretations will be abandoned, new ones will develop, and those finally accepted may bear little resemblance to hypotheses considered early during the field work.

Nothing is more naïve than to believe that a field geologist should gather only "facts," the interpretation of which is to be made at a later date. Because of his numerous tentative interpretations, the field geologist will know how to evaluate the facts; these hypotheses, moreover, will lead him to critical outcrops that might otherwise never have been visited. On the other hand, the field geologist should never let his temporary hypotheses become ruling theories, thus making him incapable of seeing contradictory facts.

Although much structural information in the past has been gathered from direct observation, either on the surface of the earth, in open pits, or in mines, a progressively greater proportion of our data is gleaned from the depths of the earth by indirect means. The petroleum geologists, in particular, have obtained vast amounts of structural data from the study of drill holes and from geophysical data. Subsurface geology[8] not only involves structural geology, but also paleontology, stratigraphy, sedimentation, and geophysical methods.

Aerial photographs[9] are not only of great value as base maps, but they often display unsuspected structural features. Moreover, geological maps, with a small amount of geological ground control, may be prepared from aerial photographs where the vegetative cover is slight and the geologic structure simple.* Spectacular aerial photographs have been obtained from orbiting satellites.[10] *Remote sensing* employs aerial photographic techniques that record gamma radiation and permit the taking of infrared and radar imagery photographs.[11] Geological maps of the moon, primarily structural in nature, have been prepared from telescopic and satellite photographs.[12] Special field techniques are necessary in studying the geology of the ocean floor[13, 14] and the moon.[15]

A second objective of the structural geologist is to relate the structure to some chronology. One phase of this study is to determine the sequence in which the structural features developed. For example, he may find an anticline, a fault, and a dike. What are their relative ages? The anticline may be the oldest and the dike may be the youngest. It is also possible that the fault is the oldest and that the anticline is the youngest. There are also other possibilities. In some areas the sequence may be exceedingly complex.

The structural geologist is interested not only in the sequence of events in the area in which he is studying but he also wants to fit them into the geological history of the whole earth. This can be done by paleontological methods[16] or by radiogenic dating.[17]

A third objective of the structural geologist is to determine the physical processes that produced the observed structure. What was the temperature and pressure at the time the structural feature formed, and what was the stress distribution? It is desirable to answer these questions before we try to deduce the ultimate causes. Without knowing the stress distribution at the time the structural feature formed, it is difficult to decide whether a given fold was the result of contraction of the earth, subcrustal convection currents, or the forceful injection of magma.

Experimental geology provides significant data for the understanding of tectonic processes. The physical properties of many rocks have been investigated.[18] It is difficult to simulate natural conditions and consider all the variables involved, but much has already been accomplished by the use of ingeniously conceived apparatus.[19]

In another type of experiment, attempts have been made to reproduce geological structures in small models, or to observe the structures that result from the application of known forces. A classic example is the formation of folds when layers of suitable material are slowly compressed by a moving piston (see also p. 127). But the significance of many of these experiments is questionable because in many cases the investigator repeatedly changed either

---

* Many such maps of $7\frac{1}{2}$-minute quadrangles, on a scale of 1: 24,000, were published by the U.S. Geol. Survey for some of the southwestern states between 1955 and 1965 as part of the "Miscellaneous Geological Investigations Maps."

the materials or the conditions of the experiment until he obtained the results he desired. It is possible, however, by the use of sound engineering principles, to construct small-scale models that will simulate natural conditions.[20,21]

Scope of this Book

Before it is possible to analyze the structure of entire mountain ranges, it is essential to have precise information on the many small separate areas that comprise each range. These small areas may cover 50 to 200 square miles, or they may be single mines or oil fields. This investigation of the structure of relatively local structure is the first and inevitable approach to the problem.[22]

Equally important, and perhaps in some ways more fascinating, is the synthesis or weaving together of the many facts obtained from local areas into a unified picture of the structure and tectonic history of the outer shell of the whole earth. Such studies are necessarily based in large part upon an intimate knowledge of the literature of structural geology because it is manifestly impossible for one man to investigate many areas in detail. But in order that he may more judiciously evaluate the reliability and importance of the published information, such an investigator must have made detailed studies of his own. One of the old classics in this field of synthesis is that by Ed. Suess, published in German and translated into French and English.[23] Excellent more recent books are those by Beloussov, Bucher, and Umgrove.[24,25,26] Others deal with the structure of the sea floor.[13,14]

Only local structural features will be considered at any length in this book. Synthesis of the structure of large areas are important and fascinating, but they are more advanced studies that cannot be comprehended until the principles of local structures are fully mastered. Moreover, to expand the text to include regional tectonics of the continents and oceans would occupy far more space than could possibly be made available in an elementary text. Fortunately, superb texts and tectonic maps are available for many areas.[27,28,29,30,31,32,33]

A study of geologic structures would be quite barren and fruitless if unaccompanied by a discussion of the forces involved. In the natural course of events, the structural geologist makes his observations first, then deduces the geological structure, and, finally, considers the nature of the causative forces. Normally, observation and description precede interpretation. It might seem logical, therefore, in a book such as this, to reserve a discussion of mechanics until the end. But it is far more satisfactory to treat each structural feature as a unit, describing it first and then considering the forces involved. It is essential, therefore, that a chapter on mechanical principles be given first in order that the origin of the various geological structures may be intelligently discussed.

References

[1] Bove, Albert N., ed., 1969, *Earth Sciences Newsletter* (NAS-NAE-NRC), No. 5, p. 4.

[2] Wilson, Gilbert, 1961, The tectonic significance of small-scale structures and their importance to the geologist in the field, *Annales de la Société Géologique de Belgique*, Tome LXXXIV, pp. 423–548.

[3] Goguel, Jean, 1962, *Tectonics*, translated from French edition of 1952 by Hans E. Thalmann, 384 pp., San Francisco and London: W. H. Freeman & Co.

[4] Compton, Robert R., 1962, *Manual of field geology*, 378 pp., New York: John Wiley & Sons.

[5] Lahee, Fredric H., 1961, *Field geology*, 6th ed., 926 pp., New York: McGraw-Hill Book Company.

[6] Chamberlin, T. C., 1897, The method of working multiple hypotheses, *J. Geol.* 5: 837–48; also reprinted in K. F. Mather and S. L. Mason, 1939, *A source book in geology*, New York: McGraw-Hill Book Company, pp. 604–12.

[7] Gilbert, G. K., 1886, The inculcation of scientific method, with an illustration drawn from Quaternary geology of Utah, *Amer. J. Sci.*, 3rd series, 31: 284–99.

[8] Bishop, Margaret S., 1960, *Subsurface mapping*, 198 pp., New York: John Wiley & Sons.

[9] Ray, Richard G., 1960, *Aerial photographs in geologic interpretation and mapping*, U. S. Geol. Survey, Prof. Paper 373, 230 pp.

[10] Pesa, Angelo, 1968, *Gemini space photographs of Libya and Tibesti*, Petroleum Exploration Society of Libya, Tripoli, 82 pp.

[11] Rydstrom, H. P., 1967, Interpreting geology from radar imagery, *Geol. Soc. Amer. Bull.* 78: 429–36.

[12] Wilhelms, Don E., 1968, Geological map of *Mare Vaporum* quadrangle, *Geological atlas of the moon*, I-54g (LAC 59), U. S. Geol. Surv.

[13] Menard, H. W., 1964, *Marine geology of the Pacific*, 271 pp., New York: McGraw-Hill Book Company.

[14] Shepard, Francis P., 1959, *The earth beneath the sea*, 275 pp., Baltimore: The Johns Hopkins Press.

[15] Abelson, Philip H., ed., 1970, The moon issue, *Science* 167: 480–792.

[16] Kummel, Bernard, 1970, *History of the earth*, 2d ed., 707 pp., San Francisco and London: W. H. Freeman and Company.

[17] Hamilton, E. I., 1965, *Applied geochronology*, 267 pp., London and New York: Academic Press.

[18] Clark, Sydney P., Jr., 1966, *Handbook of physical constants*, Geological Society of America, Memoir 97, 587 pp., New York.

[19] Donath, Fred A., 1970, Some information squeezed out of rock, *American Scientist*, 58: 54–72.

[20] Hubbert, M. K., 1937, Theory of scale models as applied to the study of geologic structures, *Geol. Soc. Amer. Bull.* 48: 1459–1520.

[21] Ramberg, Hans, 1967, *Gravity, deformation, and the earth's crust*, 214 pp., London and New York: Academic Press.

[22] Bucher, W. H., 1950, Megatectonics and geophysics, *Trans. Amer. Geophys. Union* 31: 495–507.

[23] Suess, Ed., 1904–1924, *The face of the earth*, 5 vols., Oxford: Clarendon Press.

[24] Beloussov, V. V., 1962, *Basic problems in geotectonics*, 816 pp., New York: McGraw-Hill Book Company.

[25] Bucher, W. H., 1933, *The deformation of the earth's crust*, 518 pp., Princeton: Princeton University Press. reprinted 1957, New York: Hafner Publishing Company.

[26] Umgrove, J. H. F., 1947, *The pulse of the earth*, 2d ed., 358 pp., The Hague: Martinus Nijhoff.

[27] King, Philip B., 1969, *The tectonics of North America—a discussion to accompany the tectonic map of North America, scale 1: 5,000,000*, U. S. Geol. Surv., Prof. Paper 628, 95 pp.

[28] Eardley, Armand J., 1962, *Structural geology of North America*, 2d ed., 743 pp., New York: Harper and Row.

[29] Markovskiy, editor-in-chief, 1958, *Geologic structure of the U.S.S.R.*, 3 vols., Moscow (in Russian): Ministry of Geology and Conservation of Mineral Resources; French edition, *Structure géologique de l'U.R.S.S.*, 1959, 3 vols., Paris: Centre National de la Recherche Scientifique.

[30] Schatsky, N. S., chief editor, 1964, *Tectonics of Europe, explanatory note to the international tectonic map of Europe, scale 1: 2,500,000*, International Geological Congress, 360 pp., Moscow: Publishing House Nedra.

[31] Yanshin, A. L., chief editor, 1966, *Tectonics of Eurasia* (in Russian), The explanatory notes to tectonic map of Eurasia, scale 1: 5,000,000, with legend in Russian and in French.

[32] United Nations, 1968, *International tectonic map of Africa, scale 1: 5,000,000*. Unesco Publications Center, New York.

[33] Haller, John, 1970, Tectonic map of East Greenland, *Meddelelser om Grønland*, Bd 171, No. 5, 286 pp.

2

MECHANICAL PRINCIPLES

Materials of Structural Geology

Structural geology is concerned primarily with solids, but also with liquids and to some extent with gases.

In a *gas* the atoms are in rapid motion, move independently of one another, and have no orderly arrangement. The forces of mutual attraction are less than the forces of movement. Gases have high mobility.

In a *liquid* the atomic forces are strong enough to keep the atoms together, but there is either no orderly arrangement or only a limited orderly arrangement.

In *crystalline solids* the atoms have an orderly arrangement. Common salt, for example, is composed of sodium and chlorine atoms in a ratio of one to one, and forming a cubic lattice (Fig. 2-1); the external shape of a salt crystal is a cube or a related form. A rock such as basalt is composed primarily of two minerals, plagioclase and augite. The sodium and aluminum atoms,

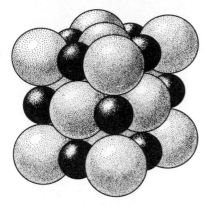

Fig. 2-1. Atomic structure of common salt. The large spheres represent chlorine atoms; the small spheres represent sodium atoms. (From R. W. G. Wyckoff's *Structure of crystals*, 2d ed., The Chemical Catalog Co., 1931.)

as well as some of the calcium and silicon atoms, combine to form crystals of plagioclase, characterized by its own lattice. Similarly, the atoms of iron and magnesium, as well as some of the atoms of calcium and silicon, combine to form crystals of augite, also characterized by its own lattice.

In *noncrystalline solids* there is no such systematic arrangement of the atoms into crystals. The *vitreous solids* or *glasses* are the product of liquids that cooled so rapidly that the atoms could not organize into crystals. Window glass is a vitreous solid. A basaltic glass is composed of the same elements as basalt, but is composed of glass rather than crystals of plagioclase and augite. *Amorphous solids* are a special group that lack crystal structure, but are not the product of rapidly cooled liquids.

The outer shell of the earth consists predominantly of solids, but gases and liquids are also present; their importance varies with time and space. Groundwater and active volcanoes attest to the importance of liquids at the present time, and the igneous rocks of intrusive bodies indicate the abundance of liquids in the past. Gases, present in the outer shell of the earth, are strikingly manifested in regions where petroleum is found; vast quantities of gas are sometimes expelled by active volcanoes. Never, however, does the gas occupy great underground chambers. The natural gas associated with petroleum occupies small pore spaces and fractures in solid rock, and the gas of volcanoes effervesces from magma.

In this section of the book we are concerned primarily with solids. Gases and liquids are important only if their presence in pore spaces modifies the behavior of the solids.

The outer shell of the earth consists of sedimentary, igneous, and metamorphic rocks. The structural geologist, however, is interested primarily in the mechanical properties of the rocks with which he deals rather than in their origin.[1,2] Is the rock well consolidated or not? A poorly cemented sandstone will be weaker than a well-cemented one, and quartzite will have greater strength than lava full of gas bubbles. Is the rock massive or not? Thin-bedded strata are weaker than thick-bedded formations. A thick, massive limestone will be stronger than a series of thin lava flows, although in laboratory tests of individual specimens, the lava may be the stronger of the two. A thick, massive sandstone can be stronger than a highly fractured granite. Is the composition such that the fractures may be readily healed? Specimens of quartzite may be stronger than a limestone. But fractures in quartzite heal less readily than those in limestone.

Force

FORCE AND ACCELERATION

Force is an explicitly definable vector quantity that changes or tends to produce a change in the motion of a body. The locomotive of a train exerts the force that moves the cars. Force is defined by its magnitude and direction, hence it may be expressed by an arrow, the length of which is proportional to the magnitude of the force, and the direction of which indicates the direction in which the force is acting.

An *unbalanced force* is one that causes a change in the motion of a body. The *acceleration* is the rate of change of velocity. If a train starts from rest and acquires a velocity of 20 miles per hour at the end of 10 minutes, the acceleration is two miles per hour per minute. A body dropped from a high building is subjected to an unbalanced force because of the gravitational pull of the earth, and the body accelerates at the rate of approximately 32 feet per second per second.

Balanced forces exist where no change in motion occurs. If a train is moving at a constant velocity, the frictional resistance of the tracks and the air equals the force exerted by the locomotive. If a man pushes against a wall that he cannot move, the wall is exerting a force equal and opposite to that exerted by the man.

Most problems confronting the structural geologist may be analyzed by assuming balanced forces because the velocity of rock bodies is so small that acceleration is negligible. Along faults, however, the motion causing earthquakes may be so rapid that acceleration is important.

UNITS OF MEASUREMENT

Two different systems of measurement are commonly used in English-speaking countries, one, the so-called absolute or C.G.S. system, the other the English or Engineer's system.

In the C.G.S. system the principal units are length (centimeters), mass (grams), time (seconds), and force (dynes).

$$F = ma \qquad (1)$$

where F is force, m is mass, and a is acceleration.

A dyne is the unbalanced force that will give a mass of one gram an acceleration of one centimeter per second per second. The force of gravity at sea level is approximately 980 dynes per square centimeter, producing an acceleration of 980 centimeters per second per second.

In the English system the units are length (feet or inches), mass (pounds), time (seconds), and force (poundal). A *poundal* is the unbalanced force that will give a mass of one pound an acceleration of one foot per second per second. The force of gravity at sea level is approximately 32.2 poundals per square foot, producing an acceleration of 32.2 feet per second per second.

In practice, force is often expressed in terms of grams or grams per square centimeter (or kilograms, that is, thousands of grams, per square centimeter) or in pounds or pounds per square inch.

Other units are also used, such as atmospheres and bars. Table 2-1 shows how the various units may be converted into one another.

Table 2-1. Conversion Table for Units of Mass and Force (Clark, 1967)

Pounds per Square Inch	Atmosphere	Bars	Kilograms per Square Centimeter	Dynes per Square Centimeter*
1	0.0680458	0.0689474	0.0703070	68,947
14.6960	1	1.01325	1.03323	1,013,250
14.5038	0.986924	1	1.01972	1,000,000
14.2234	0.967842	0.980665	1	980,665

*Assuming "normal" acceleration of gravity of 980.665 cm/sec.^{2}

COMPOSITION AND RESOLUTION OF FORCES

Force may be represented by a *vector*, that is, a line oriented in the direction in which the force is operating and proportional in length to the intensity of the force. Two or more forces may act in different directions at a point, as in Fig. 2-2, where OX (8 pounds) and OY (12 pounds) act at O. The same result would be produced by the force OZ ($14\frac{1}{4}$ pounds) acting in the direction indicated; OZ is the resultant of OX and OY. A *resultant* is the single force that produces the same result as two or more forces, and it may be represented by the diagonal of a parallelogram constructed on two arrows that represent the two forces. The *equilibrant* is the force necessary to balance

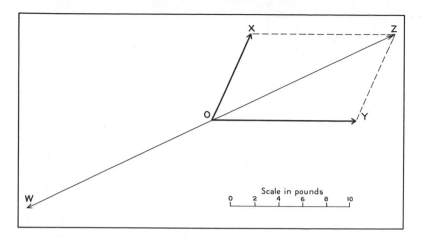

Fig. 2-2. Composition of forces.

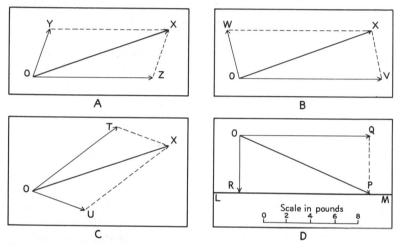

Fig. 2-3. Resolution of forces.

two or more forces. In Fig. 2-2, *OW* is the force necessary to balance *OX* and *OY*; it is equal to the resultant of the two forces, but acts in the opposite direction. The process of finding the resultant of two or more forces is called the *composition of forces*.

Conversely, the effect of a single force may be considered in terms of two or more forces that would produce the same result. Thus, in Fig. 2-3A, *OY* and *OZ* would produce the same result as *OX*; in Fig. 2-3B, *OW* and *OV* would produce the same result as *OX*; in Fig. 2-3C, *OT* and *OU* would produce the same result as *OX*. A single force may thus be resolved into two *components*, acting in defined directions, by constructing a parallelogram, the diagonal of which represents the given force, and the sides of which have the directions of the components. The process of finding the components of a single force is called the *resolution of forces*.

13

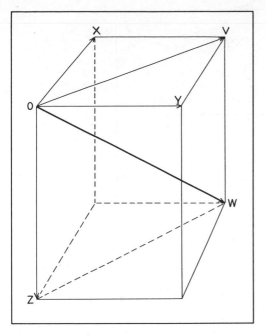

Fig. 2-4. Resolution of forces in three dimensions.

In Fig. 2-3D the force OP (12 lb) impinges on the line LM, and it is necessary to find the value of the component parallel to LM. This component OQ has a value of about 11 lb, as can be determined from the scale in the figure. OR, which is the component perpendicular to LM, has a value of about 5 lb.

The preceding discussion of the composition and resolution of forces has been confined to two dimensions, but geology is concerned with three dimensions. In Fig. 2-4 an inclined force OW lies in the vertical plane $OZVW$. This force may be resolved into two components, one of which, OZ, is vertical; the other, OV, lies in the horizontal plane $OXVY$. The OV may in turn be resolved into OX and OY, which lie in the horizontal plane and at right angles to each other. Moreover, any force, regardless of its value and its angle of inclination, may be similarly resolved into three components parallel to the X, Y, and Z axes of Fig. 2-4.

LITHOSTATIC OR CONFINING PRESSURE

The pressure on a small body immersed in a liquid is described as *hydrostatic pressure*. For example, at a depth of a mile in the ocean, the pressure is equal to the weight of a column of salt water one mile high. The pressure is 337,900 pounds per square foot, or 2346 pounds per square inch. Every square inch of the surface of a small sphere at this depth would be under a pressure of 2346 pounds per square inch. Such an undirected, all-sided pressure is called hydrostatic pressure.

Rocks in the lithosphere, because of the weight of whatever rocks lie above them, are subjected to a similar but not identical kind of pressure. The weight of a column of rock one mile high will be several times that of an equally high column of water, because rocks have a higher specific gravity. The weight of a column of granite one mile high and one inch square would be 6178 pounds. A small imaginary sphere at a depth of one mile in the granite would be subjected to an all-sided pressure that would simulate hydrostatic pressure. This type of pressure may be called *lithostatic pressure*, but in experimental work this equal, all-sided pressure on solids is called the *confining pressure*.

Obviously, the lithostatic pressure increases with depth in the earth and reaches tremendous values in the interior. It is equal to the weight of the overlying column of rocks, but near the surface this is only approximately true.

An increase in confining or lithostatic pressure causes a decrease in the volume of rocks but an increase in the density. A decrease in confining pressure causes an increase in volume but a decrease in density.

DIFFERENTIAL FORCES

In many instances the forces acting on a body are not equal on all sides. A body is said to be under *tension* when it is subjected to external forces that tend to pull it apart. Tension may be represented, as in Fig. 2-5A, by two arrows that are on the same straight line and are directed away from each other; the arrows represent the forces, whereas the rectangle represents the body or part of a body upon which the forces act. The rectangle may be omitted.

A body is said to be under *compression* when it is subjected to external forces that tend to compress it. Compression may be represented, as in Fig. 2-5B, by two arrows that are on the same straight line and are directed toward each other; the arrows represent the forces, whereas the rectangle represents the body or part of the body acted upon. The rectangle may be omitted.

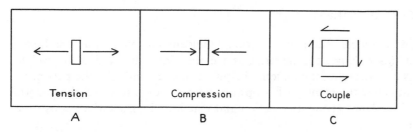

Tension	Compression	Couple
A	B	C

Fig. 2-5. Arrows representing tension, compression, and a couple.

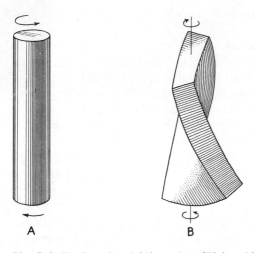

A B

Fig. 2-6. Torsion. A rod (A) or plate (B) is subjected to torsion when the ends are twisted in opposite directions.

A *couple* consists of two equal forces that act in opposite directions in the same plane, but not along the same line. In Fig. 2-5C a couple is represented by the upper and lower arrows, which are not on the same straight line and which are directed away from each other. To prevent rotation and preserve equilibrium a second couple is necessary, as shown by the vertical arrows. The rectangle, which represents the body or part of the body acted upon, may be omitted.

Torsion results from twisting. If the two ends of a rod are turned in opposite directions, the rod is subjected to torsion (Fig. 2-6A). A plate undergoes torsion, as in Fig. 2-6B, if two diagonally opposite corners are subjected to forces acting in one direction while the other two corners are subjected to forces acting in the opposite direction.

Stress

CONCEPT OF STRESS

Imagine a vertical column of material. Along any imaginary horizontal plane within this column, the material above the plane, because of its weight, pushes downward on the material below the plane. Similarly, the part of the column below the plane pushes upward with an equal force on the material above the plane. The mutual action and reaction along a surface constitutes a stress.

Moreover, along any imaginary plane within the column there are similar actions and reactions. The imaginary plane may be horizontal, vertical, or inclined at any angle. The force, due to the weight of that part of the column that lies above the plane, acts in a vertical direction. This force would be directed normally to a horizontal plane. Along an inclined plane, however, the vertically directed force would be resolved into a normal component and a tangential component. If Fig. 2-3D were turned so that *OP* were vertical, *LM* would represent the inclined plane, *OR* the normal component, and *OQ* the tangential component.

The normal component is a *compressive stress* if it tends to push together the material on opposite sides of the plane. The normal component is a *tensile stress* if it tends to pull apart the material on opposite sides of the plane. The tangential component is generally called a *shearing stress* or *shear*.

In this book, in accordance with common geological practice, a compressive stress will be considered positive, and a tensile stress will be considered negative. In engineering and physics the opposite convention is often followed.

The *stress-difference* at any point in a body is the algebraic difference between the greatest stress and the least stress at that point. This concept is more fully developed in Chap. 7.

Physicists measure stress as the force per unit area; it is stated in pounds per square inch, tons per square foot, kilograms per square centimeter, or similar convenient units. Engineers prefer to use *unit stress* for the force per unit area.

It is essential to distinguish between the external force that is applied to a body and the resulting internal actions and reactions that constitute the stress.

CALCULATION OF STRESS

There is no direct way to measure the stresses in a body, but they may be calculated if the external forces are known. If a body is compressed or stretched, the stress is referred to a plane perpendicular to the direction in which the external forces are acting. Thus, if a vertical square column 10 inches on a side supports a load of 5000 pounds, every horizontal plane in the column is subjected to a compressive force of 5000 pounds if we neglect the weight of the column itself. Each square inch of these horizontal planes supports a load of 50 pounds per square inch. The *compressive stress* is said to be 50 pounds per square inch. If a vertical rod with a cross-sectional area of 10 square inches carries a weight of 5000 pounds at its lower end, every horizontal plane in the rod is subjected to a pull of 500 pounds per square inch. The *tensile stress* is said to be 500 pounds per square inch.

Various techniques have been devised for deducing the stresses in open cuts, bore holes, tunnels, and other underground openings. These devices are attached to the wall of the opening or placed in a bore hole. They actually measure the strain, from which the stresses are deduced. One method employs photoelastic techniques.[3] Others use signals derived from electrical, mechanical, or hydraulic components.[4,5] Such measurements are important in planning large underground openings.

Strain

DEFINITION

Strain is the deformation caused by stress; strain may be *dilation*, which is a change in volume, or *distortion*, which is a change in form, or both.

When there is a change in the confining pressure, an isotropic body—that is, a body whose mechanical properties are uniform in all directions—will change in volume, but not in shape. With increasing confining pressure, the volume of the body decreases and the dilation is negative. With decreasing confining pressure, the volume of the body increases and the dilation is positive.

Under directed forces distortion occurs. For example, a steel rod 10 inches long with a cross section of one square inch is subjected to tension. A pull of 20,000 pounds stretches the rod 0.007 inch. The stress is 20,000 pounds per square inch and the strain is 0.0007 inch per inch.

THREE STAGES OF DEFORMATION

If a body is subjected to directed forces lasting over a short period of time—minutes or hours—it usually passes through three stages of deformation, although in brittle substances the intermediate stage may be omitted. At first, the deformation is *elastic;* that is, if the stress is withdrawn, the body returns to its original shape and size. There is always a limiting stress, called the *elastic limit;* if this is exceeded, the body does not return to its original shape. Below the elastic limit, the deformation obeys *Hooke's law*, which states that strain is proportional to stress.

If the stress exceeds the elastic limit, the deformation is *plastic*; that is, the specimen only partially returns to its original shape even if the stress is removed. Steel rods under tension, for example, begin to get thinner or "neck" in the middle, and, even after the stress is released, the constriction remains in the rod.

When there is a continued increase in the stress, one or more fractures develop, and the specimen eventually fails by *rupture*. The arrangement and form of the fractures depend upon several factors which are fully discussed in Chapter 7.

Brittle substances are those that rupture before any significant plastic deformation takes place. *Ductile substances* are those that undergo a large plastic deformation before rupture. After the elastic limit has been exceeded, ductile substances undergo a long interval of plastic deformation, and in some instances they may never rupture.

ELASTIC DEFORMATION

At room temperature and pressures, and under stresses applied for a short period of time, most rocks are brittle. They behave elastically until they fail by rupture. For such rocks the elastic limit or yield point is the stress at rupture. Ideally an elastic substance will return to the original shape after the deforming stress has been removed, although there may be a slight delay as unloading occurs.

If a solid cylinder of rock is subjected to stress parallel to its long axis, it will lengthen under tension and shorten under compression. The ratio of the stress to the deformation is a measure of the property of the rock to resist deformation.

$$E = \frac{\sigma}{\epsilon} \tag{2}$$

where E is *Young's modulus* (also called *modulus of elasticity*), σ is stress, and ϵ is strain.

$$\epsilon = \frac{\Delta l}{l_o} \tag{3}$$

where Δl is change in length, l_o is original length. Thus, if a cylinder 10 centimeters long is stretched 0.001 centimeters under tension by a stress of 10^8 dynes per square centimeter, Young's modulus is 10^{12} dynes per square centimeter. This is an average value for rocks; in the English system it is 1.45×10^7 lb/in.^{2}.

Under tension the diameter of a cylinder subjected to tension parallel to the axis becomes smaller; under compression parallel to the axis the diameter becomes greater. Poisson's ratio is the ratio of transverse strain to axial strain.

$$v = \frac{\Delta d/d_o}{\Delta l/l_o} \tag{4}$$

where v is Poisson's ratio, Δd is change in diameter, d_o is original diameter, and Δl and l_o are as above.

If in the preceding example the diameter decreased by 0.00025 centimeters, Poisson's ratio is 0.25; this is a good average for rocks.

Rigidity measures the resistance to change in shape (Fig. 2-7).

$$G = \frac{\tau}{\gamma} \tag{5}$$

where G is rigidity modulus, τ is shear stress and γ is the shear strain.

$$\gamma = \frac{ab}{ac} \tag{6}$$

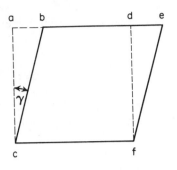

Fig. 2-7. Rigidity. Square *acdf* is deformed into parallelogram *bcef*.

If *ac* were 10 centimeters, *ab* were 0.001 centimeters, and τ were 10^7 dynes per square centimeter, the rigidity modulus would be 10^{11} dynes/cm². In rocks the rigidity averages 2 to 5×10^{11} dynes per square centimeter or 2.9 to 7.25×10^6 lb/in.².

G may also be expressed in terms of Young's modulus and Poisson's ratio.

$$G = \frac{E}{2(1 + v)} \tag{7}$$

The bulk modulus or incompressibility is

$$K = \frac{\Delta h}{\Delta V/V_o} \tag{8}$$

where K is the bulk modulus, Δh is change in hydrostatic pressure, ΔV is change in volume, and V_o is original volume. Δh may refer to lithostatic pressure instead of hydrostatic pressure. Bulk modulus may be expressed in dynes per square centimeter or some equivalent unit.

A rock at a depth of 10 kilometers (6.214 miles) is under a lithostatic pressure of 2.7×10^6 g/cm². If it is raised to a depth of 4 km, Δh becomes 1.62×10^6 g/cm². If the original volume were 1000 cubic centimeters and the new volume is 999.000 cubic centimeters, $\Delta V/V_o$ is 10^{-3}. The bulk modulus is 1.62×10^9 g/cm² or 1.59×10^{10} dynes/cm².

The compressibility is the reciprocal of the bulk modulus

$$\beta = \frac{1}{K} \tag{9}$$

where β is compressibility.

Elastic deformation is primarily of importance in analyzing tidal deformation of the solid earth and in investigating the transmission of seismic waves through the earth. It is of even more direct significance to structural geology in studying elastic rebound associated with earthquakes, in the fracturing of rocks to produce joints and faults, and in certain aspects of folding. Although most structural features observed by structural geologists are the result of deformation beyond the elastic limit, the same parameters of length, mass, time, and force are involved.

PLASTIC DEFORMATION

Although most rocks at room temperatures and pressures fail by rupture before attaining a stage of plastic deformation, most rocks, at sufficiently high temperatures and confining pressures deform plastically even in experiments lasting for a short time. This plastic deformation is not recoverable or is only partially recoverable. That is, if the stress is removed, the material does not return to its original shape.

In much experimental work the deformation of rocks results from internal adjustments within the mineral grains, notably gliding and dislocations (Chap. 20). Hence metallurgists tend to confine the concept of plasticity to such deformation. But in nature other factors are also important, notably the rotation of grains and recrystallization. But the internal changes that take place in a rock undergoing plastic deformation are discussed more fully in Chap. 20.

Stress-Strain Diagrams

The relation existing between stress and strain is commonly expressed in graphs known as stress-strain diagrams (Fig. 2-8). The stress is plotted on the ordinate (vertical axis), whereas the strain is plotted on the abscissa (horizontal axis). In Fig. 2-8 the material is under compression and the compressive stress parallel to the axis of the cylinder is in pounds per square inch. With increasing stress the specimen becomes shorter and the strain is plotted in terms of the percentage of the shortening of the specimen.

Curve A is the stress-strain diagram of a brittle substance. It deforms elastically up to a stress of 20,000 lb/in.^{2} and has shortened one-half of one percent; it then fails by rupture.

Curve B is an ideal plastic substance. It behaves elastically at first. At a stress of about 24,000 lb/in.^{2} it reaches the *proportional elastic limit*, which is the point at which the curve departs from the straight line. The shortening is slightly less than one percent. Thereafter the specimen deforms continuously without any added stress.

Curve C represents a more normal type of plastic behavior. At a stress of about 28,000 lb/in.^{2} and a strain of somewhat over 1 percent, the specimen reaches the proportional elastic limit and thereafter deforms plastically. But for every increment of strain an increase in stress is necessary. This is the result of what is called *work hardening*; that is, the specimen becomes progressively more difficult to deform.

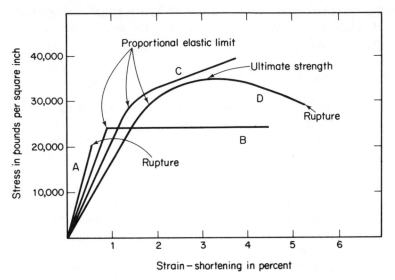

Fig. 2-8. Stress-strain diagrams.

Curve *D* represents a very common type of plastic deformation. The specimen deforms elastically up to a stress of about 28,000 lb/in.² and a shortening of somewhat less than 2 percent. At first an increase in stress is necessary for continued deformation. But when the shortening is somewhat over 3 percent, progressively less stress is necessary to continue the deformation. This high point on the curve is the *ultimate strength*. However, the ultimate strength of a rock is a function of many variables, such as confining pressure and temperature.

The term "strength" is a rather meaningless term unless all the environmental conditions are specified. The value of the breaking strength—normally applied to brittle materials under temperature and pressures at the surface of the earth—is of use in engineering projects. Under compression, diabase, basalt, felsite, and quartzite are strong, with an average breaking strength of about 1800–2600 kg/cm². Marlstone, limestone, marble, and sandstone are weak, with an average breaking strength 700–1000 kg/cm². Under tension, the breaking strengths are far less, averaging only 5 percent to 10 percent of the breaking strength under compression.

More precise values are not given intentionally, because the figures may be misleading. The shape of the test specimen may control the results. Moreover, the strength of the same kind of rock may vary greatly, depending on the locality from which it comes. Finally, the published data usually represent "intact" rocks, that is, small specimens that are not marred by flaws. Larger units, such as those significant to regional tectonics, are generally characterized by joints and other planes of weakness.

Factors Controlling Behavior
of Materials

CONFINING PRESSURE

The mechanical behavior of rocks is controlled not only by their inherent properties—mineralogy, grain size, porosity, fractures, etc.—but also by factors that are of little or no concern in planning man-made structures at the surface of the earth. These factors are confining pressure, temperature, time, and solutions. Tremendous ingenuity has been required to design the experimental apparatus that has been utilized in obtaining the data presented in the following pages.

Cylinders of rock are prepared for the experiments; usually the length is several times the diameter. In some experiments the cylinders are small, the length not exceeding one or a few inches. By using small cylinders it is possible to employ both high confining pressures and high temperatures. When high temperatures are not a factor, much larger specimens, several feet long, may be used. The confining pressure—in this case a hydrostatic pressure —is obtained by a fluid under pressure. The compressive (or tensile) stress is applied by a plunger (or gripping device) at the end of the cylinders. Often the specimens are "jacketed" by aluminum foil or some similar material to prevent fluid from escaping into the specimen and thus weakening it.

In this experimental work various systems of units have been used. Some experiments use the C.G.S. system exclusively. Others use the English system. Others use a combination of the two. Table 2-1 is a conversion table to show how the various systems are related.

Figure 2-9 illustrates the behavior of Solenhofen Limestone under such conditions.[6] The compressive stress on the ends of the cylinder is given on the ordinate in kilograms per square centimeter. The percentage of shortening of the cylinder is given on the abscissa. Seven separate experiments are shown at confining pressures of 1, 300, 700, 1000, 2000, 3000, and 4000 kilograms per square centimeter. Separate curves are given for the behavior at each of these confining pressures. Below a compressive stress of 3700 kg/cm² the curves run together and appear as one. One experiment was run in air so that the confining pressure was equal to 1 kg/cm², that is, 14.7 lb/in.², or 1 atmosphere. This specimen behaved elastically up to a compressive stress of 2800 kg/cm², when it failed by rupture. The specimens tested under confining pressures of 300 and 700 kg/cm² deformed elastically, went through a short stage of plastic deformation—that portion of the lines that is bending— and then failed by rupture. The specimens tested under confining pressures of 1000 or more kg/cm² began to deform plastically at a compressive stress of about 4000 kg/cm² and continued to deform plastically. The specimen tested under a confining pressure of 2000 kg/cm² had shortened 30 percent

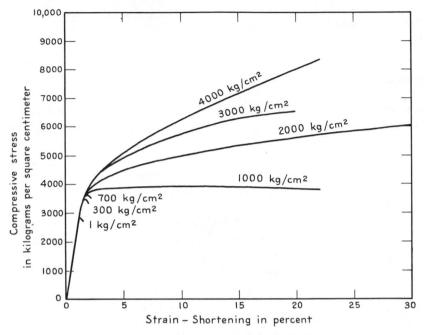

Fig. 2-9. Effect of confining pressure on behavior of Solenhofen Limestone under compression. (After E. Robertson.[6])

when the test was terminated. The curves representing the tests at a confining pressure of 1000, 2000, 3000, and 4000 kg/cm² end, not because of failure by rupture, but because the tests were not carried any further.

It is also readily apparent that the strength increases with the confining pressure. Whereas the specimen tested at a confining pressure of 1 kg/cm² fails by rupture at a compressive stress of 2800 kg/cm² and the specimen tested at a confining pressure of 1000 kg/cm² cannot support a compressive stress greater than 3900 kg/cm², the specimen tested at a confining pressure of 4000 kg/cm² can support a compressive stress in excess of 8000 kg/cm².

Such experiments indicate that rocks exhibiting very little plastic deformation near the surface of the earth may be very plastic under high confining pressure. Thus under a confining pressure of 1000 kg/cm² or greater, Solenhofen limestone will deform plastically. This means that at a depth of 2.5 miles Solenhofen limestone will deform plastically if sufficient compressive stress is applied; as will be shown later, this figure may be still less because of other factors.

Different rocks, of course, behave differently. Figure 2-10 shows the stress-strain diagram for several rocks and one mineral.[6] The results are not strictly comparable because, as the figure shows, the confining pressure was not the same in all experiments, ranging from 300 to 500 kg/cm². Pyrite, Cambridge

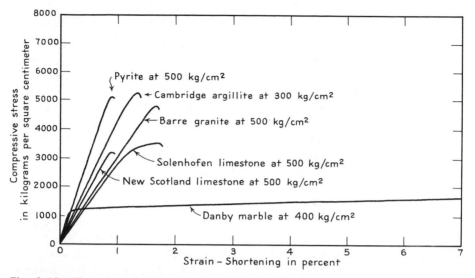

Fig. 2-10. Effect of confining pressure on behavior of various rocks and pyrite under compression. (After E. Robertson, "An experimental study of flow and fracture in rocks," doctoral thesis, Harvard University, 1952.)

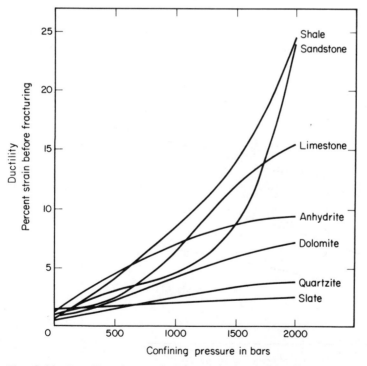

Fig. 2-11. Ductility of several common rocks under differing confining pressures. (After Donath.[7] Reprinted by permission, *American Scientist*, Journal of The Society of Sigma Xi.)

Argillite, and Barre Granite are relatively brittle rocks, which behave elastically up to a compressive stress of over 4500 kg/cm². Above the elastic limit there is a small zone of plastic deformation, and then rupture takes place. New Scotland limestone was elastic up to a compressive stress of about 3000 kg/cm², deformed plastically for a short interval, and then ruptured at 3200 kg/cm². Solenhofen Limestone shows a still larger range of plastic deformation. Danby Marble is much weaker. It deforms elastically up to a compressive stress of 1000 kg/cm² and then deforms plastically. Although the curve scale ends at 7 percent, the original data show that the specimen shortened 14 percent before the test was ended.

Figure 2-11 illustrates the effect of confining pressure on the breaking strength of several different rocks.[7] At atmospheric pressure—confining pressure of one bar—the rocks deform only a few percent before fracturing. Under a confining pressure of 1000 bars the sandstone and shale deform more than 5 percent before rupturing. Under a confining pressure of 2000 bars the limestone deforms nearly 15 percent and shale and sandstone over 20 percent before rupturing.

TEMPERATURE

Changes in temperature modify the strength of rocks.[8] Hot steel, for example, undergoes plastic deformation much more readily than does cold steel. Figure 2-12 shows two tests run on Yule Marble. Conditions were identical

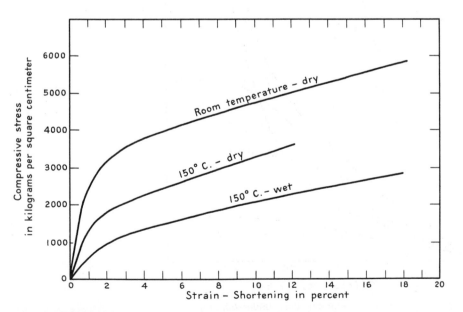

Fig. 2-12. Effect of temperature and solutions on deformation of marble. Yule Marble, confining pressure of 10,000 atmospheres, cylindrical specimen cut perpendicular to foliation. (After D. T. Griggs, et al.[8])

except for temperature; the axes of the cylinders were perpendicular to the foliation, the confining pressure was 10,000 atmospheres, and the deformation was produced by compressive stress. The uppermost curve is that obtained at room temperatures, whereas the intermediate curve is that obtained at a temperature of 150°C. At room temperature the elastic limit is at a compressive stress of about 2000 kg/cm², at 150°C the elastic limit is at about 1000 kg/cm². Moreover, to produce a given strain far less stress is necessary when the specimen is hot than when it is cold. For example, to produce a strain of 10 percent at 150°C the compressive stress is 3000 kg/cm², but at room temperature the stress necessary to produce a similar deformation is 4500 kg/cm².

It is apparent that plastic deformation is far less common near the surface of the earth, where the confining pressure and the temperature are low, than it is at greater depths, where higher temperatures and greater confining pressure increase the possibility of plastic deformation.

TIME

Geological processes have great lengths of time in which to operate. Although geologic time is impossible to duplicate experimentally, it is possible from experiments to make some deductions concerning the influence of time. An analysis of the effects of time is concerned with such subjects as creep, strain-rate, and viscosity.

Creep refers to the slow continuous deformation with the passage of time.[9] The stresses may be above or below the elastic limit, but we are especially interested in creep caused by stresses below the elastic limit.

Solenhofen Limestone under atmospheric pressure and at room temperature has a strength of 2560 kg/cm². In a long-time experiment, Solenhofen Limestone subjected to a compressive stress of 1400 kg/cm²—half the value of the strength—deforms rapidly at first, then more slowly (Fig. 2-13). At the end of one day, it has been shortened about 0.006 percent; after 10 days about 0.011 percent; after 100 days about 0.016 percent; and after 400 days a little more than 0.019 percent.

The general form of a creep curve is shown in Fig. 2-14. The ordinate is the total strain and the abscissa is time. The intercept A on the ordinate represents the instantaneous strain when the load is added. The first part of the curve, B, represents primary creep, when the strain decreases with time. The main part of the curve, C, represents secondary or steady-state creep. Finally, in tertiary creep, D, the curve sharply rises just before rupture.

This curve may be expressed as an equation

$$S = A + B \log t + Ct + D \tag{10}$$

where S is total strain, t is time, A, B, and C are constants, and D is strain during tertiary creep.

Strain rate is the amount of strain (pp. 18 and 21) divided by the time.

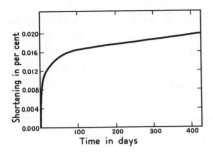

Fig. 2-13. Creep curve for Solenhofen Limestone under a stress of 1400 kg/cm². (After D. T. Griggs.[9])

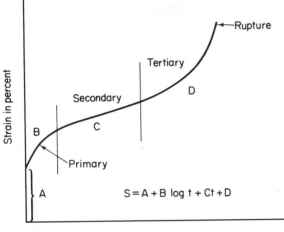

Fig. 2-14. Ideal creep curve. (A) Instantaneous deformation. (B) Primary creep. (C) Secondary creep. (D) Tertiary creep. *S* is the total strain; *t* is time.

$$\dot{\epsilon} = \frac{\epsilon}{t} \tag{11}$$

where $\dot{\epsilon}$ is strain rate, ϵ is strain, and t is time, usually given in seconds. In Fig. 2-13 the instantaneous deformation and the primary and secondary creep are shown well. For that part of the curve in Figure 2-13 in which the shortening exceeds 0.016, the strain rate is

$$\dot{\epsilon} = \frac{3.2 \times 10^{-5}}{325 \times 24 \times 60 \times 60} = \frac{3.2 \times 10^{-5}}{2.81 \times 10^{7}} = 1.139 \times 10^{-12} \text{ sec} \tag{12}$$

Experimental apparatus has been devised so that the strain rate can be kept constant.[10] Figures 2-15 and 2-16 show two sets of graphs for Yule Marble

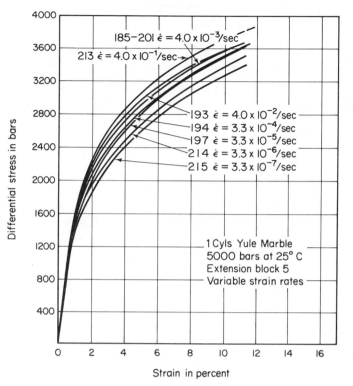

Fig. 2-15. Effect of strain rate on stress-strain curve at 25°C, confining pressure of 5000 bars. (Hugh C. Heard.[10]) Permission University of Chicago Press.

subjected to tension under a confining pressure of 5000 bars. The cylinders were cut perpendicular to the foliation. Figure 2-15 gives the results for experiments conducted at 25°C, Fig. 2-16 for experiments at a temperature of 500°C. In Fig. 2-16 eight experiments were run, at strain rates of 4.0×10^{-1}/sec, 4.0×10^{-2}/sec, etc., up to a minimum strain rate of 3.3×10^{-8}/sec. Curve 153, $\dot{\epsilon} = 3.3 \times 10^{-7}$/sec shows that the strain is very small until the differential stress reaches 500 bars, when the strain increases relatively rapidly, attaining a value of 10 percent at a differential stress of about 725 bars. The time to reach this strain can be calculated from equation (11) as 3.03×10^4 sec or $8\frac{1}{2}$ hr.

From Figs. 2-15 and 2-16 two important principles may be established. (1) The slower the strain rate, the less the differential stress to attain a given strain. In Fig. 2-16 a differential stress of 1820 bars is necessary to attain 10 percent lengthening if the strain rate is 4.0×10^{-1}/sec, but only 450 bars are necessary if the strain rate is 3.3×10^{-8} sec. (2) The higher the temperature, the less the required differential stress for a given strain; compare Fig. 2-15 (25°C) and Fig. 2-16 (500°C).

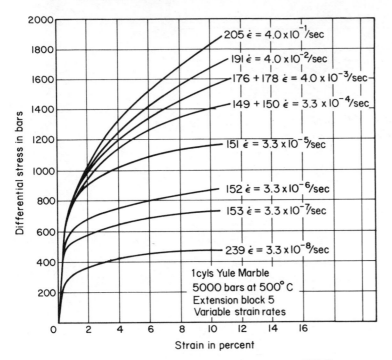

Fig. 2-16. Effect of strain rate on stress-strain curves at 500°C, confining pressure of 5000 bars. (Hugh C. Heard.[10]) Permission University of Chicago Press.

Normally we think of *viscosity* in relation to liquids and more specifically the ease with which they flow. Water is much less viscous than syrup, and syrup much less viscous than tar. But the concept of viscosity, or, more precisely, apparent viscosity, may be applied to solids.

In a perfectly viscous substance, characterized by Newtonian viscosity, viscosity is defined as the ratio of shearing stress to the rate of shear. Referring to Fig. 2-7, if shear stress is applied the rapidity with which the square *adcf* is deformed into the parallelogram *becf* is a function of the viscosity. The rate of shear is measured by the change in angle γ per unit of time. If the stress is measured in dynes/cm² and the rate of shear in seconds, the viscosity is given in dynes-sec/cm²; this unit is called the *poise*.

$$\dot{\gamma} = \frac{\gamma}{t} \tag{13}$$

and

$$\eta = \frac{\tau}{\dot{\gamma}} \tag{14}$$

where $\dot{\gamma}$ is rate of shear strain, γ is shear strain, t is time, η is viscosity, and τ is shearing stress.

Calculations show that the viscosity of Yule Marble, used in the experiments referred to in Figs. 2-15 and 2-16, ranges from 10^{23} poises at 25°C to 10^{16} poises at 500°C. Additional values of viscosity are given in Table 2-2.

Table 2-2. Some Values of Viscosity in Poises

Water at 100°C and one atmosphere	0.00284
Water at 30°C and one atmosphere	0.00801
Water at 0°C and one atmosphere	0.01792
Corn syrup, room temperature and pressure	7×10^2
Roofing tar, ready to apply	3×10^7
Lava, Mt. Vesuvius, 1400°C	2.56×10^2
Lava, Mt. Vesuvius, 1100°C	2.83×10^4
Rock salt, near surface	10^{17}
Rocks in general	10^{17} to 10^{22}
Mantle of earth	10^{23}

Creep is the combined effect of an elastic strain and a permanent strain. It is viscoelastic deformation rather than pure viscous deformation or elastic deformation. The specimen recovers from that portion of the deformation that is caused by the elastic strain, whereas that portion of the deformation that is the result of permanent strain is unrecoverable.

The *fundamental strength* of any material is defined as the stress which that material is able to withstand, regardless of time, under given physical conditions—temperature, pressure, solutions—without rupturing or deforming continously. The fundamental strength, which is always less than the breaking strength and the ultimate strength, is much more significant to the geologist. Unfortunately, at the present time we have few data on the value of the fundamental strength of rocks.

The mechanics of plastic deformation and creep—that is, the changes that take place in the rock during these processes—is discussed in Chap. 20.

SOLUTIONS

Much rock deformation takes place while solutions capable of reacting chemically with the rock are present in the pore spaces. This is notably true of metamorphic rocks, in which extensive or complete recrystallization occurs. The solutions dissolve old minerals and precipitate new ones. Under such conditions the mechanical properties of rock are greatly modified.

Creep experiments have been performed[11] on alabaster (a variety of gypsum) with solutions present (Fig. 2-17). In all cases the compressive stress was 205 kg/cm² (less than half the normal elastic limit of 480 kg/cm²), and the temperature 24°C. The lowest curve represents the deformation of a dry specimen. Within a few days the specimen had shortened about 0.03 percent,

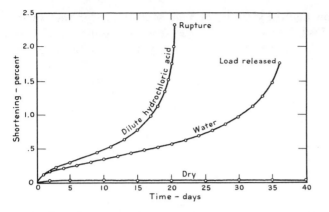

Fig. 2-17. Effect of solutions on deformation of alabaster. (After D. T. Griggs.[9])

but there was no further detectable deformation even after 40 days. A specimen deformed under such conditions that water had access to the alabaster (intermediate curve) had shortened 1 percent at the end of 30 days and 1.75 percent by the end of 36 days, when the load was released. A specimen deformed with access of dilute hydrochloric acid had deformed more than 2 percent before rupturing at the end of 20 days. Whereas the strength of the dry alabaster under room temperature and a confining pressure of 1 atmosphere is 480 kg/cm², and the ultimate strength is 520 kg/cm², the fundamental strength under similar conditions, but with the specimen free to react with water, is estimated to be only 92 kg/cm². In this particular case, therefore, the fundamental strength is less than 20 percent of the strength and the ultimate strength.

The lowest curve in Fig. 2-12 shows the effect of water on the strength of Yule Marble. At a temperature of 150°C the elastic limit and strength of the wet specimen is much less than the strength of the dry specimen at the same temperature.

PORE PRESSURE

In recent years it has been suggested that high fluid pressure in the pore spaces of rocks partially balances the lithostatic pressure. The reality of such high pressures has been demonstrated in some oil fields. Imagine an open bore hole 1 km deep filled with water. The hydrostatic pressure at the bottom of the hole is 10^5 g/cm². If the rock is sufficiently permeable for water to penetrate into all the pore spaces, this pressure in part balances the lithostatic pressure.

Let λ be the ratio of pore pressure to lithostatic pressure. In the case just cited $\lambda = 1/2.3 = 0.435$. In some deep wells λ is much greater, and may approach 1 in value.

This pore pressure weakens the rock. Normally the strength of a rock increases at depth because of the increase in confining pressure. But with increasing pore pressure the effective confining pressure decreases. Moreover, with increasing pore pressure the rocks are less coherent.

ANISOTROPY AND INHOMOGENEITY

Most of the tests described in the preceding sections were made on isotropic materials, that is, rocks whose mechanical properties were uniform in all directions. Rocks that show bedding, banding, or foliation are not isotropic. The strength of such rocks would depend upon the orientation of the applied forces to the planar structures of the rock. This point is well illustrated in Fig. 2-18. The rock was Yule Marble, confining pressure was 10,000 kg/cm², and the tests were run at room temperature. All the specimens show great plastic deformation. The solid lines represent experiments under compression; in this case the stress is compressive and the strain is shortening parallel to the axis of the cylinder. Under compression the cylinder perpendicular to the foliation is stronger than the cylinder parallel to the foliation. The broken lines represent tests under tension; in this case the stress is tensile and the strain is lengthening parallel to the cylinders. Under tension the cylinder

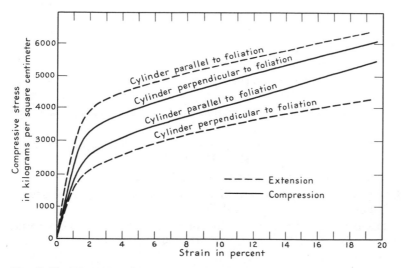

Fig. 2-18. Effect of anisotropy on deformation of marble. Yule Marble at confining pressure of 10,000 kg/cm² and room temperature. (After D. T. Griggs, et al.[8])

parallel to the foliation is much stronger than the cylinder perpendicular to the foliation.

SUMMARY

It is clear that the mechanical properties of rocks are profoundly modified by confining pressure, temperature, the time factor, and the presence of reacting solutions. The combined effect of these factors is so great that it is impossible in the present state of our knowledge to treat rock deformation in a quantitative way. Increase in confining pressure increases the elastic limit and the ultimate strength. Increase in the temperature weakens the rocks. After long continued stress the rocks become much weaker. The fundamental strength is of more interest to the structural geologist than is the strength or ultimate strength. Reacting solutions lower the strength, the ultimate strength, and the fundamental strength of rocks.

References

[1] Clark, Sydney P., Jr., ed., 1966, *Handbook of physical constants*, rev. ed., Geol. Soc. Amer. Memoir 97, 587 pp.

[2] Griggs, David T., and Handin, J., eds., 1960, *Rock deformation*, Geol. Soc. Amer. Memoir 79, 382 pp.

[3] Voight, Barry, 1967, On photoelastic techniques, in situ stress and strain measurement, and the field geologist, *J. Geol.* 75: 46–58.

[4] Hast, N., 1958, *The measurement of rock pressure in mines*, Sveriges Geologiska Undersökning, Avhandlingar Och Uppsator, ser. C, 560, Årksbok, vol. 52, pt. 3, 183 pp.

[5] Terzaghi, K., 1962, Measurements of stresses in rock, *Geotechnique*, 12: 105–24.

[6] Robertson, E. C., 1955, Experimental study of the strength of rocks, *Geol. Soc. Amer. Bull.* 66: 1294–1314.

[7] Donath, Fred A., 1970, Some information squeezed out of rocks, *Amer. Sci.* 58: 54–72.

[8] Griggs, David T., et al., 1951, Deformation of Yule Marble: part IV, effects at 150°C, *Geol. Soc. Amer. Bull.* 62: 1385–1406.

[9] Griggs, David T., 1939, Creep of rocks, *J. Geol.* 47: 225–51.

[10] Heard, Hugh C., 1963, Effect of large changes in strain rate in the experimental deformation of Yule Marble, *J. Geol.* 71: 162–95.

[11] Griggs, David T., 1940, Experimental flow of rocks under conditions favoring recrystallization, *Geol. Soc. Amer. Bull.* 51: 1001–22.

3

DESCRIPTION
OF FOLDS

Introduction

Folds (Frontispiece) are best displayed by stratified formations such as sedimentary or volcanic rocks or their metamorphosed equivalents. But any layered or foliated rock, such as gabbro or granite gneiss, may show folds. Some folds are a few miles across. The width of others is to be measured in feet or inches or even fractions of an inch. Folds of continental proportions are hundreds of miles wide.

Folds may be observed directly or may be inferred from various kinds of data. The size of the exposure determines the size of the folds that may be observed. Folds many thousands of feet across may be observed in regions of high relief. Conversely, where the exposures are small, only folds a few feet or tens of feet across may be observed.

Attitude of Beds

Attitude refers to the three-dimensional orientation of some geological feature, such as a bed, a joint, a hornblende needle, or a fold. The attitude of planar features, such as beds or joints, is defined by their strike and dip. The *strike* (Plate 1) of a bed is its trend measured on a horizontal surface. More precisely, strike may be defined as the direction of a line formed by the intersection of the bedding and a horizontal plane. In Fig. 3-1A the

Plate 1. *Measuring strike.* Measurement is made at base of a sandstone in the Chinle Formation. Usually the geologist stands at a distance to eliminate effect of small irregularities in the bedding. Colfax County, New Mexico. Photo: J. R. Stacy, U. S. Geological Survey.

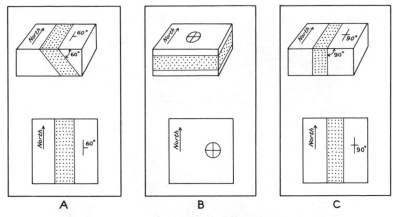

Fig. 3-1. Dip-strike symbols used for inclined, horizontal, and vertical strata. Block diagram above, map below. (A) Inclined strata. (B) Horizontal strata. (C) Vertical strata. The position of the 90° may be used to indicate the top side of the bed (see page 81).

strike is north; the upper part of the figure is a block diagram, the lower part a map or plan.

The *dip* (Plate 2) of a stratum is the angle between the bedding and a horizontal plane; it is measured in a vertical plane that strikes at right angles to the strike of the bedding. In Fig. 3-1A the dip is 60 degrees to the east.

A special *dip-strike symbol* is employed on geological maps to show the attitude of beds (Plates 3, 4, and 5). It is shaped somewhat like the capital letter *T*, but the relative lengths of the two parts are reversed (Fig. 3-1A). The longer line is parallel to the strike of the bedding, whereas the shorter line points in the direction of the dip, and a figure gives the value of the dip. For horizontal strata (Plate 6), a special symbol ⊕ may be used (Fig. 3-1B). For vertical strata a long line gives the strike, and a short crossbar extends on either side of the long line (Fig. 3-1C). Although most geological maps use symbols similar to these to give dip and strike of the bedding, the system is not standardized, and it is necessary to look at the legend accompanying the map in order to ascertain the meaning of the symbols employed.

Parts of a Fold

The *hinge* of a fold is the line of maximum curvature in a folded bed. It is characterized by orientation and position. There is a hinge for each bed. In the fold shown in Fig. 3-2A, the line *aa'* is the hinge on the top of the bed.

The hinges may be horizontal (Fig. 3-2A, D, E), inclined (Fig. 3-2B, F) or vertical (Fig. 3-2C).

The axial plane is the surface connecting all the hinges. It may be a simple

Plate 2. *Measuring dip.* Dakota Sandstone. Usually the geologist stands at a distance to eliminate effect of small irregularities in the bedding. Colfax County, New Mexico. Photo: J. R. Stacy, U. S. Geological Survey.

plane or a curved surface. In cross sections, the axial plane is represented by a line.

In some folds the axial plane is vertical (Fig. 3-2A, B, and C), in others it is inclined (Fig. 3-2D and F), and in still others it is horizontal (Fig. 3-2E). Although in many folds the axial surface is a relatively smooth plane, it may be curved. The attitude of the axial plane is defined by its strike and dip, just as the attitude of a bed is defined. In Fig. 3-2, north is toward the upper left-hand corner. In Fig. 3-2A, B, and C, the axial plane strikes north and has a vertical dip. In Fig. 3-2D, the strike is north, the dip 45° to the west. In Fig. 3-2F, the axial plane strikes north and dips 60 degrees to the west; in

Plate 3. *Gently dipping strata.* Precambrian Eleanore Bay Group. Devil's Castle (4396 feet high), Kejser Franz Joseph Fjord, East Greenland. Photo: Lauge Koch Expedition.

Plate 4. *Dipping strata.* Alameda Parkway. Shows how to measure dip. Jefferson County, Colorado. Photo: J. R. Stacy, U. S. Geological Survey.

Fig. 3-2E, the axial plane is horizontal. If the axial plane is curved, the dip or strike—or both—may differ from place to place, as in the case of a curved bedding plane.

The *axis* is a line parallel to the hinges. It is that straight line moving parallel to itself that generates the fold. The term *fold axis* as thus used is an abstraction. The term axis has also been used as synonomous with hinge as defined above.

The sides of a fold are called the *limbs* or *flanks*. Terms used in the past, but now obsolete, were *legs*, *shanks*, *branches*, and *slopes*. A limb extends from the axial plane in one fold to the axial plane in the next. For example,

Plate 5. *Dipping strata.* Beds strike diagonally to cliff. Cliff is 4000 feet high. Rocks are quartzites and quartz schists of Precambrian Eleanore Bay Group. Kejser Franz Joseph Fjord, East Greenland. Photo: Lauge Koch Expedition.

41

Plate 6. *Flat strata.* Height of cliff is 5000 feet. Early Tertiary basalts (black) and interbedded sedimentary rocks (light) resting unconformably on gneiss. Gaaserfjord, East Greenland. Photo: Lauge Koch Expedition.

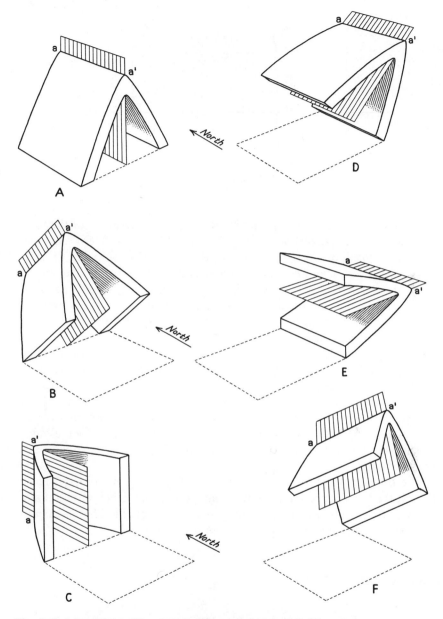

Fig. 3-2. A few of the different attitudes assumed by axial planes and hinges of folds. The axial plane is shaded in each diagram; *aa'* is hinge of fold.

in Fig. 3-3A, *a'b* is the limb of a fold. It may be considered either the east limb of the upfold or the west limb of the adjacent downfold. In other words, every limb is mutually shared by two adjacent folds.

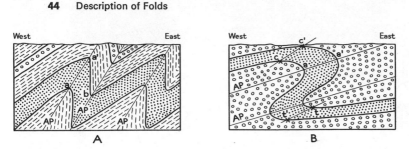

Fig. 3-3. Parts of a fold. *AP*, axial plane; *a'b*, limb of a fold; *c*, crest on one bed; *c'*, crest on another bed; *cc'*, crestal plane; *t*, trough on one bed; *t'*, trough on another bed; *tt'*, trough plane.

Although in many instances the hinge is at the highest part of the fold, as in Fig. 3-3A, this is not necessarily the case. In Fig. 3-3B, for example, *a* and *a'* are hinges, or, to be more precise, the intersection of hinges with the plane of the paper; *c* and *c'* are the highest points on the folds. The crest is a line along the highest part of the fold, or, more precisely, the line connecting the highest points on the same bed in an infinite number of cross sections. There is a separate crest for each bed. The plane or surface formed by all the crests is called the *crestal plane* (*cc'* of Fig. 3-3B).

In many phases of geology, the distinction between the crest and hinge is not important, either because the two correspond or, if there is a difference, because the distinction is of academic interest only. The same is true of the distinction between crestal plane and axial plane. In the accumulation of gas and petroleum, however, the difference is significant. The trapping of such materials is controlled by the crest and crestal plane rather than by the hinge and axial plane. In American oil fields, however, the crestal plane and axial plane are usually identical.

The *trough* is the line occupying the lowest part of the fold, or, more precisely, the line connecting the lowest parts on the same bed in an infinite number of cross sections. In Fig. 3-3B, *t* and *t'* are troughs. The plane connecting such lines may be called the *trough plane*.

Nomenclature of Folds

In general an *anticline* (Plates 7A and B and Plates 8A and B) may be defined as a fold that is convex upward; it may also be defined as a fold that has older rocks in the center. The word is from the Greek, meaning "opposite inclined," referring to the fact that in the simplest anticlines the two limbs dip away from each other (Fig. 3-4A). But the term has also been extended to folds such as that in Fig. 3-4B where the two limbs dip in the same direction at different angles. Normally this means that older rocks are in the center of the fold. Consequently

Plate 7A. *Anticline.* South Sunshine Anticline looking south across Gooseberry Creek. Park County, Wyoming. Photo: W. G. Pierce, U. S. Geological Survey.

45

Plate 7B. Overturned anticline. Panther Gap, Rockbridge County, Virginia. Photo: N. H. Darton U. S. Geological Survey.

Plate 8A. *Folds.* Gypsiferous shales. An anticline lies to the left of a syncline. Ventura County, California. Photo: H. S. Gale, U.S. Geological Survey.

Plate 8B. Overturned syncline. Lower Cambrian limestone, York, York County, Pennsylvania. Photo: C. D. Walcott, U. S. Geological Survey.

the term has been extended to any fold where older rocks are in the center (Fig. 3-4C). The definition of antiform is deferred to page 66.

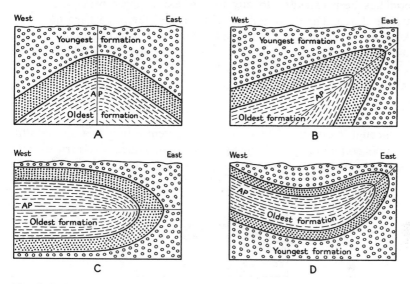

Fig. 3-4. Some varieties of anticlines. *AP*, axial planes.

In general a *syncline* (Plates 8A and B) may be defined as a fold that is convex downward. The word is from the Greek meaning "together inclined," referring to the fact that in the simplest synclines the two limbs dip toward each other (Fig. 3-5A). But the term has been extended to such folds as that in Fig.

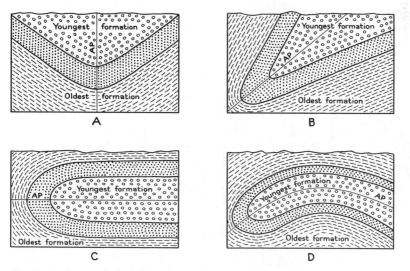

Fig. 3-5. Some varieties of synclines. *AP*, axial planes.

3-5B where the two limbs dip in the same direction at different angles. Normally this means that younger rocks are in the center of the fold. Consequently the term has been extended to any fold where younger rocks are in the center (Fig. 3-5C). The definition of synform is deferred to page 66.

A *symmetrical* fold is one in which the axial surface is essentially vertical (Fig. 3-6A); *upright* is also used. Conversely, an *asymmetrical* fold is defined as one in which the axial surface is inclined (Fig. 3-6B).

In the *overturned fold* or *overfold* (Plate 9) the axial plane is inclined, and both limbs dip in the same direction, usually at different angles (Fig. 3-6C). The *overturned, inverted,* or *reversed limb* is the one that has been rotated through more than 90° to attain its present attitude. A special dip-strike symbol is commonly used on modern maps to indicate overturned strata (Fig. 3-7). The *normal limb* is the one that is right-side-up.

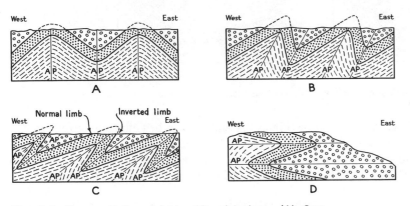

Fig. 3-6. Some varieties of folds. *AP*, axial plane. (A) Symmetrical (upright) folds. (B) Asymmetrical folds. (C) Overturned folds (overfolds). (D) Recumbent folds.

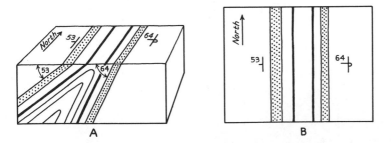

Fig. 3-7. Dip-strike symbol for overturned strata. (A) Block diagram. (B) Map. The dip-strike symbol with 53 beside it indicates beds that dip 53° to the west (left) and are not overturned. The dip-strike symbol with 64 beside it indicates beds that dip 64° to the west (left), but are overturned.

Plate 9. *Overturned syncline.* Exposure about 3000 feet high. Axial surface axis dips to right; plunge of axis is gentle to the left. Eleanore Bay Group. Blaabaerdal, Andrée Land, East Greenland. Photo: Lauge Koch Expedition.

51

A *recumbent fold* is one in which the axial plane is essentially horizontal (Fig. 3-6D). Large-scale folds of this type are especially well exposed in the Alps.[1] Consequently, a rather elaborate terminology has been evolved by European geologists to describe such folds (Fig. 3-8). The strata in the inverted limb are usually much thinner than the corresponding beds in the normal limb. The term *arch-bend* has been used for the curved part of the fold between the normal and inverted limbs. It is synonymous with hinge. Many of the recumbent folds in the Alps have Paleozoic crystalline rocks in the center and Mesozoic sedimentary rocks in the outer covering. Thus there is a distinct core of crystalline rocks within a shell of sedimentary rocks. Even in a recumbent fold composed entirely of one kind of rock, the terms *core* and *shell* may be used to refer, respectively, to the inner and outer parts of the fold. Many recumbent folds have subsidiary recumbent anticlines attached to them; these subsidiary folds may be called *digitations* because they look like great fingers extending from a hand. All recumbent folds, if satisfactory exposures are available, may be traced back to the *root* or *root zone*—that is, to the place on the surface of the earth from which they arise; in other words, recumbent folds may be traced to the place where the axial plane becomes much steeper.

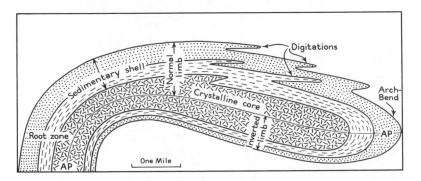

Fig. 3-8. Recumbent anticline with names of various parts.

An *isoclinal fold,* from the Greek meaning "equally inclined," refers to folds in which the two limbs dip at equal angles in the same direction (Fig. 3-9). A vertical isoclinal fold (Fig. 3-9A) is one in which the axial plane is vertical; an inclined or overturned isoclinal fold is one in which the axial plane is inclined (Fig. 3-9B). A recumbent isoclinal fold is one in which the axial plane is horizontal (Fig. 3-9C). Many recumbent folds are isoclinal.

A *chevron fold* is one in which the hinges are sharp and angular (Fig. 3-10A).

A *box fold* is one in which the crest is broad and flat; two hinges are present, one on either side of the flat crest (Fig. 3-10B).

A *fan fold* is one in which both limbs are overturned (Fig. 3-11A). In

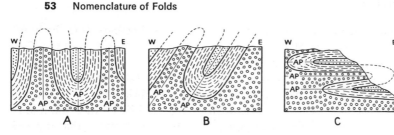

Fig. 3-9. Isoclinal folds. *AP*, Axial planes. (A) Vertical isoclinal folds. (B) Inclined isoclinal folds. (C) Recumbent isoclinal folds.

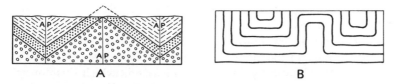

Fig. 3-10. Some varieties of folds. *AP*, axial plane. (A) Chevron fold. (B) Box fold.

the anticlinal fan fold, the two limbs dip toward each other; in the synclinal fan fold, the two limbs dip away from each other.

Kink bands (Fig. 3-11B) are narrow bands, usually only a few inches or few feet wide, in which the beds assume a dip that is steeper or gentler than that in the adjacent beds.

In plateau areas, where the bedding is relatively flat, the strata may locally assume a steeper dip (Fig. 3-12A). Such a fold is a *monocline*. The beds in a monocline may dip at angles ranging from a few degrees to 90° and the eleva-

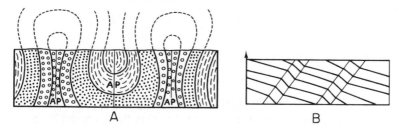

Fig. 3-11. Some varieties of folds. *AP*, axial plane. (A) Fan fold. (B) Kink bands. A fracture may separate the kink band from the rest of beds.

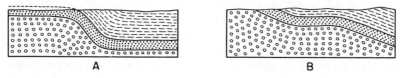

Fig. 3-12. Monocline and terrace. (A) Monocline. (B) Structural terrace.

tion of the same bed on opposite sides of the monocline may differ by hundreds or even thousands of feet.

The term *homocline*, from the Greek meaning "one inclination," may be applied to strata that dip in one direction at a relatively uniform angle. Although many homoclines are, if large areas are considered, limbs of folds, the term is useful to refer to the structure within the limits of a small area. But many geologists use the term monocline to refer to rocks that dip uniformly in one direction.

In areas where dipping strata locally assume a horizontal attitude, a *structural terrace* is formed (Fig. 3-12B). This usage should not be confused with that of the physiographer, who employs the term to refer to terraces that are structurally controlled.

A *closed* or *tight fold* is one in which the deformation has been sufficiently intense to cause flowage of the more mobile beds so that these beds thicken and thin (Fig. 3-13B). Conversely, an *open fold* is one in which this flowage has not taken place (Fig. 3-13A). Although the more extreme cases of these two types may be readily distinguished from each other, there are intermediate examples that are difficult to classify.

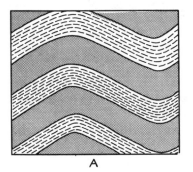 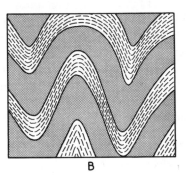

A   B

Fig. 3-13. Open and closed folds. (A) Open folds. (B) Closed folds.

Drag folds form when a competent ("strong") bed slides past an incompetent ("weak") bed (Fig. 3-14). Such minor folds may form on the limbs of larger folds because of the slipping of beds past each other, or they may develop beneath overthrust blocks (Chap. 10). The axial planes of the drag folds are not perpendicular to the bedding of the competent strata, but are inclined at an angle. Under a couple, of the type illustrated in Fig. 3-14, an imaginary circle in the incompetent bed would be deformed into an ellipse. The traces of the axial planes of the folds are parallel to the long axis of this ellipse. The acute angles between the axial planes and the main bedding plane point in the direction of the differential movement. Such structural features may form during sedimentation, when a sheet of sediment slides over a weaker bed.

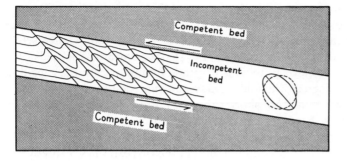

Fig. 3-14. Drag folds resulting from shearing of beds past each other.

The shape of folds may vary along the axial plane at right angles to the fold hinges. A theoretical approach sheds some light on the problem. Figure 3-15A illustrates *similar folding*. Line *a* is taken as the form of the fold shown by one bedding plane. The other lines have been drawn on the assumption that they have the same form as line *a*. In this way the form of the fold is propagated indefinitely upward and downward. Moreover, lines *b* and *c* have the same length as *a*. In this type of folding every bed is thinner on the limbs and thicker near the hinges. To produce folds of this type there is considerable plastic movement of material away from the limbs and toward the hinges. In natural folds the stronger or more competent beds preserve a relatively uniform thickness, but the weaker, less competent beds adjust themselves by flowage and drag folding.

Figure 3-15B illustrates *parallel folding (concentric folding)*. Line *a* is taken as the form of the fold shown by one bedding plane. The rest of the figure has been constructed on the assumption that the thickness of the beds

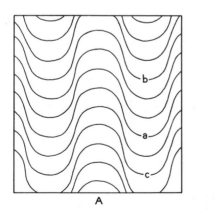
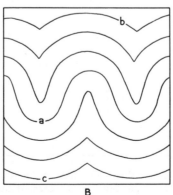

A   B

Fig. 3-15. Types of folding. (A) Similar folding. (B) Parallel folding.

has not changed during the folding. It is apparent that, under such conditions, the form of the fold must change upward and downward. The anticlines become sharper with depth, but broader and more open upward. Conversely, the synclines become broader with depth, but sharper upward. The folds die out downward and upward. In regions of gentle folding, where the dips do not exceed 10 or 20 degrees, the folding may well approach the parallel type. In Fig. 3-15B, lines b and c are shorter than a, but in the original basin of deposition, they must have had the same length as a.

The examples of similar and parallel folds in Fig. 3-15 are ideal and limiting cases. Most folds are a combination of the two extremes. Moreover, the axial planes shown in Fig. 3-15 are vertical, but, of course, may assume any attitude.

Where unusually good data are available, it is clear that most folding is disharmonic; that is, the form of the fold is not uniform throughout the stratigraphic column. Figure 3-16 shows a structure section in the Northern Anthracite Basin of Pennsylvania, based on data obtained from mines and drill holes.[2] A symmetrical anticline between two bore holes passes downward into an overturned anticline at 400 feet above sea level. At 400 feet below sea level the fold has disappeared. Disharmonic folding is well displayed on some of the great cliffs in the Alps. Figure 3-17 is one example, but inasmuch

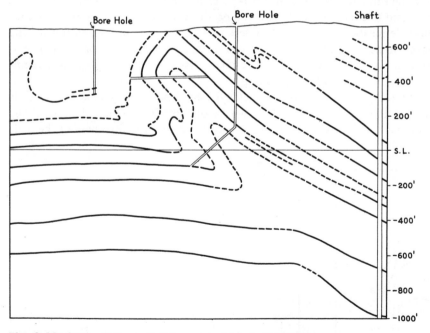

Fig. 3-16. Cross sections of disharmonic folds in the Northern Anthracite Basin of Pennsylvania. Solid lines represent beds of coal that have been mined out. Broken lines represent beds of coal based on drill records. (After N. H. Darton.[2])

as the folds are recumbent, the change of form takes place in the horizontal direction rather than with depth. The formation labeled *Jd* shows four recumbent synclines approximately equal to each other in size, whereas the top (north) of formation *Ja* shows folds of very different form.

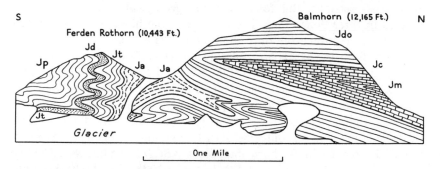

Fig. 3-17. Folds in the Ferden Rothorn and Balmhorn, Switzerland. Sketched by author from the Hockenhorn. The cliff on the east face of the Balmhorn is about 4000 feet high; that on the east face of the Ferden Rothorn is about 1600 feet high. *Jp*, Pliensbachian limestone; *Jd*, Domerian quartzite; *Jt*, Toarcian formation; *Ja*, Aalenian shale; *Jdo*, Dogger shale; *Jc*, Callovian shale; *Jm*, Malm limestone.

An extreme case of disharmonic folding is inferred for the Jura Mountains (Fig. 3-18). The Mesozoic and Tertiary strata are thrown into a series of anticlines and synclines that do not affect the underlying Paleozoic crystalline rocks. The weak shales and salt beds near the base of the Mesozoic served as a lubricant over which the higher strata slid, to form a *décollement*, that is, a "shearing off." Although most geologists agree with this interpretation, a few do not.

Piercing or diapir folds are anticlines in which a mobile core—rock salt in many cases—has broken through the more brittle overlying rocks. (See also Chap. 16.)

Fig. 3-18. *Décollement* of the Jura Mountains. The lowest formation, with nearly vertical structure, consists of Paleozoic crystalline rocks. Directly above these is a thin bed of flat-lying Triassic quartzite, left blank. The lower, solid black formation, which is very incompetent, consists of anhydrite, shale, and salt. The higher beds are Triassic, Jurassic, and Tertiary sedimentary rocks. (After A. Buxtorf.)

Plunge of Folds

In the preceding section, emphasis has been placed upon the appearance of folds in cross sections. But folds, like any geological structure, must be considered in three dimensions. The attitude of the hinge is of the greatest importance in describing the third dimension (Plate 10).[3,4]

In some folds the axis is horizontal (Plate 11 and Fig. 3-2A, D, and E); in other folds the hinge is inclined (Fig. 3-2B and F); in one (Fig. 3-2C) it is vertical. The attitude of the hinge of a fold is defined by two measurements: the *bearing* or *strike of its horizontal projection* and the *plunge*. It must be remembered that a hinge is a line, such as *FD* in Fig. 3-19. Of all the possible vertical planes in the figure, only one, *ADFG*, contains the line *FD*. The intersection of this plane with the horizontal plane *ABCD* is the line *AD*. The line *AD* is the horizontal projection of *FD*. In Fig. 3-19 the line *AD* bears northwest, and this is therefore the bearing of the horizontal projection of *FD*. The plunge of *FD* is the angle *P*, which is the angle between *AD* and *FD* measured in the vertical plane *ADFG*.

Although the larger plunging folds cannot be directly observed, they are easily recognized from their outcrop pattern. Figure 3-20 is a block diagram of a fold that does not plunge. On the map, the beds on the opposite

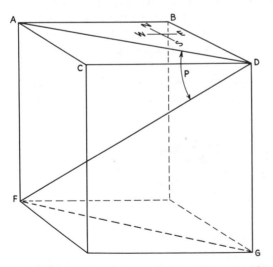

Fig. 3-19. Attitude of hinge of a fold. If *FD* is the hinge, *AD* is the bearing of the horizontal projection of the hinge; the angle of plunge is *P*.

Plate 10. *Plunging anticline.* Axial surface dips to right, hinge plunges to left. Park County, Wyoming. Photo: W. G. Pierce, U. S. Geological Survey.

Plate 11. *Minor folds.* Interbedded impure marbles and quartzofeldspathic schists of Lower Paleozoic Age. Godøstraumen, Saltfjord, Northern Norway. Photo: J. Haller.

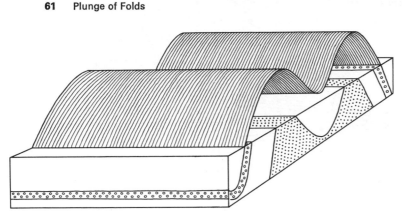

Fig. 3-20. Nonplunging folds.

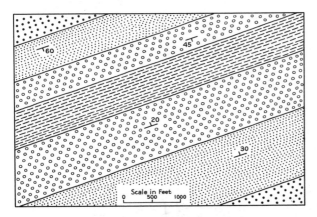

Fig. 3-21. Geological map of nonplunging fold.

limbs strike parallel to each other; they do not converge. Figure 3-21 is a geological map of a nonplunging syncline. Figure 3-22 is a block diagram of plunging folds and shows that on the map the beds converge; the formations have a zigzag pattern. Figures 3-23 and 3-24 are geological maps of plunging folds; the beds on the opposite limbs strike toward each other and the formations converge. Figure 3-23 shows a plunging anticline; Fig. 3-24 shows a plunging syncline. The *axial trace* of a fold connects the points where, on the map, each bed shows the maximum curvature (Figs. 3-23 and 3-24). For symmetrical folds or nonplunging folds, the axial trace and the horizontal projection of the hinge coincide, but this is not true if the axial plane is inclined and the fold plunges. A plunging fold is shown in Plate 10.

In the preceding paragraphs it has been tacitly assumed that the plunge is constant. In most instances, however, the value of the plunge changes along the bearing and the direction of plunge may even be reversed. Figure 3-25A

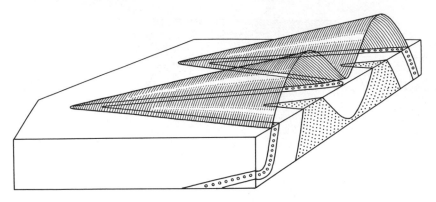

Fig. 3-22. Plunging folds. Plunge is about 10° to the left. One bed is shown by open circles; the part of this bed that has been removed by erosion is shown by lining.

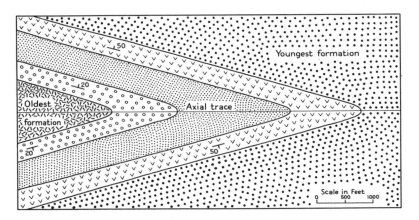

Fig. 3-23. Geological map of an anticline plunging east (to the right).

is a geological map of an anticline, the axial trace of which trends northeast. In the northeast corner of the map, the fold plunges 10 degrees to the northeast, and the strata converge northeastward. Toward the southwest, the value of the plunge decreases, and in the center of the map it is zero; here the strata on opposite limbs are parallel in strike. In the southwest corner, the anticline plunges 15 degrees to the southwest. Figure 3-25B shows a syncline that plunges southwest at the northeast corner of the map, and northeast at the southwest corner.

A *doubly plunging fold* is one that reverses its direction of plunge within the limits of the area under discussion. Most folds, if followed far enough, are doubly plunging.

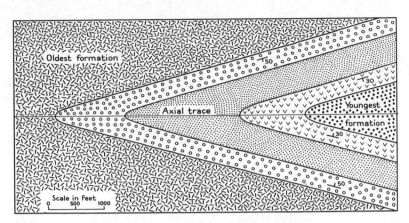

Fig. 3-24. Geological map of a syncline plunging east (to the right).

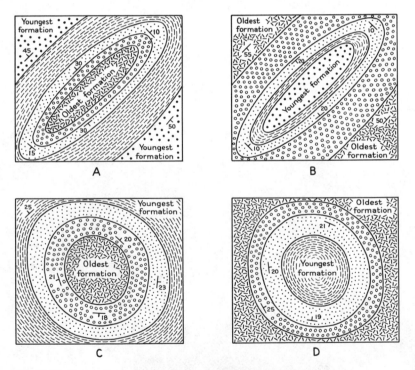

Fig. 3-25. Maps of plunging folds. (A) Doubly plunging anticline. (B) Doubly plunging syncline. (C) Dome. (D) Basin.

A *dome* is an anticlinal uplift that has no distinct trend (Fig. 3-25C). A *basin* is a synclinal depression that has no distinct trend (Fig. 3-25D).

Plate 12. *Refolded fold.* An older fold, the axial surface of which is horizontal on right side of photograph, has been refolded by a fold of which the axial surface dips steeply to the left. Loch Hourn, Scotland. Photo: J. Haller.

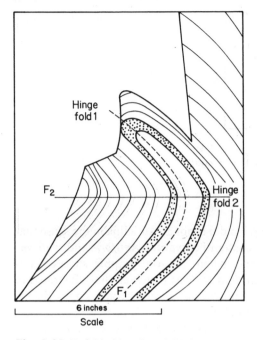

Fig. 3-26. Refolded isoclinal fold. (After G. E. Moore.[6])

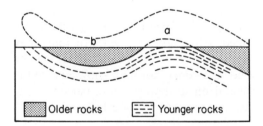

Fig. 3-27. Antiform and synform. *a*, Antiform; *b*, Synform.

Refolding

Folds may be refolded (Plate 12). One of many possible examples is illustrated in Fig. 3-26. A vertical isoclinal fold has been refolded by an open fold with a horizontal axial plane.[6]

Figure 3-27 illustrates the complexity that may occur in highly folded areas. Fold *a* appears to be an anticline, fold *b* appears to be a syncline. But it is possible, of course, that the surface of the earth here truncates the refolded inverted limb of a recumbent fold with the normal limb removed by erosion;

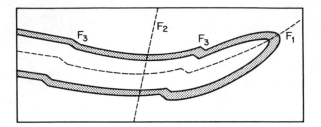

Fig. 3-28. Three stages of folding. Isoclinal recumbent fold, F_1, is oldest; broad open syncline, F_2, is intermediate in age; kink bands, F_3, are youngest.

that is, the shaded rock is the older. Fold *a* is not an anticline in the sense that older rocks are exposed toward the center of curvature. *Antiform* and *synform*[5] are often used as geometrical terms when the stratigraphic succession is unknown. The terms are also often used for cases such as that illustrated in Fig. 3-27, when the stratigraphy is known, but younger rocks appear in the center of what appears to be an anticline.

A *reclined fold* is one in which the axes plunge directly down the dip of the axial surface. Many reclined folds form when steeply dipping beds are subjected to shearing parallel to the strike of the beds.

In some areas the refolding may be deduced from the map pattern.[7] Figure 3-28 shows three stages of folding. F_1 is the axial surface of a recumbent fold. This was then refolded into an open synform (F_2) with gentle limbs. Finally, the kink bands (F_3) developed. Although a sequence of folds may be observed and deduced in the field, the dating of the various phases may be very difficult. Often the dating must be based on a thorough knowledge of the regional geology, which in turn is based on stratigraphy and paleontology. Theoretically radiogenic dating may be used, but to date little of this has been done in cases such as this.

Fold Systems

The preceding discussion has not been concerned with the structure of entire mountain ranges or large parts of continents. But folds are generally part of a larger system of folds, and this subject will be treated briefly here.

The wavelength of folds is the distance from one anticlinal hinge to the next anticlinal hinge (or between synclinal hinges). The wave lengths may range from a fraction of an inch to many miles. In many instances folds of various sizes are superimposed on one another. The amplitude of folds is the distance between a bed on an anticlinal hinge and the same bed in an adjacent synclinal hinge, the distance to be measured parallel to the axial surface and perpendicular to the fold axes.

A major anticline that is composed of many smaller folds is called an *anticlinorium*. The term is restricted to large folds that are at least several miles across. Similarly, a *synclinorium* is a large syncline composed of many smaller folds; it should be at least several miles across.

Geosyncline literally means an "earth syncline," but should not be used for large synclines. It is a large depression, hundreds of miles long and tens of miles wide, in which many thousands or tens of thousands of feet of sediments accumulate.[8, 9] Although the sediments deposited in the Appalachian geosyncline during the Paleozoic Era were 40,000 feet thick in places, the water was usually not very deep and at times deposition went on above sea level. The floor of the depression sank while the sediments were being deposited. In other geosynclines the water may have been deep at times. But the evolution of geosynclines has been so complex and varied that further generalizations here would be misleading.

A *geanticline* is a broad uplift, comparable in size to a geosyncline. It may lie either outside or inside a geosyncline.

The interior of the United States, between the Appalachian Highlands on the east and the Rocky Mountains on the west, is characterized by broad folds. The limbs of these folds have very low dips, rarely exceeding one or two degrees. The wavelengths are hundreds of miles and the amplitudes are many thousands of feet.

In orogenic belts, such as the Valley and Ridge Province of the Appalachian Highlands, the California Coast Range, and the Jura Mountains, the limbs of the folds dip more steeply, and may even be vertical or even overturned. The axial traces of the folds are generally parallel in any one part of the belt. But, over the folded belt as a whole the axial traces sweep in broad arcs or garlands. In Pennsylvania, the axial traces swing from almost north-south in the southern part of the state, through northeast-southwest, to nearly east-west in the northeast part of the state.

Figure 3-29 is a geological map of an area in which the axial traces are curved. Along the line *cc'* they are convex toward the north, and they diverge toward the east and west. In a *salient* the axial traces of the folds are convex toward the outer edge of the folded belt. Along the line *cc'* the folds are in a salient. In a *recess* the axial traces of the folds are concave toward the outer edge of the folded belt. Along the lines *dd'* the folds are in a recess.

In many areas all the folds plunge in the same direction. In Fig. 3-30 the folds between *cc'* and *ee'* plunge toward the east. East of *ee'* the folds plunge to the west. West of *cc'* but east of *dd'* the folds plunge west. West of *dd'* the folds plunge to the east. The line *cc'* is on a *culmination*, whereas *dd'* and *ee'* are on *depressions*. The folds plunge away from a culmination and toward depressions.

Salient and recess refer to the plan. Culmination and depression refer to

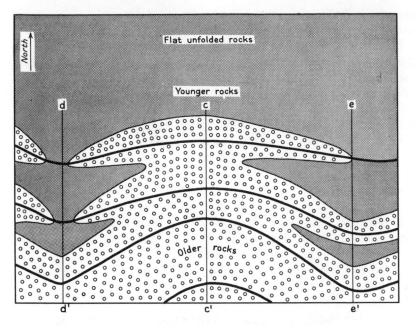

Fig. 3-29. Convergence and divergence of axial traces of folds. The axial traces (axes) are shown as heavy black lines. These lines diverge from one another on either side of *cc'*, and converge toward *cc'* if approached from *dd'* or *ee'*.

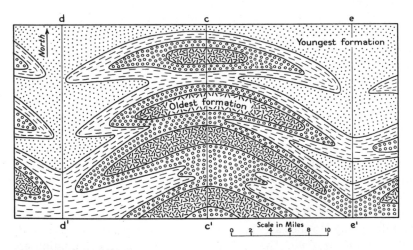

Fig. 3-30. Culmination and depression. The line *cc'* is a culmination, away from which the folds plunge; *dd'* and *ee'* are depressions, toward which the folds plunge.

the plunge in longitudinal section. Salients generally correspond to culminations. Along a line of maximum compression at right angles to the fold belt the folds are pushed forward the greatest amount and the anticlines are uplifted the most.

In some localities individual folds do not extend any great distance, but overlap one another, *en échelon*, as shown in Fig. 3-31.

An entire orogenic belt may show a sharp bend, which is called an orocline.[10] Some believe that these oroclines result from the bending of the whole folded belt.

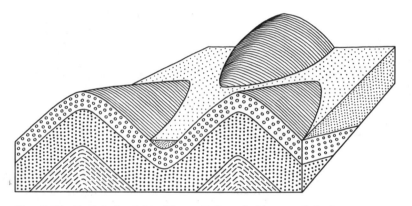

Fig. 3-31. *En échelon* folds. The anticlines hold up anticlinal mountains rising above a flat plain. The axial trace of the fold in the background is *en échelon* relative to the folds in the foreground.

References

[1] Collet, L. W., 1935, *The structure of the Alps*, 2d ed., 289 pp., London: Edward Arnold and Company.

[2] Darton, N. H., 1940, *Some structural features of the Northern Anthracite Coal Basin, Pennsylvania*, U.S. Geol. Survey, Prof. Paper 193, pp. 69–81.

[3] Mackin, J. Hoover, 1950, The down-structure method of viewing geologic maps, *J. Geol.* 58: 58–72.

[4] Stockwell, C. H., 1950, The use of plunge in the construction of cross-sections of folds, *Proc. Geol. Assoc. Canada*, 3: 97–121.

[5] Bailey, E. B., 1934, West Highland tectonics: Loch Leven to Glen Roy, *Quart. J., Geol. Soc. London*, 90: 462–525.

[6] Moore, George E., 1949 Structure and metamorphism of the Keene-Brattleboro area, New Hampshire-Vermont, *Geol. Soc. Amer. Bull.* 60: 1613–70.

[7] Stauffer, Mel R., 1968, The tracing of hinge-line ore bodies in areas of repeated folding, *Can. J. Earth Sci.* 5: 69–79.

[8] Kay, Marshall, 1951, North American geosynclines, *Geol. Soc. Amer. Memoir* 48, 143 pp.

[9] Murray, Grover E., et al., 1952 Sedimentary volumes in Coastal Plain of United States and Mexico, *Geol. Soc. Amer. Bull.* 63: 1157–1228.

[10] Carey, S. Warren, 1955, The orocline concept in geotectonics, part I, *Roy. Soc. Tasmania, Papers and Proceedings* 89: 255–88.

4

FIELD STUDY
OF FOLDS

Recognition of Folds

DIRECT OBSERVATION

Folds may be recognized in many ways. The easiest and most satisfactory method is to observe the fold, but this can be done in comparatively few regions. Folds may be readily seen on some of the great cliffs in the Alps (Fig. 3-17). Folds are exposed in some parts of the Appalachian Mountains in large highway cuts (Fig. 4-1), railroad cuts, or in natural exposures (Fig. 4-2). Folds may be observed also in the Rocky Mountains and in other parts of the North American Cordillera, as well as in many other places in the world. Far more commonly, however, folds must be deduced from other data, and detailed studies show that most visible folds, even in such places as the Alps, are relatively minor features associated with much larger folds.

Wherever small folds are observed in single outcrops, it is desirable to

SE   NW

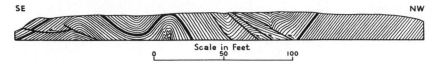

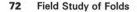

Scale in Feet

0   50   100

Fig. 4-1. Folds formerly exposed in highway cut one mile south of Alexandria, Pennsylvania. Strata belong to the Wills Creek formation, which is of Upper Silurian age. The bed represented by solid black is a red calcareous shale.

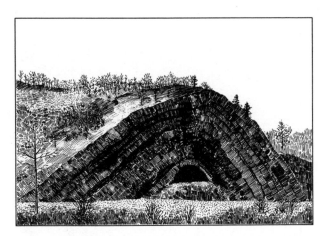

Fig. 4-2. Anticline one-quarter mile east of Roundtop, Maryland. View looking across Potomac River from West Virginia. Strata belong to Upper Silurian Bloomsburg formation.

record their attitude. To do this one must measure the attitude of the hinges and the attitude of the axial planes.

The attitude of the hinge of a fold may be readily measured if it is possible to look down on the fold. One edge of the compass is held in such a way as to cover the hinge of the fold. In this way the bearing of the horizontal projection of the hinge is obtained (see also Fig. 3-19). The plunge is measured by means of the clinometer attached to the compass. The attitude of the hinge can then be recorded by an arrow. The symbols are given in the upper row of Fig. 4-3. Figure 4-3(a) means that the hinge of the fold plunges west at an angle of 40 degrees; the semicircle convex toward the arrowhead means the fold is an anticline. Figure 4-3(b) means that the hinge of the fold plunges northwest at an angle of 50 degrees; the semicircle concave toward the arrowhead means that the fold is a syncline. Figure 4-3(c) means that the hinge of an anticline bears northeast-southwest and is horizontal. Figure 4-3(d) indicates the hinge of a syncline that bears east-west and is horizontal. The symbol in Fig. 4-3 (e) represents a vertical hinge.

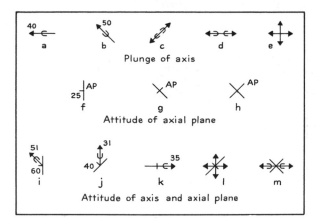

Fig. 4-3. Symbols used to represent attitude of folds.

A notebook or sheet of paper may be held parallel to the axial plane of a fold in order to measure its attitude. A second person measures the strike and dip of the notebook, just as the attitude of a bedding plane is measured. An experienced field geologist can measure the attitude of the axial plane without the help of a second person. Symbols similar to those for bedding may be used, except that the letters *AP* are placed beside the symbol. Figure 4-3(f) indicates an axial plane striking north and dipping 25°W. Figure 4-3(g) represents a vertical axial plane striking northwest. Figure 4-3(h) is a horizontal axial plane.

Normally, of course, the symbols for axial plane and hinge will be combined to form the symbols shown in the lowest row of Fig. 4-3. Figure 4-3(i) indicates an anticline plunging 51°NNW, with the axial plane striking north and dipping 60°W. It is not necessary to add the letters *AP* because the symbol for the hinge shows that the entire symbol refers to a fold. Figure 4-3(j) is a syncline plunging 31°N, with the axial plane striking northeast and dipping 40°NW. Figure 4-3(k) is a fold plunging 35°E and a vertical axial plane striking east. Figure 4-3(l) is a vertically plunging fold, the vertical axial plane of which strikes northeast. Figure 4-3(m) is the symbol for a syncline with a horizontal axial plane and a horizontal hinge that bears east-west.

The pattern shown by the minor folds in plan should also be recorded, as it is of great use, in conjunction with the plunge of folds, in deducing larger structures. The patterns (Fig. 4-4) may be called left-handed, right-handed, or neutral. The pattern is left-handed if one turns to the left to stay in the same bed. The pattern is right-handed if one turns to the right. It is neutral if one turns neither to the left nor to the right. These terms should not be used for observations made on vertical faces or projections into the vertical face, because on opposite ends of the same outcrop the appearance is reversed. The term shear-sense may be used for the appearance in vertical sections (Fig. 4-5); the shear-sense is southeast-side-up (A), northwest-side-up (B) or noncommittal (C); of course, these terms are used in a relative sense.

A

Left-handed

B

Right-handed

C

Neutral

Plans

Fig. 4-4. Terminology for pattern of minor folds in plan. Terms S, Z, and M folder are also used for patterns A, B, and C respectively.

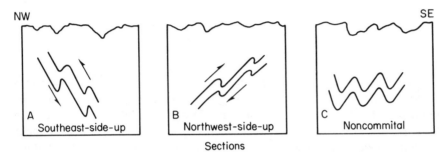

NW   SE

A
Southeast-side-up

B
Northwest-side-up

C
Noncommital

Sections

Fig. 4-5. Terminology for pattern of minor folds in cross section, showing "shear sense."

In order to visualize the three-dimensional attitude of folds, take a sheet of paper and bend it back on itself to make an isoclinal fold. The crease is the hinge of the fold, and the plane of the paper not only represents the two limbs of the fold, but the axial plane as well. The fold may then be held in any position desired to show the distinction between the attitude of the axial plane and the attitude of the hinge.

In a larger fold that cannot be observed in a single outcrop or in a series of closely adjacent outcrops, the attitude of the axial plane and the hinge cannot be measured directly. Nevertheless the same principles apply, and the geologist should always think of folds in three dimensions. Methods of calculating the attitude of the hinge are discussed on pages 100–108, 574.

INFERRED FOLDS

Folds larger than an outcrop are based on inference. Moreover, the part of the fold that was above the present surface of the earth has been removed by erosion. One or more of the following pieces of information are commonly used to deduce folds: (1) differences in attitude of some planar feature at different localities; (2) areal map pattern; (3) subsurface exploration by drilling, mining, and tunneling; and (4) subsurface studies by geophysical methods. Drag folds, pi diagrams and beta diagrams may also be employed.

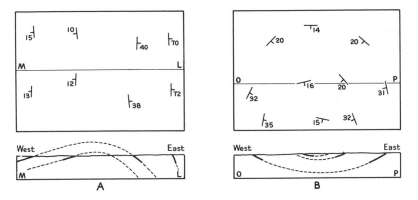

Fig. 4-6. Fold shown by dip-strike symbols. Maps are above, structure sections are below. (A) Asymmetric, nonplunging anticline. (B) Symmetrical (upright) syncline plunging south.

Inference concerning the geological structure should be made as field work progresses. Constantly changing working hypotheses can thus be tested. Nothing is intellectually more sterile than to gather data for future analysis without considering the significance of those data. Conversely, the field geologist should never expect to get the final answer immediately, and should not become too firmly attached to any one interpretation in the early stages of his investigation.

PLOTTING ATTITUDE OF BEDS ON A MAP

Bedding is the planar feature that is most commonly employed, but in some instances the geologist may be concerned with the folds shown by deformed foliation, flow banding in igneous rocks, folded dikes, etc.

The common procedure is to plot on a map or aerial photograph the strike and dip of the bedding, each observation being placed in its proper geographical position. Figure 4-6A is the map of an anticline, the axial surface of which strikes north and dips to the west. The hinge bears north and the plunge is zero. Figure 4-6B is the map of a syncline, the hinge of which plunges 15°S.

The attitude of beds may be determined qualitatively and even quantitatively from a geologic map that has topographic contours. If the contact between two formations is rigorously parallel to the contours, the strata are horizontal (Fig. 4-7A). If, regardless of the topography, a contact maintains a uniform trend, the strata are vertical (Fig. 4-7B). Dipping strata have an outcrop pattern that is partially controlled by the contours. The strike and dip of the beds may be calculated, as is shown by Fig. 4-8. The southern

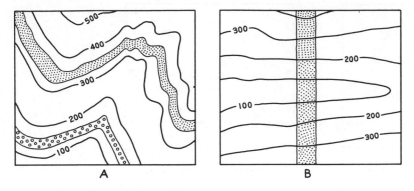

Fig. 4-7. Relation of outcrop pattern to topography. Every hundred-foot topographic contour is shown; special beds are shown by dots and circles. (A) Horizontal beds. (B) Vertical beds.

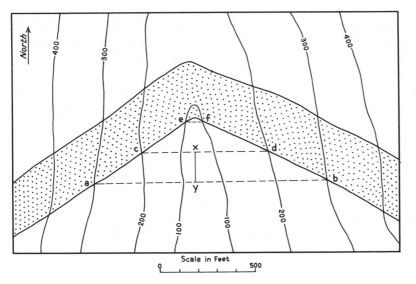

Fig. 4-8. Relation to topography of outcrop pattern of a dipping bed.

boundary of the stippled bed crosses the 300-foot contour at *a* and *b* on opposite sides of the valley. Inasmuch as those two points are at the same altitude, line *ab* is horizontal; moreover, *ab* lies in the plane of the bedding. It is apparent, therefore, that *ab* gives the strike of the bedding; in this instance it is east-west. Similarly, lines *cd* and *ef* are parallel to the strike. The dip may also be readily determined. Line *xy* is perpendicular to *ab* and *cd*, and is 160 feet long; that is, the bed drops 100 feet vertically in 160 feet horizontally.

The dip is found by the equation

$$\tan \delta = \frac{100}{160} \qquad (1)$$

and

$$\delta = 32°$$

AREAL MAP PATTERN

The areal map pattern refers to the pattern that results from mapping the different lithologic units. This is based, of course, on the data obtained at individual outcrops. However, the map pattern by itself is not sufficient; the topography must be known and additional stratigraphic or structural data must be available. Thus the structure illustrated by Fig. 3-21, without the strike-dip symbols, could be either an anticline, a syncline, or a homocline dipping northwest or southeast. Moreover, the beds could be flat-lying with the topographic surface sloping either northwest or southeast. Similarly, in Fig. 4-9 various alternate interpretations are possible if the strikes and dips, relative ages, and topography are unknown. But with a few additional pieces of information the map pattern may be exceedingly useful.

Topography is often useful in the study of folds. In heavily forested or deeply weathered regions, it may be possible to trace key units for long distances by means of the topography. A resistant formation will stand up in ridges, an easily eroded bed will be followed by valleys, and a limestone may be traced by karst topography. In reconnaissance studies, particularly by airplane, the topography may give important clues to the geological structure.

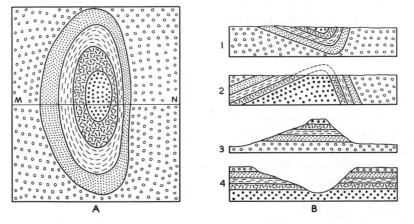

Fig. 4-9. Geological map with alternate interpretations of structure. (A) Geological map. (B) Alternate interpretations.

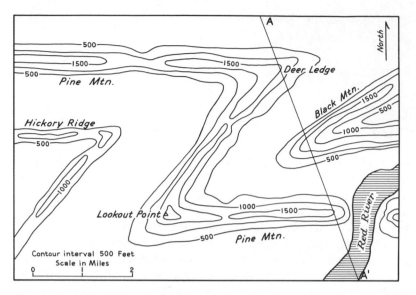

Fig. 4-10. Zigzag ridges that indicate plunging folds.

Figure 4-10 is a topographic map of an area in which ridges rise more than 1500 feet above sea level. The ridges are heavily wooded, and exposures are scanty. But a good section is exposed at the water gap where the Red River cuts through the ridge. At this locality, hard quartzites dipping 45° to the north are right-side-up (see page 81). Apparently this same quartzite holds up Pine Mountain throughout its extent. The zigzag pattern of the ridge indicates that the strata are folded.

The observations at the water gap reveal that this portion of the ridge is on the south limb of a syncline. The hinge of the syncline must lie at Lookout Point, where the ridge makes a sharp bend. In plan, the ridge is concave toward the northeast; the syncline, therefore, plunges to the northeast because the pattern of a plunging syncline is concave in the direction of plunge (Fig. 3-24).

Another hinge must be located at Deer Ledge. This is an anticlinal hinge, and it must plunge to the northeast because the pattern of the plunging anticline is convex in the direction of plunge.

It follows that Black Mountain is held up by some resistant formation stratigraphically higher than the quartzite on Pine Mountain. Hickory Ridge is held up by a formation stratigraphically lower than the quartzite on Pine Mountain. The inferred structure is shown in Fig. 4-11. The value of the dips at Deer Ledge and Black Mountain are necessarily schematic unless actually observed in the field.

The careful field geologist would consider such an interpretation suggestive, and would feel compelled to visit as many outcrops as time permitted.

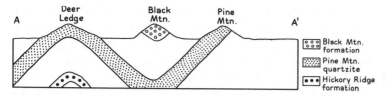

Fig. 4-11. Cross section along line *AA'* of Fig. 4-10.

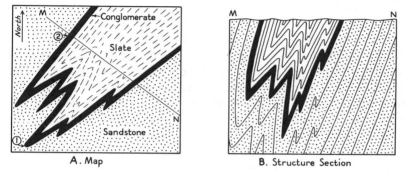

Fig. 4-12. Fold shown by pattern of formations on a map. (A) Geological map. (B) Inferred structure section.

In areas of complex geology the pattern shown by a key bed (key horizon, marker bed) may be exceedingly important. Figure 4-12A is the geological map of a region where the actual outcrops might constitute 25 percent of the area. In practically all of the exposures the strata strike northeast and dip steeply to the northwest. A traverse across the region from *M* to *N* suggests that there are five stratigraphic units: two sandstones, two conglomerates, and one slate. The map shows, however, that the two sandstones at opposite ends of *MN* belong to the same formation. Moreover, the map reveals that there is only one conglomerate. The pattern is that of a plunging fold— either an anticline that plunges southwest, or a syncline that plunges to the northeast. Without some additional data, a solution would not be possible. At some such place as locality 1, the nose of the fold might be exposed. By following the base of the conglomerate, one might be able to ascertain that at locality 1 it plunges 60° to the northeast. If this were so, the structure is synclinal. Primary features (pp. 81–90) might solve the problem. If cross-bedding in the conglomerate at locality 2 shows the top to be to the southeast, the fold is a syncline plunging northeast. Fossils might give the answer to the problems; if fossils in the slate are younger than those in the sandstone, the fold is a syncline plunging northeast.

A cross section along the line *MN* of Fig. 4-12A is given in Fig. 4-12B.

The section must satisfy the condition that the beds dip steeply to the north-west. Moreover, although no direct evidence for subsidiary folds may be obtained along the line *MN*, it is obvious from the map that such folds must exist, and, assuming harmonic folding, the number and location of these folds can be predicted from the map.[1] Every anticline shown by the map must also appear in the structure section. The depth of each anticlinal crest and synclinal trough depends upon the value of the plunge. In this case the plunge is about 60 degrees to the north-northeast.

In subsurface studies, notably in petroleum geology and mining geology, every effort is made to identify one or more key beds. The key bed or beds may be identified by lithology, fossils, or distinctive minerals.

In many metamorphic areas, such as New England[2] or the Highlands of Scotland[3], the pattern of one or more key beds, along with knowledge of the direction of plunge, is indispensible in deducing the structure. Over large areas the bedding may be vertical and thus suggest a homocline. But the complex pattern of the key bed may indicate a complicated structure. Flat bedding may suggest a simple structure, but the complex pattern of the key bed may demonstrate the presence of isoclinal recumbent folds. The correct structural interpretation must be consistent with the attitude of the bedding and minor folds, but, paradoxical as it may seem, these features by themselves may not give the solution.

The areal map pattern may be very conspicuous on aerial photographs, especially in regions of scanty vegetation. Moreover, from the relation of contacts to topography (Figs. 4-7 and 4-8) it may be possible to calculate the dip and strike of the strata. Thus, with some field checking, it is possible to map very rapidly large areas of simple structure. Photogeologic maps of the Colorado Plateau by the United States Geological Survey are excellent examples of such maps. In recent years infrared photography from planes has proved very valuable.

DRILLING

Where exposures are rare or absent, the structure may be deduced from drilling. Moreover, vast amounts of information on the subsurface geology have been obtained by drilling in those areas with sufficient economic incentive. Data obtained in the search for and exploitation of petroleum resources cover large areas. Conversely, the areas involved in the search for other types of mineral deposits, notably metallic ores, tend to be very restricted.

If some bed is sufficiently distinctive, either because of lithology or because of fossil content, its altitude in various drill holes can be recorded and the structure determined. If drill cores are obtained, the angle of dip of the bedding can be determined, but not the strike.[4] The more complex the structure, the greater should be the number of drill holes per unit area. This method is expensive, however, and has been used only where the possibility of financial return has justified the cost. A thorough discussion of the use of drill records is given elsewhere.[5, 6, 7]

MINING

Mining operations give the most complete information concerning underground geological structure. Coal mining, especially, furnishes valuable data because individual beds are followed for long distances. The Anthracite Basins of Pennsylvania are unusually well known.[8, 9] A cross section from the Northern Anthracite Basin is reproduced in Fig. 3-16. It is obvious that this method can be used only where there is economic incentive. The structure of bodies exploited in metal mines and nonmetal mines is exceptionally well known.[10]

GEOPHYSICAL METHODS

During the last five decades, under the impetus of the exploration for petroleum and metals, various geophysical methods have been utilized to determine geological structure. The principal methods may be classified as *gravimetric, magnetic, seismic,* and *electrical.* These methods will be discussed in Chaps. 22 and 23.

Determination of Top of Beds by Primary Features

NATURE OF THE PROBLEM

The features developed in sedimentary rocks during their deposition is of great interest to geologists.[11] Much information may be derived concerning the source of the sediments and, if marine, the depth at which they were deposited. The direction in which sediments were being transported may also be determined. Obviously such studies are of inestimable value in analyzing the larger tectonic movements of the crust, but this subject is beyond the scope of this book. However, many of these primary sedimentary features may be confused with secondary features produced during deformation, and are thus of great concern to the structural geologist.

In overturned folds and in recumbent folds the strata on one limb are overturned. Obviously, it would greatly facilitate the solution of structural problems if methods were available to determine whether the beds are right-side-up or overturned.[12] Where the folds are exposed on the face of a great cliff, the whole structure may be clearly observed, and special methods are unnecessary. In some instances, even in regions of low relief, the exposures may be sufficiently continuous to show a progressive change from beds that

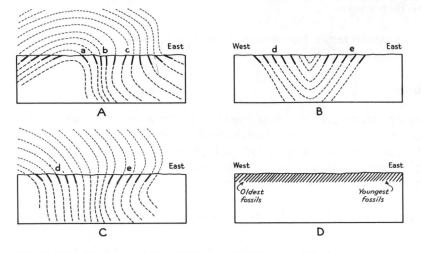

Fig. 4-13. Importance of determining top and bottom of beds. (A) The gradual change from beds that are right-side-up at *a* through vertical beds at *b* to overturned beds at *c* may be observed. (B) Beds at *d* and *e* dip toward each other, suggesting a syncline. (C) If beds at *e* are overturned, the structure cannot be a simple syncline such as shown in diagram (B). (D) Beds dip west; fossils indicate that the strata are overturned.

are right-side-up to those that are overturned. Figure 4-13A illustrates such a case. At *a* the beds are in normal position, but toward the east they become progressively steeper and are vertical at *b*; still farther east, as at *c*, they dip to the west and must be overturned. In many areas, however, exposures are not sufficiently numerous to show the gradual change. In Fig. 4-13B for example, the beds at *d* and *e* dip toward each other, and at first it might be supposed that the structure is a simple syncline. If, however, the strata at *e* are overturned, such an interpretation is impossible, and the structure would be that shown in Fig. 4-13C.

In recent years the words "top," "young," and "face" have been used as verbs, so that geologists talk of the direction in which the beds *top, young,* or *face. Downward facing* beds are overturned beds. In areas where such information is pertinent, special symbols can be added to show the direction in which the beds face and the kind of information used.

PALEONTOLOGICAL METHODS

Paleontological methods,[13] of course, may be of great aid in indicating whether beds are right-side-up or not. In Fig. 4-13D the beds dip 48 degrees to the west. If the youngest fossils are at the east end of the section, it is apparent that the strata are overturned.

Plate 13. *Ripple marks.* Dakota Sandstone. Jefferson County, Colorado. Photo: J. R. Stacy, U. S. Geological Survey.

RIPPLE MARKS

Ripple marks (Plate 13) may be aqueous or eolian in origin; that is, ripple marks may form on the bottom of a body of water or, by wind action, at the surface of the earth. The origin and formation of ripple marks is a subject that cannot be fully discussed here, and only those phases of the subject that are significant to the structural geologist will be considered.

Oscillation ripples, as shown in Fig. 4-14A, are symmetrical, and they consist of broad troughs that are convex downward and of sharp crests that point upward. Ideally, oscillation ripples form in bodies of standing water.

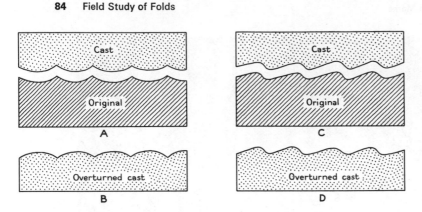

Fig. 4-14. Ripple marks. (A) Original oscillation ripple marks and the cast. (B) Overturned cast of oscillation ripple marks. (C) Original current ripple marks and the cast. (D) Overturned cast of current ripple marks.

Whenever waves disturb the upper surface of the body of water, the individual water particles move in vertical orbits that are nearly circular. Although the wave form moves across the water, the individual particles do not. The motion of the particles is transmitted downward with decreasing intensity. The sand or mud on the bottom is affected by the same motion and is thrown into ripples.

Current ripples, as shown in Fig. 4-14C, are asymmetrical, and both the crest and trough are rounded. Such ripples develop when a current, either of water or of air, moves across sand or mud. In Fig. 4-14C, the current moved from left to right.

Forms transitional between oscillation ripples and current ripples are not uncommon; although these forms are asymmetrical, they have sharp crests that point upward. There are other forms of ripple marks, but they do not concern the structural geologist.

Either the original ripple mark itself or its cast may be preserved. In Fig. 4-14A original oscillation ripple marks are represented on the lower block; the upper block is the cast. Figure 4-14B shows the cast after it has been removed from the original and has been turned over. In Fig. 4-14C, original current ripple marks are represented on the lower block; the upper block is the cast. Figure 4-14D shows the cast after it has been removed from the original and has been turned over.

Oscillation ripple marks can readily be used to tell whether a bed is right-side-up or overturned. The sharp crest points toward the younger beds, whereas the rounded trough is convex toward the older beds. This is true whether the specimen is an original or a cast. In Fig. 4-15 the beds at outcrop *I* dip to the west at an angle of 30°. At *a* the originals of ripple marks are preserved; the crests point upward to the left, indicating that the beds are

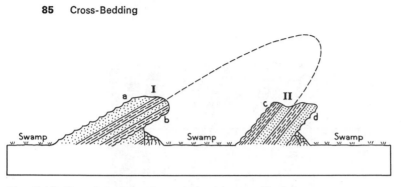

Fig. 4-15. Use of ripple marks to determine top of beds.

right-side-up. At *b* there is an overhanging cliff on which the casts of ripple marks are preserved; here, also, the crests point upward to the left, confirming the conclusion that the beds are right-side-up. At outcrop *II*, the beds dip 50 degrees to the west. On the sloping surface of the outcrop at *c*, the casts of ripple marks are preserved; the crests point downward to the right, indicating that the beds are overturned. On the face of the overhanging cliff at *d*, the originals are preserved, and again the crests point downward to the right. The inferred structure is indicated by a broken line.

A brief consideration of Fig. 4-14 shows that current ripples cannot be used to determine top from bottom. An overturned current ripple has the same form as one that is right-side-up.

CROSS-BEDDING

Cross-bedding, (Plate 14) which is also known as *cross-stratification*, *cross-lamination*, or *false-bedding*, is illustrated by Fig. 4-16A. Whereas the true bedding in this figure is horizontal, the cross-bedding is inclined at varying angles. Cross-bedding develops wherever sand has dropped over the edge of a growing sand bar, over the front of a sand dune, or over the edge of a small delta.[14] The upper extremity of each cross-bed is commonly inclined at a considerable angle to the true bedding, whereas the lower extremity is essentially parallel to the true bedding. The cross-beds thus are sharply truncated above and are tangential to the true bedding below. The cross-beds in *planar cross-bedding* are inclined to the true bedding at a considerable angle at both their upper and lower extremities (Fig. 4-16B).

The use of cross-bedding to distinguish the top from the bottom of beds is apparent from Fig. 4-16A. The cross-beds are tangential downward, but are truncated upward. Thus in Fig. 4-17A, the beds, which dip 45 degrees to the left, are right-side-up. The top of the vertical beds in Fig. 4-17B is to the right. The beds in Fig. 4-17C, which dip 45 degrees to the left, are overturned. In order to use cross-bedding properly, the tangential portion must be observed; it is insufficient to observe that the cross-beds are sharply truncated. If the

Plate 14. *Cross-bedding.* Coarse fluviatile sands in Glenns Ferry Formation (Late Pliocene), 1½ miles east of Hammett. Note pencil lower left-hand corner for scale. Elmore County, Idaho. Photo: H. E. Malde, U. S. Geological Survey.

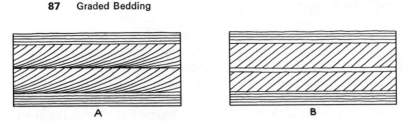

Fig. 4-16. Cross-bedding. (A) Normal cross-bedding. (B) Planar cross-bedding.

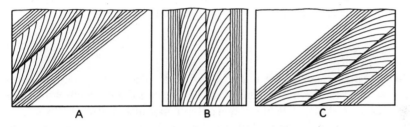

Fig. 4-17. Use of cross-bedding to determine attitude of beds. (A) Beds are right-side-up. (B) Top is to the right. (C) Beds are overturned.

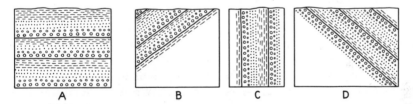

Fig. 4-18. Graded bedding. (A) As deposited in horizontal beds; each bed gets finer toward the top. (B) Dipping beds, right-side-up. (C) Vertical beds, top toward the right. (D) Dipping beds, overturned.

beds in Fig. 4-16B were folded to a vertical position, the cross-bedding could not be used to tell top and bottom.

GRADED BEDDING

In many instances the grains in a thin bed are progressively finer from bottom to top (Fig. 4-18A). This feature is known as *graded bedding*.[15] The materials comprising a sediment are transported when the currents are swifter than usual. As the velocity subsides, the largest particles are dropped first, and then progressively finer particles are deposited. *Turbidity currents*, loaded with sediments, may sweep into a sedimentary basin. When the current comes to rest, the larger particles settle first; the consolidated rocks resulting from this process are known as *turbidites*. Field experience shows that the grading may be reversed, especially among such coarse sedimentary rocks as conglomerate. On the other hand, in the finer sedimentary rocks—notably shales and siltstones, where the individual beds are a fraction of an inch thick—the method is more reliable, but by no means infallible. Pyroclastic volcanic rocks (page 308) may show graded bedding.

If this method is applied to the examples shown in Fig. 4-18, the beds in Fig. 4-18B are right-side-up; the tops of the beds in Fig. 4-18C are to the right; and the beds in Fig. 4-18D are overturned.

SOLE MARKINGS

Sole markings are casts on the underside of beds. The casts, generally composed of sandstone, are more readily preserved than the original, which are usually made in shale. The term hieroglyph has also been used for any markings on bedding planes, but has generally been restricted to sole markings.[16] The principal sole markings used in determining top and bottom of beds are load casts, groove casts, and flute casts.

LOCAL UNCONFORMITIES, CHANNELING, AND RELATED FEATURES

During the accumulation of sediments, particularly those laid down by rivers, erosion may alternate with deposition. In Fig. 4-19A, for example, conglomerate occupies a channel in shale. After the original mud had been deposited, swiftly flowing streams in flood carved a channel. When the flood was subsiding, or at some later time, gravel was deposited in the channel. The base of the conglomerate truncates the bedding of the shale.

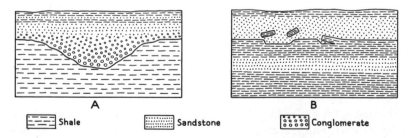

A

B

⫶⫶⫶⫶ Shale   ⫶⫶⫶⫶ Sandstone   ⫶⫶⫶⫶ Conglomerate

Fig. 4-19. Channeling and local unconformity. (A) Channel cut into shale has been filled by conglomerate. (B) Fragments of shale deposited in overlying sandstone.

A related feature is illustrated by Fig. 4-19B where sandstone lies on top of shale. The currents that transported the sand ripped up pieces of mud, fragments of which are preserved as shale in the sandstone. Similarly, fragments of lava may be found in the sedimentary rocks directly above a lava flow.

Features that are the results of short intervals of erosion during a period of sedimentation are known as local unconformities (Fig. 4-19A).

In Fig. 4-20 the beds dip to the west. By following the contact of the conglomerate and shale at *a*, it becomes obvious that the conglomerate truncates the shale and fills a channel in it. Also, at *b*, the sandstone contains pebbles of the lava to the west, and the sandstone at *c* contains fragments of the shale directly west of it. All the evidence thus indicates that the beds are overturned.

PILLOW STRUCTURE

Some lavas, particularly those of basaltic composition, are characterized by pillow structure[17, 18] (Fig. 4-21A). The individual pillows are roughly ellipsoidal, and they range from a foot to several feet in diameter. In flat lavas, the tops and bottoms of the pillows are generally convex upward. This method

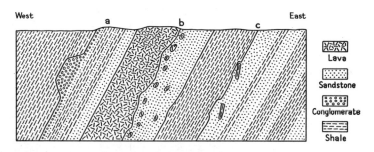

Fig. 4-20. Overturned beds, shown by channeling, pebbles of lava in sandstone, and fragments of shale in sandstone.

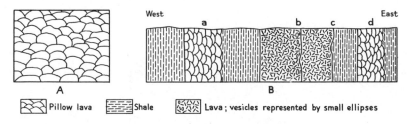

Fig. 4-21. Top of lava flows. (A) Pillow structure, right-side-up. (B) Pillow structure and vesicular structure show that the top is to the right.

of distinguishing top from bottom has been used with particular success in the Precambrian rocks of the Canadian Shield (Fig. 4-21B). Gas bubbles, also called vesicles, tend to concentrate at the top of flows.

Drag Folds

RELATION OF DRAG FOLDS
TO MAJOR FOLDS

Drag folds, in the restricted sense of the term, are those folds that develop in an incompetent bed lying between two competent beds that shear past one another (Fig. 3-14). But drag folds may form beneath an overthrust block or in a layer of mud over which other layers of mud slide. In the broader sense of the term, drag folds refer to any minor folds genetically associated with major folds.

Ideally, the drag folds are systematically related to the contemporaneous major folds. The upper beds slide away from the synclinal hinges relative to the lower beds, as is shown by the arrows in Fig. 4-22. The acute angles between the axial planes of the drag folds and the more competent bed point in the direction of differential movement.

The use of drag folds in deducing the larger structures is illustrated by Fig. 4-23. In this example it is assumed that the major folds plunge at a low angle. At *a* the strata are vertical and the drag folds show that the beds slipped past each other in the manner indicated by the arrows. The synclinal hinge must lie to the east. The beds at *b* dip to the west, and the drag folds show that the bed to the left moved upward relative to the bed to the right; the synclinal axis must lie to the west, and the strata are right-side-up. At *c* the beds

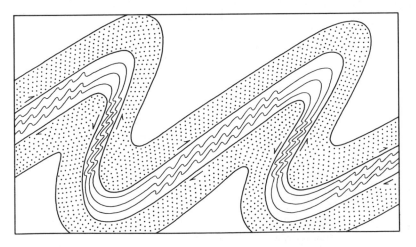

Fig. 4-22. Structure section of overturned folds showing relation of drag folds and direction of shearing.

also dip to the west, but the drag folds reveal that the beds to the right moved upward relative to the beds to the left. The synclinal hinge must lie to the right. Assuming that the beds at *a*, *b*, and *c* are the same, the probable structure is indicated by the broken lines.

In a nonplunging fold the younger beds shear directly up the dip (Fig. 4-24A). Drag folds would be displayed on a vertical cross section at right

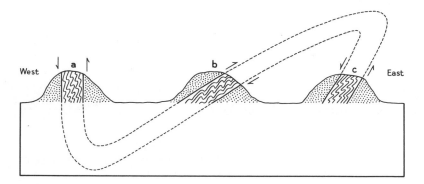

Fig. 4-23. Use of drag folds in determining major structure. Letters are referred to in text.

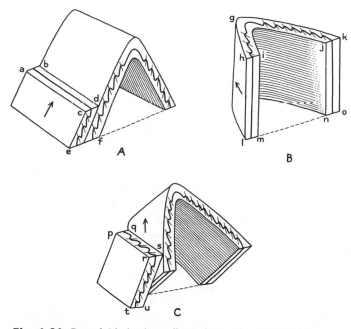

Fig. 4-24. Drag folds in three dimensions. The small blocks on the left side of diagrams (A) and (C) show the appearance of the drag folds on a map and on a vertical section that strikes perpendicular to the axial plane of the fold.

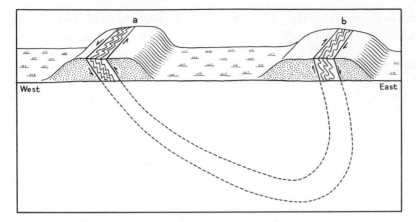

Fig. 4-25. Use of drag folds in determining major structure.

angles to the fold axes (*cdef*), but would not show on a horizontal surface (*abcd*). But, if the major fold has a vertical plunge, the drag fold would be displayed on a horizontal surface (*ghijk*) of Fig. 4-24B), but would not show on the vertical surfaces (*hilm* and *jkno*). The most general case is illustrated by Fig. 4-24C, where the drag folds are displayed on both the horizontal and vertical surfaces.

A specific example is shown in Fig. 4-25. The evidence in the vertical sections is analyzed first. It shows that a synclinal hinge lies between *a* and *b*. In plan, the relative component of shear of the beds nearer the synclinal hinge is northward. Hence the major fold also plunges northward. This may be inferred from a comparison with Fig. 4-24C.

STRATIGRAPHIC SEQUENCE FROM DRAG FOLDS

In many instances drag folds can be used to deduce the stratigraphic sequence. It has already been pointed out that the strata nearer the synclinal hinge shear upward relative to the beds nearer the synclinal hinge (Fig. 4-22). Normally the younger beds will be in the synclines, the older beds in the anticlines. In Fig. 4-23 the shear sense at locality *c* shows that the beds "face" to the east, despite the fact that they dip to the west.

LIMITATIONS IN USE OF DRAG FOLDS

These conclusions concerning the stratigraphy would be incorrect, of course, if the strata had been overturned by an earlier folding. In this sense, therefore, drag folds by themselves cannot give an unambiguous solution to the stratigraphic sequence. On the other hand, some other type of evidence, such as paleontology or primary features, may give a clue to the stratigraphic sequence at some locality in the area. If this confirms the conclusions also reached from the drag folds at this locality, the drag folds may be used throughout the area to deduce stratigraphic relationships.

It should be emphasized that drag folds are generally consistent with the contemporaneous major structural features. Thus, if the strata were completely overturned in an F_1 phase of folding, and then refolded by F_2 folds, the F_2 drag folds would be consistent with the major F_2 folds.

But it may be difficult to distinguish the drag folds and minor folds of different phases. Assume an area in which the strata are thrown into large vertical isoclinal folds with horizontal north-south axes (F_1). Then let the vertical strata be subjected to a gigantic couple, east side moving north relative to the west side. Minor drag folds (F_2) developing during this phase will be vertically plunging and left-handed. Using them to deduce the major structure would be quite disastrous.

Deducing the successive phases of folding in highly deformed areas is one of the modern challenges of structural geology.[3] Where several phases of folding are involved, it may be very difficult to correlate the drag folds with the proper phase of major folding. On the other hand, drag folds are usually consistent with the contemporaneous major phase of folding.

References

[1] Mackin, J. Hoover, 1950, The down-structure method of viewing geologic maps, *J. Geol.* 58: 58–72.

[2] Zen, E-an, et al., 1968, *Studies of Appalachian geology: northern and maritime*, 475 pp., New York: John Wiley & Sons.

[3] Craig, Gordon Y., 1965, *The geology of Scotland*, 556 pp., Hamden, Connecticut: Archon Books.

[4] Lahee, F. H., 1960, *Field geology*, 6th ed., 926 pp., New York: McGraw-Hill Book Company.

[5] LeRoy, L. W., 1950, *Subsurface geologic methods*, 2d ed., 1166 pp., Golden, Colorado: Colorado School of Mines.

[6] Bishop, Margaret S., 1960, *Subsurface mapping*, 198 pp., New York: John Wiley & Sons.

[7] Lyons, Mark S., 1964, *Interpretation of planar structure in drill-hole core*, Geol. Soc. Amer., Spec. Paper 78, 67 pp.

[8] Darton, N. H., 1940, *Some structural features of the Northern Anthracite Coal Basin, Pennsylvania*, U.S. Geol. Surv., pp. 69–81.

[9] Wood, Gordon H., Jr., Trexler, J. Peter, and Kehn, Thomas, 1969, *Geology of the west-central part of the Southern Anthracite field and adjoining areas, Pennsylvania*, U.S. Geol. Surv. Prof. Paper 602, 150 pp.

[10] Ridge, John D., ed., 1968, *Ore deposits of the United States 1933–1967*, Amer. Inst. Mining, Metal., and Petrol. Eng., 2 vols., 1880 pp., New York.

[11] Pettijohn, F. J., 1957, *Sedimentary rocks*, 2d edition, 718 pp., New York: Harper and Row.

[12] Shrock, Robert R., 1948, *Sequence in layered rocks: a study of the features used for determining the top and bottom for order of succession in bedded and tabular rock bodies*, 507 pp., New York: McGraw-Hill Book Company.

[13] Shrock, Robert R., and Twenfofel, William H., 1953, *Principles of invertebrate paleontology*, 2d ed., 816 pp., New York: McGraw-Hill Book Company.

[14] Joplin, Alan V., 1966, Origin of cross-laminae in a laboratory experiment, *J. Geophys. Res.* 71: 1123–33.

[15] Kuenen, P. H., 1953 Significant features of graded bedding, *Amer. Assoc. Petrol. Geol., Bull.* 37: 1044–66.

[16] Pettijohn, F. J., and Potter, Paul Edwin, 1964, *Atlas and glossary of primary sedimentary structures*, 370 pp., New York: Springer-Verlag.

[17] Snyder, George L., and Fraser, George D., 1963, *Pillowed lavas*, U.S. Geol. Surv. Prof. Paper 454, pp. B1–B23.

[18] Wilson, M. E., 1960, Origin of pillow structure in early Precambrian lavas of western Quebec, *J. Geol.* 68: 97–102.

5

OFFICE TECHNIQUES
USED IN
STUDYING FOLDS

Introduction

The emphasis in the preceding chapters has been on field observations and their analysis in the field or office. Additional analytical methods may be employed, primarily in the office or laboratory. But some of this analysis can be done in the field office as the field work is proceeding and does not necessarily have to wait until the field season is over. These methods include the use of pi diagrams and beta diagrams (page 100) plotted on the equal-area net; such diagrams may be prepared manually or by the use of computers. But before describing pi diagrams and beta diagrams it is necessary to describe the use of the equal-area and stereographic projections.

Equal-Area
and Stereographic Projections

PRINCIPLES

The basic principles of these projections are illustrated in Fig. 5-1. Figure 5-2 is a stereographic net and Fig. 5-3 is an equal-area net. The similarity is obvious; the differences are discussed on a later page. Figure 5-1A shows a sphere; the *AF* axis is vertical, the *BD* axis is horizontal east-west, and the *CE* axis is horizontal north-south. *BCDE* is a horizontal plane, *ACFE* is a vertical north-south plane, and *BFDA* is a vertical east-west plane. Let us

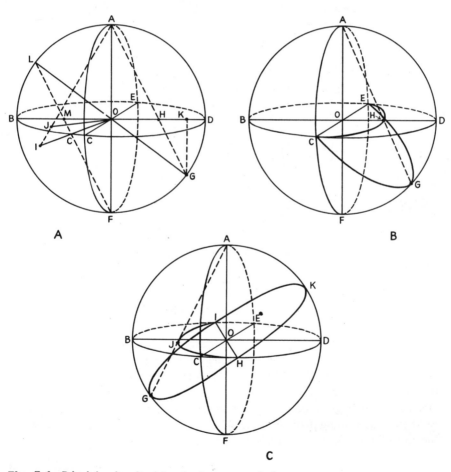

Fig. 5-1. Principles involved in plotting trace of plane on an equal-area net.

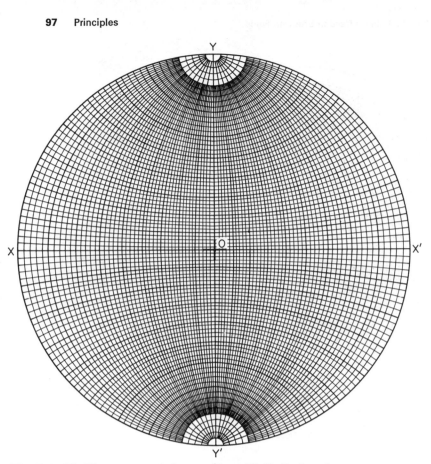

Fig. 5-2. Meridian stereograhpic net. (From W. H. Bucher; by permission of the University of Chicago Press.)

consider a line that plunges due east at an angle of 40 degrees. The point on the sphere penetrated by the line is called a pole. Now assume, as is always done in these projections, that the line passes through O, the center of the sphere. Since the line plunges due east, it must lie in the plane $BFDA$. The pole of this line is G on the lower hemisphere and L on the upper hemisphere. If G were projected vertically upward to plane $BCDE$, the point of projection would be K.

In the stereographic projection the line is drawn from G to the uppermost pole of the sphere (A), so that the point of projection is H.

$$OH = OD \times \tan \frac{\alpha}{2} \tag{1}$$

where α = angle GOF.

In the equal-area projection, for reasons explained below,

$$OH = 2 \times 0.707 \times OD \times \sin \frac{\alpha}{2} \tag{2}$$

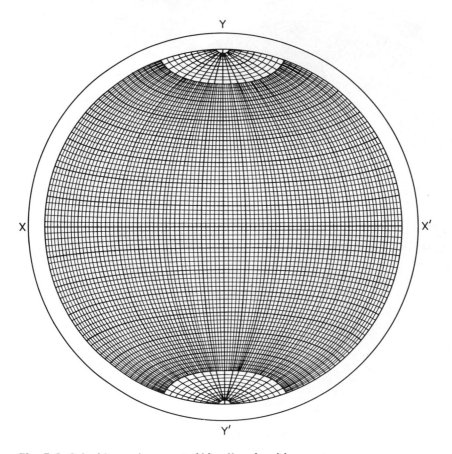

Fig. 5-3. Schmidt equal-area net. (After Knopf and Ingerson, Memoir 6, Geol. Soc. Amer., 1938.)

Thus in these projections, *HO* represents *GO* in space, which is a line plunging east at an angle of about 40° (*GOF* = 50°). Thus the distance from the center of the projection is a function of the value of the plunge. A line bearing east-west and plunging zero, would be represented by *OD* (or by *OB*). A line bearing north and plunging zero would be represented by *OE* (or by *OC*). A vertical line would be represented by a point at *O*. Fuller treatment of the methods of preparation of the equal-area projection are available.[1]

It is apparent that only one hemisphere need be used. If the upper hemisphere were used, the projection would be the mirror image of the lower hemisphere. Of course, there is ambiguity in lines with zero plunge—they can be represented by points at either end of the line. In each instance one of the two points can be chosen arbitrarily. In statistical analysis, as will be shown below, the method of counting the points automatically takes care of the problem. In most structural problems the lines have no vector properties, that is, there is no directional preference. But in paleomagnetic problems the vectoral properties may be significant. In such cases the plot can be made on the lower hemisphere, but special symbols may indicate that the point on the upper hemisphere is meant.

In Fig. 5-1A assume that a line plunges 25°SW. The pole on the lower hemisphere is at I, the projection of which is J on the plane of projection.

The method of representing planes is shown in Fig. 5-1B and 5-1C. Assume a plane that strikes north-south and dips 50°E. If such a plane is imagined to pass through the center of the sphere at O (Fig. 5-1B), its intersection with the lower hemisphere is CGE. The projection of G onto the $BCDE$ plane is H, and the line CGE is represented by CHE. If a reader were presented with this projection, he could tell that the plane strikes north. Because CHE lies east of O, the plane dips east. The distance HD gives the dip of the plane; it depends on the projection used. A horizontal plane would be represented by the circumference of the plane of projection. A vertical north-south plane would be represented by COE; a vertical east-west plane by the line BOD.

If a plane strikes northwest and dips 50°SW, its projection on the plane of projection is HJI (Fig. 5-1C).

The stereographic and equal-area nets are equally satisfactory for many purposes in structural analysis. But the equal-area net is constructed in such a way that areas of equal size on the surface of the sphere are also of equal size on the projection. For example, all areas covering one square centimeter on the surface of the sphere will cover 0.5 square centimeters on the projection, because

$$\frac{Ac}{Ah} = \frac{\pi r^2}{2\pi r^2} = \frac{1}{2} \tag{3}$$

where Ac is area of the projection circle, Ah is area of the surface of a hemisphere, and r is the radius of the sphere and circle.

The data plotted on an equal-area net may be analyzed statistically. This is discussed on pages 104, 428.

The equal-area net to be used in the problems is shown in Fig. 5-4, on page 589. There is, of course, a great similarity to the meridians of longitude and parallels of latitude on maps of the world. But the use of such terms should be avoided in referring to the equal-area net, because these lines are not compass directions.

The arcs convex toward the circumference represent great circles (Fig. 5-3). They are the projections of the trace on the surface of the sphere of the various planes that contain the line YY'. Henceforth they will be called great circles. The arcs convex toward the center of the net represent small circles. They are the projections of the trace on the surface of the sphere of the various planes perpendicular to YY'. Henceforth they will be called small circles.

METHODS OF CONSTRUCTION

Plotting a plane. Assume a plane striking N.30°W. and dipping 60°NE. An *overlay* (Fig. 5-5A) is then prepared on a piece of tracing paper showing the circumference of the net, the center (*O*) and a few principal points of the compass, N, E, S, and W. A square on this and ensuing figures is drawn with its sides perpendicular to the cardinal compass directions. The equal-area net serves as an *underlay*. The north direction on the overlay is kept toward the top of the page (Fig. 5-5B). The underlay is then rotated counterclockwise 30° so that the line *YY'* bears N.30°W. The proper arc on the underlay is chosen, the 60° being measured in from the circumference of the net. Of course, two arcs represent the two planes dipping 60°; the arc toward the northeast is chosen and traced on the overlay (*YFY'*).

If one wishes to plot the pole of the perpendicular to the plane, a distance of 90° is measured along the line *XX'* and *G* is plotted.

But in practice it is much easier to rotate the overlay rather than the underlay (Fig. 5-5C). The overlay is rotated in the *opposite* direction of the strike of the plane being plotted; but the amount of rotation is the same. That is, in the case cited above, the overlay is rotated 30° clockwise. Of the two possible arcs representing a 60° dip, the one convex toward the NE on the overlay is chosen. With experience, it is not necessary to draw the strike-line of the plane.

Plotting a line. A line is plotted as follows. Assume a line plunging 30° N.20°E. A line striking N.20°E. (*OD*) is drawn on the overlay (Fig. 5-5D). The overlay is rotated until *OD* is parallel to *XX'* or *YY'* (Fig. 5-5E) and 30° is measured from the circumference to establish point *F*, the projection of the pole of the line plunging 30°NE.

Pi Diagrams

Pi diagrams and *beta diagrams* are used extensively in analyzing folds. The strike and dip of the bedding is recorded at many points of observation. If the attitude of the bedding is relatively uniform at an outcrop, one observation is sufficient. But if the outcrop displays folds, several observations on dip and strike will be necessary.

The projection of the planes representing all bedding planes will be arcs except for those beds that are vertical or horizontal. The projection of a vertical bed will be a straight line striking in the same direction as the bed. The projection of a horizontal bed will coincide with the circumference of the net.

A few examples are shown in Fig. 5-6. Arc *a* is the projection of a plane striking N.20°E. and dipping 70°SE. Arc *b* is the projection of a plane striking N.80°W. and dipping 20°SW. Line *c* is the projection of a vertical bed striking

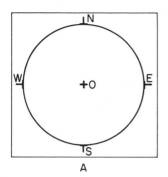

A

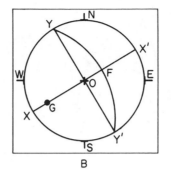

B

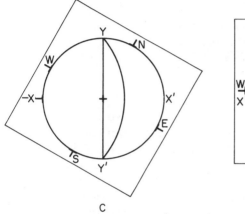

C

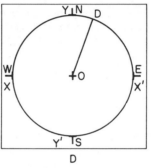

D

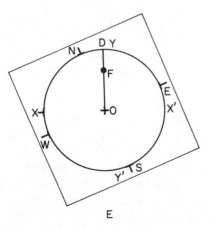

E

Fig. 5-5. Method of plotting planes and lines on an equal-area net.

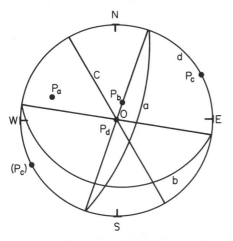

Fig. 5-6. Equal-area plot of several planes and perpendiculars to them. Trace of planes marked *a*, *b*, *c*, and *d*. Projection of poles of perpendiculars to these planes are marked *Pa*, *Pb*, *Pc*, and *Pd*.

N.30°W. Circle *d* is the projection of a horizontal bed. Normally—except for vertical beds—the strike lines are not shown on the projection.

Pi diagrams are the projection of the poles of the perpendiculars to the planar feature, in this case the bedding. In Fig. 5-6 the perpendiculars plot as *Pa*, *Pb*, *Pc*, and *Pd*. The projection of the perpendicular to a horizontal plane will coincide with the center of the projection. A vertical plane will be represented by two diametrically opposed poles. In statistical studies each such pole counts as a half pole; however, either point may be plotted and the method of contouring (page 104) makes the necessary adjustments.

Figure 5-7 is a pi diagram of Fig. 4-6A. Each point in Fig. 5-7 represents a strike-dip symbol in Fig. 4-6A; each point in Fig. 5-7 is identified by a numeral corresponding to the dip as given in Fig. 4-6A. To avoid crowding, some of the points have been moved slightly. The points lie on a *girdle* that is a straight line trending east-west. This girdle is the projection of the plane containing all the perpendiculars to the bedding. Similarly, Fig. 5-8 is the pi diagram for Fig. 4-6B. The points lie on a girdle that is convex toward the north. This girdle is the trace of the projection of the plane containing all the perpendiculars to the bedding. Neither Fig. 5-7 nor 5-8 has points on the periphery of the circle, because none of the beds are vertical.

The pi diagram is analyzed on the assumption that the folds are cylindrical. A *cylindrical fold* is the surface generated by a line moving through space parallel to itself. In such folds the axis is perpendicular to the plane represented by the girdle. In Fig. 5-7, since the girdle is the projection of a vertical plane striking east-west, the axis of the fold strikes north-south and

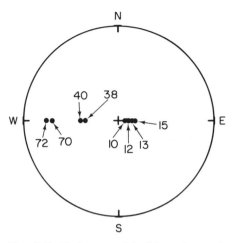

Fig. 5-7. Pi diagram of bedding planes shown in Fig. 4-6A.

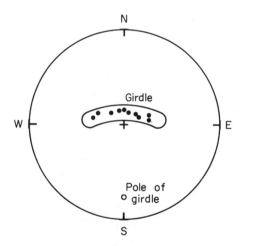

Fig. 5-8. Pi diagram of bedding planes shown in Fig. 4-6B.

is horizontal. In Fig. 5-8 the girdle is the projection of a plane dipping 75°N.; the axis of the fold therefore plunges 15°S.

This preparation of pi diagrams may thus be used to determine the direction and value of the plunge of folds. They may also indicate more than one orientation of the folds. Figure 5-9 suggests two sets of folds. One set, shown by dots, is relatively open, because the points are evenly distributed on the east-west line; the axes are horizontal and trend north-south. A second set consists of chevron folds, shown by crosses, because only steeply dipping limbs are shown; the axes are horizontal and trend east-west.

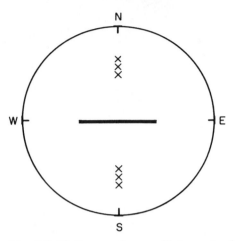

Fig. 5-9. Pi diagram of area with two sets of folds. Axis of open fold is horizontal and bears north-south. Axis of chevron fold is horizontal and bears east-west.

But there are many limitations to the use of pi diagrams. Anticlines and synclines give the same diagram. Figures 5-7 and 5-8 can represent either anticlines or synclines. Moreover, many folds may be present instead of one. In Fig. 5-9 it is impossible to tell whether the two sets of folds are in different parts of the area, or intimately associated. Finally, if the value of the plunge changes, the folds are no longer cylindrical; the greater the variation in the value of the plunge, the greater the complexity of the diagram.

It is thus apparent much of the interpretation must be based on the geographic distribution of the observations. Thus, after preparing the pi diagrams, it is usually necessary to go back to the field notes and maps; if these are inadequate it may be necessary to go back into the field to make additional observations. But the best procedure is to prepare and analyze the pi diagrams as the field work is proceeding.

Contour Diagrams

The data are often shown by *contour diagrams.* Figure 5-10 is a point diagram based on the perpendiculars to 152 bedding planes in Devonian schists in the Peterborough Quadrangle of New Hampshire.[2] From this diagram it is apparent that most of the bedding strikes east, northeast, and north, and dips 10° to 60° north, northwest, and west. Figure 5-11 is a contour diagram based on Fig. 5-10. The solid black area, labeled 12-14 percent, means that 12 to 14 percent of all the points shown in Fig. 5-10 lie within an area equal to one percent of the total area of the diagram. That is, if a small circle, cover-

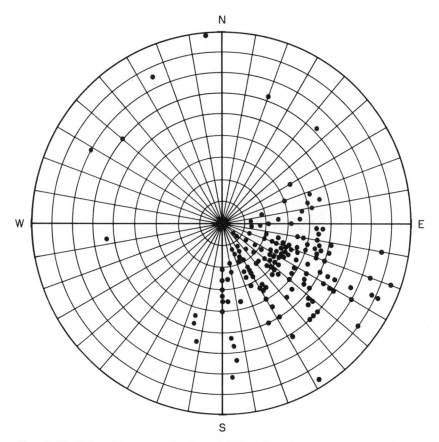

Fig. 5-10. Point diagram, projection of 152 poles of perpendiculars to 152 bedding plane measurements, area in Peterborough quadrangle, New Hampshire. (After Robert Greene.[2])

ing an area equal to one percent of the area of the large circle, were placed over this solid black area, it would contain 12 to 14 percent of the points. This is a so-called *maximum* in the figure. A maximum is not necessarily an average or mean. But the diagram taken as a whole shows that the beds strike northeast and dip northwest. The beds could be on the southeast limb of a syncline and face consistently northwest; they could be on the northwest limb of a syncline and face consistently southeast, that is, be overturned; or there could be many isoclinal and overturned folds, the axial planes of which dip northwest.

The preparation of such a contour diagram from a point diagram is illustrated by Fig. 5-12. A piece of tracing paper is placed over the point diagram. The center counter, *CC* of Fig. 5-12, consists of a circular hole in the center of a piece of paper, cardboard, celluloid, or any satisfactory material.

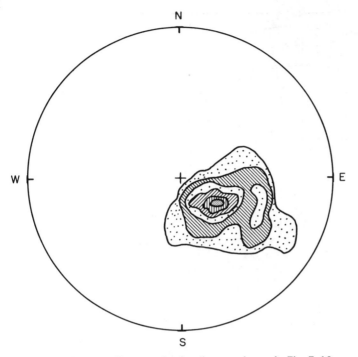

Fig. 5-11. Contour diagram of point diagram shown in Fig. 5-10.
Contours are 2, 4, 6, 8, 10, and 12 percent.

The area of this circle is equal to one percent of the area of the large circle; if the large circle has a radius of 10 centimeters, the small circle has a radius of 1 centimeter. Two hundred points are plotted on Fig. 5-12. (Some of the 200 points are covered by the counters.) Six of these points lie within the center counter; six points are three percent of the total number of points in the large circle, and the figure 3 is written in the center of the center counter. The center counter is moved over the whole diagram, and the percentage of points at each place is recorded. In order that the sampling may be systematic, a grid system is placed on the point diagram—or beneath it if the point diagram is on tracing paper—and the center counter is moved from left to right one centimeter at a time. After a traverse from left to right has been completed, the counter is moved down one centimeter, and a second traverse is run. It should be noted that a single point in the point diagram may lie within the center counter several times in its successive positions. The point is counted each time.

For points closer to the circumference than a distance equal to the diameter of the center counter—one centimeter if the large circle is 10 centimeters—a special technique is required. The peripheral counter (*PC* of Fig. 5-12) is used for such points; it is made of paper, cardboard, celluloid, or any satisfactory material. Half of each of the two circles at either end extends

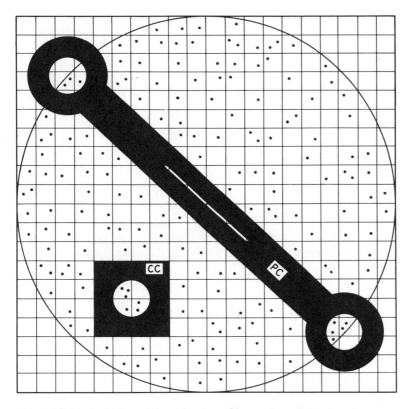

Fig. 5-12. Method of counting points in making contour diagram.
CC, center counter. *PC*, peripheral counter.

beyond the circumference of the large circle. The points in each circle are added together. In Fig. 5-12 they total 8, which is 4 percent of the total 200. The figure 4 is then entered on the diagram in the center of *both* circles at the ends of the peripheral counter.

After the diagram has been covered with percentage figures, contours are drawn in the same manner by which topographic contours are prepared from points of known altitude. Figures 7-5 and 7-6 show how point and contour diagrams are related.

Beta Diagrams

Beta diagrams involve the projection of the planes rather than the perpendiculars to the planes. The plunge of a fold may be calculated if the attitudes of the bedding on the opposite limbs are known. The same principle may be applied by projecting the bedding onto an equal-area net, using, as before, the lower hemisphere.

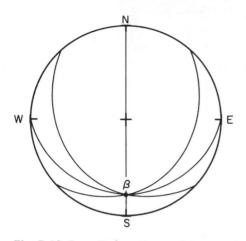

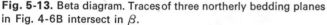

Fig. 5-13. Beta diagram. Traces of three northerly bedding planes in Fig. 4-6B intersect in β.

Figure 5-13 is the projection of the three northerly bedding planes shown in Fig. 4-6B. The three arcs intersect at a common point called β (beta). This intersection shows that the fold plunges south. The angle of plunge is determined by rotating the line connecting β with the center of the diagram until it is parallel with the vertical line of the net. The value of the plunge, 26°, is the distance, in degrees, between the periphery of the circle and β.

The number of intersections in such diagrams is

$$\frac{n(n-1)}{2} \tag{1}$$

where n is the number of planes. If 25 planes are plotted, the number of intersections is 300. But such a diagram is so cluttered up with lines that it may be difficult to interpret.

Figure 5-14 is an example from an area in New Zealand.[3] Figure 5-14A is a pi diagram of the perpendiculars to 22 bedding and schistosity planes. The girdle trends northeasterly through the center of the diagram. This suggests horizontal fold axes trending northwest. Figure 5-14B shows the trace of the 22 planes on the equal-area net. As one would expect, there is a spread in the 231 intersections. Figure 5-14C is the contour diagram of the intersections. It shows a strong concentration of 25 percent in the southeast corner, meaning that the fold axes bear northwest-southeast and are horizontal.

The same limitations apply to the interpretation of beta diagrams as to the interpretation of pi diagrams.

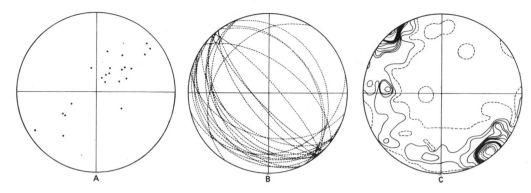

Fig. 5-14. Twenty-two bedding and foliation planes from East Otago, New Zealand. (A) Pi diagram. (B) Plot of 22 planes with 231 intersections. (C) Contour diagram of the 231 intersections. Contours are 1 to 10, 15 (dashed), 20, and 25 (dashed) percent. (After Peter Robinson et al.[3]) Permission *American Journal of Science* and Peter Robinson.

Use of Computers in Geology

Many problems in geology are now analyzed by the use of computers.[4,5] Four major computer-oriented techniques have been utilized in geology:[6] statistics, correlation and classification, trend analysis, and simulation. Many of these techniques, although known before 1960, were too involved or complicated before computers became readily available.

Statistics,[7] formerly largely descriptive, are now more analytical; in structural geology statistics are largely concerned with the extent of preferred orientation and its significance. Correlation and classification is of concern to structural geology because stratigraphy is, after all, the fundamental basis of much structural analysis. Trend analysis is concerned, as far as structure is concerned, primarily with the shape of surfaces, both before and after deformation. Simulation is the mathematical analysis of scale models by computer techniques; such analyses are far more rapid and complete than the use of physical scale models.

We shall be concerned only with the use of computers in structural geology. Moreover, a discussion of the preparation of programs is beyond the scope of this book. We can only give some examples of the results obtained by the use of computers.

The main objective in using a computer is to save time in analyzing rather simple problems or to solve problems that otherwise could not be undertaken because the time involved would be prohibitive. It takes time to prepare a program. It may be quicker to analyze a problem by simple algebraic or graphical methods. On the other hand, although any one individual in a large organization may not have enough problems of a certain type to justify a program, the group as a whole may find that a program is highly desirable. Of course, for a highly sophisticated research project a very elaborate program may be fully justified.

The raw data or input is placed on cards or tape. A code system is used, as will be shown below. The results or output may be expressed as numerals, maps, sections, or even as motion pictures on television screens. Examples of the use of computers will be given in the appropriate place in the text.

Preparation of Pi Diagrams and Beta Diagrams by Computer

Pi diagrams may be prepared by computer.[8] The raw material consists of the strikes and dips of the planar feature; this is the input. In the specific example cited (Fig. 5-15), from a portion of the Buckfield quadrangle in Maine, measurements on 162 bedding planes were utilized. The problem was to discover whether there is any preferred orientation of these bedding planes. It is a statistical problem. As in the manual construction of pi diagrams, the plot is based on the perpendiculars to these planes. The output may be expressed in one of the following ways:

1. The equal-area net is divided into one hundred or more compartments of equal area. The computer prints out in tabular form the number of poles in each compartment. These figures are manually transferred to the net and contoured by hand.
2. The same procedure is followed, but the number of poles in each compartment is converted to a percentage.
3. The results, in percentages, are printed on a net by the computer. Figure 5-15 is such a plot. The capital letters A to F represent percentages from 10 to 15. Contouring is done by hand.
4. The contours are printed directly by the computer.

Figure 5-15 shows that the average strike is north, but the dip shows considerable range, with the greatest concentration around 60°W.

Beta diagrams can also be prepared by computer.[3,9] Since the number of intersections increases astronomically with an increase in the number of planes, the manual method is very difficult to use if the number of planes exceeds 25. Figure 5-16 contains two beta diagrams for two adjacent areas in north-central Massachusetts.

Figure 5-16A is a beta diagram prepared from 162 bedding-foliation planes giving 13,041 intersections. In this case the number of intersections per compartment was printed out in tabular form. These data were converted to

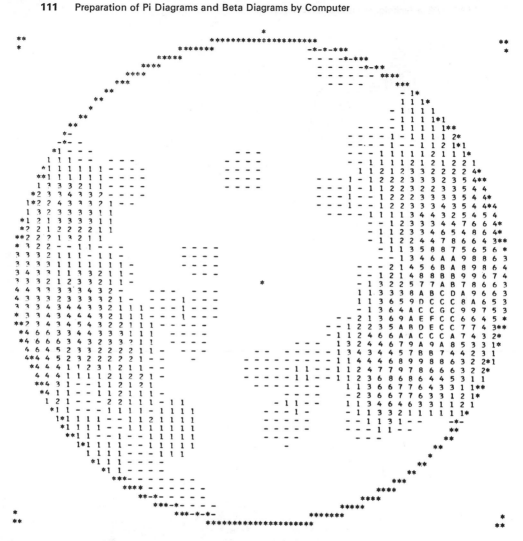

Fig. 5-15. Pi diagram prepared by computer. Perpendiculars to 162 bedding planes, part of Buckfield quadrangle, Maine, given as percentage of points falling within 1-percent areas. Capital letters A to F represent numbers 10 to 15. (After Jeffrey Warner.[8]) Permission University of Kansas Press.

percentages, entered into each compartment on the equal-area net, and manually contoured. The fold axes plunge, on the average, 20°N. 10°E. Figure 5-16B, based on 120 planes (7140 intersections) indicates fold axes with an average plunge of 10° in a direction S.10°E.

Other uses of computers in structural analysis are discussed on later pages.

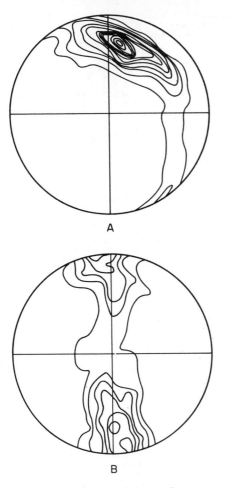

Fig. 5-16. Contoured beta diagram prepared by computer. Shows percentages of beta intersections within 1-percent areas. (A) 162 planes, contours are 1 to 12 percent. (B) 120 planes, contour interval 1 to 6 percent. (After Peter Robinson et al.[3]) Permission *American Journal of Science* and Peter Robinson.

Structure Contour Maps

Structure contours, an example of which is given in Fig. 5-17, provide the most precise way to represent folds in three dimensions. Such a map is read in the same way as a topographic contour map. The contours are based on a single horizon, such as the bottom or top of some particular bed. The position of the bed is given in reference to some datum plane, usually mean sea level. Inasmuch as the key horizon may go below sea level, negative

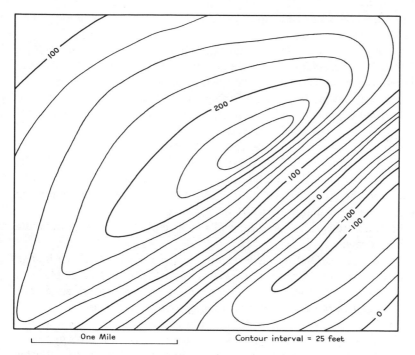

One Mile   Contour interval = 25 feet

Fig. 5-17. Structure contour map.

contours are not uncommon. For a given contour interval, the dip is steepest where the contours are closest together. In Fig. 5-17 the contour interval is 25 feet, and every 100-foot contour is labeled. The structure shown is a doubly plunging anticline, bounded on the southeast by a syncline that plunges to the northeast. The southeast limb of the anticline, where the contours are close together, is much steeper than the northwest limb, where the contours are far apart. The southwestward plunge of the southwestern end of the anti-cline is gentler than the northeastward plunge of the northeastern end. If one knows the scale of the map, one can determine the dip at any place in feet per mile, and, if desirable, one can then readily convert it into degrees. On the northwestern limb of the anticline, the key bed drops from 200 feet to 100 feet in one mile. The dip on this limb is thus 100 feet per mile, or approximately one degree.

Structure contour maps are particularly useful in regions where the dips are low. Obviously such maps are difficult to use where strata are overturned, although broken lines or some other device could be used for overturned beds.

Structure contour maps are based on two kinds of data. In areas of relatively simple geology they can be prepared entirely from surface observations. From the strike and dip of the beds it is possible to predict the altitude relative to sea level of a key bed (pp. 75–77, 507–519, 527–533) at any location on the map.

But many structure contour maps are based on subsurface geology (pp. 80–81). The necessary data include the location of the wells and the altitude relative to sea level of the key bed in these wells. This last piece of information may have to be calculated from the altitude of the top of the well, the depth to the key bed, or perhaps, if the key bed is not reached, the vertical distance of the key bed below some recognizable bed.

Obviously a computer may be very useful or indispensible in preparing such maps. The results could be presented in one of several ways: (1) the altitude of the key bed could be tabulated; (2) the altitude could be printed on a map and the structure contours drawn manually; or (3) the structure contours can be drawn automatically on the map.

If the structure is simple and the wells sufficiently close, a very accurate map can be drawn.

Trend surface analysis[10,11] as used in geology is a statistical method to represent the variation of certain properties of rocks from place to place. It might, for example, deal with the size of sand grains in a sheet of sand. The average diameter of the grains might decrease away from the source and also upward in the sheet. A surface could be constructed through all the points where the average grain size is, let us say, 0.1 mm. In structural geology we are primarily concerned with the shape of some surface, such as a stratigraphic horizon or an unconformity. The problem is to find the mathematically defined geometrical surface—the trend surface—that most nearly fits the actual surface. A series of folds may consist of two components, which, in two dimensions, may be represented by two sine curves (Fig. 5-18A). Similarly, in three dimensions, the structure may result from three sinusoidal components (Fig. 5-18B).

One problem is to recognize these various components in a structure contour map.[12] Figure 5-19A is a structure contour map on the so-called pre-Cretaceous unconformity in Alberta. In Fig. 5-19B the regional strike and dip have been filtered out, so that only features with a minimum size between 10 and 40 miles and a relief greater than 100 feet are shown. That is, the plane surface that conforms best to the unconformity dips southwest at a rate of one mile per hundred miles—about 0.05 degree. The solid black areas are bumps on this plane, the light-gray areas are depressions. The bumps could be traps for petroleum.

Calculating the Depth of Folding

The depth of folding can be calculated under certain conditions. In Fig. 5-20A it is assumed that the vertical rectangle dl is changed into the rectangle $b(d + h)$, with no change in area. Then:

$$dl = b(d + h)$$

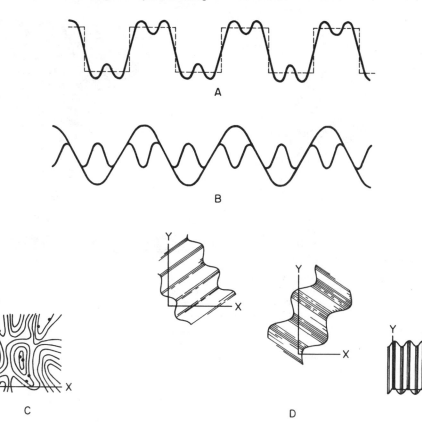

Fig. 5-18. Components comprising a surface. (A) The curve representing a fold in (A) consists of the two sine curves shown in (B). Similarly, the structure contour map shown in (C) may consist of the three components shown in (D). (After J. E. Robinson and H. A. K. Charlesworth.[12]) Permission University of Kansas Press.

With certain modifications, the same concept may be extended to folded belts. It is assumed that there is no lengthening or shortening parallel to the axes of the folds and that the rocks do not change in volume. The term b is the present breadth of the folded area; the l, which is the original width before folding, is measured along some convenient bed in the folded belt. The term h is the amount of uplift due to folding.

In Fig. 5-20B the heavy black line represents a single bed. At the left end of the section it is flat and has not been affected by the folding. In the folded area it has been uplifted from the position of the broken line to the position shown by the heavy solid line. The average uplift h can be determined in several ways. The simplest is to measure the actual uplift at stated intervals—such as at every millimeter in the figure shown—and to compute the

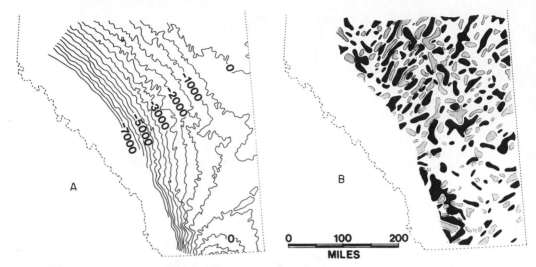

Fig. 5-19. Filtering out of regional components. Area is in Alberta. (A) Structure contour map on pre-Cretaceous unconformity; structure contour interval 500 feet. (B) Minor irregularities on hypothetical surface striking north-northwest and dipping southwest. Solid black areas are bumps on a hypothetical plane representing the unconformity; the light gray areas are depressions. Permission University of Kansas Press. (After Robinson and Charlesworth[12])

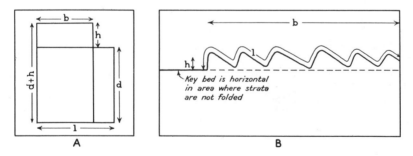

Fig. 5-20. Method of calculating depth of folding. *b*, present breadth of folded belt; *h*, average uplift due to folding; *l*, original breadth of folded belt; *d*, depth of folding. (A) Square deformed into a rectangle, no change in area. (B) Folded strata.

average. All the factors in the equation given above, except d, are known. For convenience in computation, the equation can be rewritten:

$$d = \frac{bh}{l - b}$$

The answer d gives the depth of folding measured from the key bed where it is horizontal.

Several assumptions are made in applying this method. One assumption is that there is a sharp break between the folded rocks and the unfolded rocks below; in other words, a *décollement* is assumed. If the folds gradually disappear downward, the calculations would be incorrect, and the depth of folding would be greater than the calculations indicate. Secondly, the method assumes that the base of the strata is not depressed by the folding. There are reasons for believing, however, that in many folded belts the basement is downfolded by horizontal compression. If this is so, the method employed for determining h gives too low a value, and the depth of the folded zone would be much greater than the calculated value.

The depth of folding in the Jura Mountains may be calculated. In this case the method is applicable because a relatively thin column of sedimentary rock is separated from a crystalline basement by a *décollement*. The calculated depth is comparable to that deduced by the Europeans by other means.

References

[1] Doell, Richard R., and Altenhofen, Robert E., 1960, *Preparation of an accurate equal-area projection*, U.S. Geol. Surv. Prof. Paper 400B, pp. B427–29.

[2] Greene, Robert, 1970, *The geology of the Peterborough Quadrangle, New Hampshire*, Bull. No. 4, N. H. Dept. Resources and Economic Development, 88 pp.

[3] Robinson, Peter, et al., 1963, Preparation of beta diagrams in structural geology by a digital computer, *Amer. J. Sci.* 261: 913–28.

[4] Smith, F. G., 1966, *Geological data processing using Fortran IV*, 284 pp., New York and London: Harper and Row.

[5] Merriam, Daniel F., ed., 1969, *Computer applications in earth sciences*, 282 pp., New York: Plenum Press.

[6] Merriam, Daniel F., 1969, Computer utilization by geologists, Kansas Geol. Surv., *Computer Contribution No. 40*, pp. 1–40.

[7] Krumbein, W. C., and Graybill, Franklin A., 1965, *An introduction to statistical models in geology*, 475 pp., New York: McGraw-Hill Book Company.

[8] Warner, Jeffrey, 1969, *Fortran IV program for construction of pi-diagrams with the Univac 1108 computer*, Kansas Geol. Surv., Computer Contribution No. 33, 38 pp.

[9] Lam, Peter W. H., 1969, Computer method for plotting beta diagrams, *Amer. J. Sci.* 267: 1114–17.

[10] McIntyre, Donald B., 1966, Trend-surface analysis of noisy data, Kansas Geol. Surv., *Computer Contribution No. 7*, pp. 45–56.

[11] O'Leary, Mont, Kipport, R. H., and Spitz, Owen T., 1966, *Fortran IV and map program for computation and plotting of trend surfaces for degrees 1 through 6*, Kansas Geol. Surv., Computer Contribution No. 3, 47 pp.

[12] Robinson, J. E., and Charlesworth, H. A. K., 1969, Spatial filtering illustrates relationship between tectonic structure and oil occurrence in southern and central Alberta, Kansas Geol. Surv., *Computer Contribution No. 40*, pp. 13–18.

6

MECHANICS
AND CAUSES
OF FOLDING

Introduction

The three preceding chapters have been concerned with the description, field investigation, and office analysis of folds. These chapters were primarily concerned with the geometry and kinematics of folding.[1] In the present chapter we are concerned with the dynamics and the ultimate causes.

Types of Folding

In general three principal types of folding may be recognized, but these are actually ideal limiting cases and transitions and combinations are common. The types are (1) flexure folding, (2) shear folding, and (3) flow folding.

Flexure folding is sometimes referred to as *true folding*. For purposes of analysis the behavior of flat beds under a compressive force acting parallel

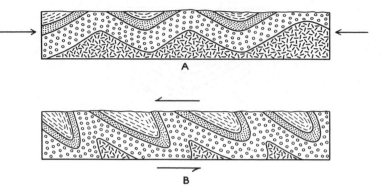

Fig. 6-1. Strata deformed by flexure folding. (A) Folds resulting from compression. (B) Folds developed by (1) a couple or (2) compression and a couple combined.

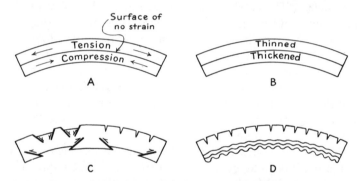

Fig. 6-2. Principles of flexure folding. (A) Stresses in bent beam. (B) Deformation entirely plastic. (C) Deformation by rupture. (D) Deformation by rupture and by crumpling.

to the bedding may be considered (Fig. 6-1). If a sheet is bent (Fig. 6-2), the convex side is subject to tension, whereas the concave side is subjected to compression. There is an intermediate *surface of no strain*. Usually the rocks are sufficiently plastic so that rupture does not occur. The convex side will lengthen and thin, whereas the concave side will shorten and thicken.

The unique feature of sedimentary rock is the presence of bedding planes. Usually, therefore, the folding is analogous to the bending of a thick package of paper, and a very important factor is the sliding of beds past one another, as illustrated in Fig. 6-3. Of two adjacent beds, the upper one moves away from the synclinal axis relative to the lower bed (Fig. 6-3). Some geologists refer to this type of folding as *flexural-slip folding*. This concept is of great importance in interpreting certain types of drag folds (pp. 90–93).

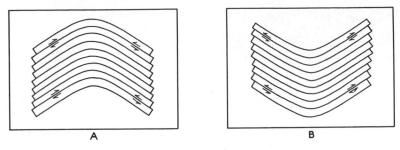

Fig. 6-3. Cross sections illustrating flexure folding. (A) Anticline. (B) Syncline.

In the folding of sedimentary rocks, some formations are competent, whereas others are incompetent. Competency is a relative property. A *competent formation* is strong and can transmit the compressive force much farther than a weak, *incompetent formation*. Many factors determine whether or not a formation is competent. The crushing strength is one of these factors. If specimens from two different formations are tested in the laboratory, the one with the greater crushing strength will be the more competent in folding, provided, of course, that all other factors are equal. Quartzite and marble are stronger than sandstone and limestone, and shale is the weakest of all.

The massiveness of the formation is an important factor. If in two formations composed of the same kind of limestone the beds of one formation are a foot thick, whereas in the other the beds are 100 feet thick, the thick-bedded formation will be the more competent.

The ease with which fractures heal may be an important factor. If a stratum fails by rupture, it is no longer competent to transmit a compressive force. A sandstone may be inherently stronger than an adjacent limestone. But once the sandstone has broken, the fracture may heal with difficulty, whereas the rupture in the limestone may heal relatively rapidly.

If the column of sedimentary rocks is composed of materials of greatly differing competency, the competent beds transmit the force, whereas the incompetent beds behave more or less passively; they are either lifted by the rising arch of competent rock or flow into potential cavities beneath the arch.

In summary, flexure folding involves the bending or buckling of the more competent layers under compressive force, the more passive behavior of the incompetent beds, and the sliding of beds past one another.

Shear folding, also known as *slip folding*, results from minute displacements along closely spaced fractures. In Fig. 6-4A original horizontal strata are broken into blocks by fractures that dip 60 degrees to the left. In Fig. 6-4B blocks 1 and 11 remain undisturbed. Block 6 has moved upward the greatest amount; the blocks on either side of it have moved upward in amounts that decrease progressively. Each fracture is actually a minute fault. If, however, the fractures are only a fraction of an inch apart, and the beds,

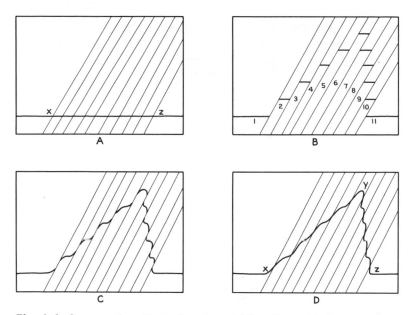

Fig. 6-4. Cross sections illustrating shear folding. Heavy black line, *xz*, is a bedding plane. Inclined light lines are fractures. (A) Before displacement on fractures. (B) After displacement. (C) Because of friction, beds tend to parallel the fractures. (D) Fold results if bed maintains continuity.

because of friction, tend to parallel the fractures, as is shown in Fig. 6-4C, the resulting structure is a major fold with many associated minor folds (Fig. 6-4D). In the simplest case, such shear folds should always be accompanied by visible fractures, usually cleavage. It is conceivable, however, that such fractures could be eliminated by later recrystallization of the rocks.

It is evident from Fig. 6-4 that in shear folding the beds are thinned, but are never thickened. Inasmuch as blocks 1 and 11 have not moved toward each other during the folding, the beds that originally had a length *xz* have been stretched so that they now have a length of *xyz*.

In the example cited, the beds were assumed to have been horizontal at the beginning of the deformation, and all of the folding is of the shear type. After earlier folds develop by some other mechanism, however, their form may be modified by shear folding. The presence of visible fractures with conspicuous, even though minute, offsets, is no proof that the folding was entirely of the shear type. Under conditions of extreme deformation, after an initial phase of flexure folding, closely spaced fractures may develop and slippage may take place. In the experience of the author, most shear folds evolve in this way.

Very complex map patterns may develop if shear folds are superimposed

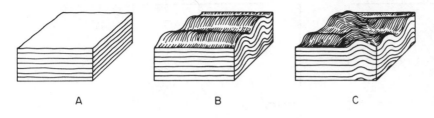

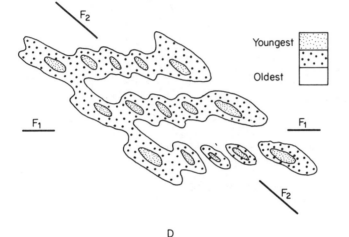

Fig. 6-5. Two stages of folding. An earlier set of upright east-west folds is deformed by a set of later shear folds trending northwest. (After E. S. O'Driscoll.[2]) Permission Alberta Society Petroleum Geologists and E. S. O'Driscoll.

on flexural folds or on older shear folds; this happens if the slip planes of the later shear folds are not parallel to the axial planes of the earlier folds.[2] An example is given in Fig. 6-5. Initially flat sediments (Fig. 6-5A) are deformed into a series of symmetrical folds with horizontal hinges and vertical axial planes, both of which strike east-west (Fig. 6-5B). These could be either shear folds or flexure folds. Shear folding then takes place by vertical slip, along vertical slip planes striking northwest (Fig. 6-5C). The map pattern, after some erosion, is shown in Fig. 6-5D. However, complex map patterns may develop in other ways.

Flow folding is similar to shear folding, except that the slip planes are infinitesimally close; that is, the deformation is anologous to the lamellar flow of liquids. Of course, flow and shear folding may be combined so that visible offsets of the beds occur in places but are absent elsewhere.

Flexure, shear, and flow folds may be distinguished from one another by various criteria, recognizing of course, that combinations may occur.

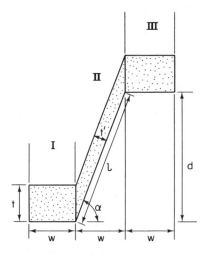

Fig. 6-6. Thinning of bed as result of shear folding.

In flexure folds the beds, especially the competent beds, tend to be the same thickness regardless of their position on the fold (Fig. 3-13). The incompetent beds may thin at the limbs and thicken at the hinges. Moreover, fossils and sedimentary structural features tend to be undistorted and maintain their original shape.

In the simplest cases of shear folding the slip planes are generally visible (Fig. 6-4) and the displaced beds have a uniform thickness in all the blocks. But because of drag the beds may be stretched and become thinner near the shear planes (Fig. 6-4).

Flow folding may usually be distinguished from flexure folding because the thickness of each bed is not constant. The thickness is a function of several variables, including the angle between the axial surface and the bed. One of many possible examples is shown in Fig. 6-6. Block I is assumed to remain stationary, whereas block III has been displaced a distance d. Block II deforms by lamellar flow. The planes of lamellar flow are vertical in this case. The width of each block is w. The thickness of the undeformed bed is t. In Block II the bed is stretched to a length l. Assuming no volume change, the thickness is reduced to t'. The angle between the normal to lamellar flow and the bedding in block II is α. The greater d, the greater will be l and the smaller will be t'.

$$\cos \alpha = \frac{w}{l} \tag{1}$$

$$l = \frac{w}{\cos \alpha} \tag{2}$$

$$tw = t' \frac{w}{\cos \alpha} \tag{3}$$

$$t' = t \cos \alpha \tag{4}$$

Thus if α is 0°, $t' = t$; if $\alpha = 45°$, $t' = 0.71t$; if $\alpha = 60°$, $t' = 0.5t$; and if $\alpha = 90°$, $t' = 0$. That is, in a flow fold formed in accordance with the

123

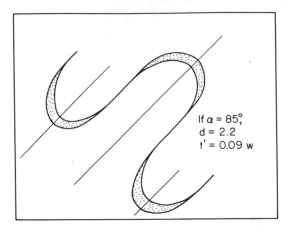

If $\alpha = 85°$,
$d = 2.2$
$t' = 0.09\, w$

Fig. 6-7. Bed deformed by shear or flow folding. The maximum thickness of the bed is at the hinge; the thickness is greatly reduced on the limbs.

stipulations given above, the beds become progressively thinner as the beds become more nearly parallel to the axial plane of the fold.

The greater the angle α, the greater the required displacement. If α is 0°, d is zero; if α is 30°, $d = 0.58w$; if $\alpha = 60°$, $d = 1.7w$; and if $\alpha = 90°$, $d = \infty$.

Figure 6-7 is a hypothetical flow fold that is not quite isoclinal; α is 85° on the limbs, where the thickness of the bed is reduced to 0.09 of its original value.

Many geologists have assumed that similar folds are necessarily shear or flow folds. But this is not necessarily so. In flexure folding the competent beds may deform into similar folds, but the incompetent beds thicken at the hinges and thin on the limbs (Fig. 3-13).

Dynamics of Folding

We are concerned here with the problem of the temperature and confining pressure at which folding takes place, as well as the stresses and time involved.[3,4] The geologist should be fully aware of the parameters that must be considered, despite the fact that so many variables are involved that precise calculations may be impossible.

Folding takes place under a wide range of temperatures and confining pressures. Folded moraines on mountain glaciers demonstrate that folding may occur at low temperature and low confining pressure.

The rocks of the Jura Mountains (Fig. 3-18) were folded under very little overburden and at comparatively low temperatures. The highly deformed strata of the Valley and Ridge Province of the Appalachians, including some very competent quartzites and limestones, may have been folded under a cover of one or two miles, but if such a cover existed, it has been removed by erosion. On the other hand, experimental data on the aluminum silicates (the system andalusite-sillimanite-kyanite) indicates that the Paleozoic metamorphic rocks of eastern Vermont and western New Hampshire were under a lithostatic pressure equivalent to at least eight miles at a temperature of 500°C.[5]

Several studies have suggested that the compressive stress to cause flexure folds in the Jura, Appalachians, and elsewhere may be in the range of 100 to 1000 kg/cm².[6, 7, 8]

Most of the studies of the dynamics of folding have utilized basic engineering principles supplemented by models constructed in accordance with proper principles of similarity.[9] In all instances the initial material is assumed to consist of one or more horizontal layers with mechanical properties differing from those of the medium in which it is embedded.

As a first approximation we may consider folding to be an elastic phenomenon. For a plate embedded in an elastic medium of infinite thickness

$$\lambda = 2\pi h \sqrt[3]{\frac{E}{6E_o}} \tag{5}$$

$$\sigma = \frac{3}{2} \sqrt[3]{EE_o^2 \frac{1}{6}} \tag{6}$$

where λ is the wavelength of the fold, h is the thickness of the plate, E is Young's modulus of the plate, E_o is Young's modulus of the medium, σ is the stress to cause folding.

For example, if a bed one meter thick with a Young's modulus of 5×10^{11} dynes were embedded in an elastic medium with a Young's modulus of 10^{10} dynes, the wavelength of the folds would be 12.7 meters and the compressive stress to cause folding would be 2×10^7 dynes/cm².

Similar equations consider cases where the enclosing medium is of finite thickness and other cases where several competent layers are present.

Equation (5) may be rewritten as

$$\frac{\lambda}{h} = 2\pi \sqrt[3]{\frac{E}{6E_o}} \tag{7}$$

That is, the ratio of the wavelength of the folds to the thickness of the beds is constant if the ratio of the moduli of elasticity is a constant. Field studies in one part of the Appalachians shows that the ratio of wavelength to thickness is 27, regardless of the size of the folds.[10]

The basic assumption that folding can be treated as an elastic phenomenon is generally unjustified. Folding is the result of permanent deformation. But it is conceivable that the wavelength of folds is initiated very early in the deformation under essentially elastic conditions.

Undoubtedly, time should be a factor in folding and the subject should be treated as an elasticoviscous phenomenon, that is, a process involving both elastic and viscous behavior.[11,12] Viscosity, in the narrowest sense, refers to the deformation of liquids. It is defined as the ratio of shearing stress to the rate of shear (page 30). Generally the stress is measured in dynes/cm². The rate of shear is the amount of shear divided by time, which is measured in seconds. Viscosity is expressed in poises, that is, in dynes/cm²sec. True viscosity applies only to fluids with Newtonian flow, that is, fluids in which the rate of flow is proportional to force. For those substances lacking true Newtonian flow, the term equivalent viscosity is often used. Some values of viscosity and equivalent viscosity are given in Table 2-2. Viscosity depends upon temperature and confining pressure as well as the material itself.

The dominant wavelength of folds in a viscous medium embedded in a viscous medium is

$$\lambda = 2\pi h \sqrt[3]{\frac{\eta}{6\eta_1}} \tag{8}$$

where λ is the wave length, h is the thickness of the bed in centimeters, η is viscosity of bed, η_1 is viscosity of enclosing medium.

The percent of strain is given by

$$S = \frac{\sigma}{4\eta} t \tag{9}$$

where S is the percent of strain, σ is the compressive stress in dynes, η is viscosity of the bed in poises, and t is time in seconds. The strain can take place both by folding and thickening of the plate.

The so-called amplification factor is the factor by which the amplitude of the folds is increased after a given time. For example, if the amplitude of the folds was one meter at time zero, but was 1000 meters at time t_1, the amplification factor at time t_1 is 1000.

The time to attain a given amplification factor is

$$t = \frac{\eta}{\sigma} \left[\frac{6\eta_1}{\eta} \right]^{2/3} \log_e A \tag{10}$$

where A is the amplification factor.

For an amplification factor of 1000, the equation becomes

$$t_1 = \frac{\eta}{\sigma} \left[\frac{6\eta_1}{\eta} \right]^{2/3} \times 6.91 \tag{11}$$

A hypothetical case will illustrate the preceeding equations. Assume that a bed 100 meters thick with a viscosity of 10^{21} is immersed in a medium with a viscosity of 10^{18} poises. Let the compressive stress, σ, be 10^8 dynes/cm². Then, from equation (8)

$$\lambda = 2\pi \times 10^4 \sqrt[3]{\frac{10^{21}}{6 \times 10^{18}}} = 3450 \text{ meters} \tag{12}$$

The time necessary for the amplitude of the folds to increase by a thousandfold is given by equation (11).

$$t = \frac{10^{21}}{10^8} \left[\frac{6 \times 10^{18}}{10^{21}} \right]^{2/3} \times 6.91 = 2.28 \times 10^{12} \text{ seconds} = 138,000 \text{ years}. \quad (13)$$

The amount of strain at this time is given by equation (9).

$$S = \frac{\sigma}{4\eta} t = \frac{10^8}{4 \times 10^{21}} \times 2.28 \times 10^{12} = 5.7 \times 10^{-2} = 5.7 \text{ percent}. \quad (14)$$

Experiments with layers of roofing tar (viscosity 3.0×10^7) immersed in two grades of corn syrup (viscosities 7.0×10^2 and 1.35×10^4) were in agreement with equation (8).

The preceding analysis indicates the parameters involved in folding. On the other hand, such mathematical analysis is difficult to apply with any great precision in natural occurrences. Strata are generally interbedded in a much more complex manner than the equations assume. For many reasons precise values cannot be given to such terms as Young's modulus and viscosity. In equation (9) it is not clear how much the "strain" is represented by folding and how much by a change in thickness of the plate. Obviously there are no experimental data concerning the time factor. As more data become available and as the equations become more refined, it will be possible to deal with these factors more precisely.[13, 14] But the field geologist should be fully cognizant of the parameters and problems involved.

Ultimate Causes of Folding

INTRODUCTION

Folds may be classified as tectonic or nontectonic. Those of *tectonic* origin result more or less directly from forces operating within the earth. Those of *nontectonic* origin are largely the result of superficial processes, often associated with erosion or deposition.

TECTONIC PROCESSES

By *horizontal compression* we mean a compressive force acting parallel to the surface of the earth. In other words, the greatest principal stress axis is horizontal. The principle of horizontal compression is illustrated by Fig. 6-1. In most cases the active force, analogous to a moving piston, operates on one side of the folded belt, whereas a resisting force is induced by a stationary block on the other side.

Laboratory experiments have been performed to produce folds by compression. The most exhaustive study was made 80 years ago[15] in order to understand more thoroughly the folds of the Appalachian Province. The experiments were performed in a pressure box (Fig. 6-8), the interior of which was $39\frac{3}{8}$ inches long, 5 inches wide, and 20 inches deep. A movable piston was at one end. Alternating layers, representing sedimentary rocks, were laid in the box. The various layers differed in thickness from $\frac{1}{16}$ of an inch to $2\frac{1}{2}$ inches. Each layer consisted of wax, plaster, and turpentine, the proportions having been varied in order to produce strata of different strength. A layer of BB shot more than a foot thick was placed over the artificial sedimentary rocks. The piston was then slowly moved from right to left to produce folds such as those shown in Fig. 6-9. A represents the first stage, D represents the last stage. The resulting folds, very similar to those in the Appalachian and

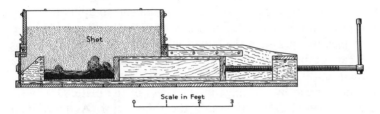

Fig. 6-8. Pressure box for producing folds. (After B. Willis.[15])

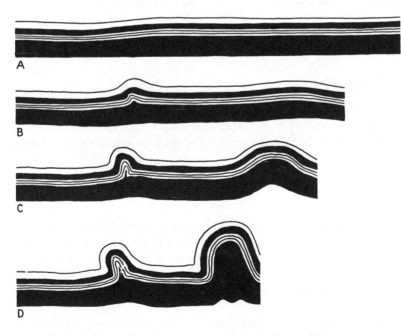

Fig. 6-9. Folds produced in pressure box. Piston active on right-hand end. (After B. Willis.[15])

Jura Mountains, indicate the effectiveness of horizontal compression to produce folds.

It is supposed, in applying this principle to orogenic belts, that the primary force acted at right angles to the trend of the folds.

If the intensity of the compressive force diminishes downward, the uppermost layers moving more than the lower layers, a couple is superimposed upon the simple compression (Fig. 6-1B). This causes asymmetry and overturning of the folds.

The primary forces in the Appalachian Province, where the folds trend northeast, were acting along northwest-southeast lines. In the North American Cordillera, where the folds trend northwest, the compressive forces acted along northeast-southwest lines.

A detailed consideration of the ultimate cause of the forces causing the horizontal compression is beyond the scope of this book, but a brief discussion should be of interest.

The *contraction theory* is classical.[16] This theory assumes that the interior of the earth has been growing progressively smaller throughout geologic time. This is supposedly due to one or more of the following reasons: (1) cooling of the earth; (2) formation of denser minerals within the earth; and (3) the extrusion of magma. The outer shell of the earth, compelled to accomodate itself to this smaller interior, has been subjected to strong compressive forces. Several objections have been made to this theory. One is that the apparent amount of crustal shortening throughout geologic time greatly exceeds the amount that can be reasonably expected. A second objection has been that the theory fails to explain the concentration of deformation in relatively narrow belts, such as the Appalachians or the Rocky Mountains, with large intermediate undeformed areas. How can the stresses be transmitted for such long distances without causing folds in the intervening areas? The old analogy with a dried apple suggests that during a single period of deformation many small folds should be more or less evenly distributed over the earth. A third objection is that the hypothesis fails to explain the supposed recurrent character of deformation, that is, the occurrence of short periods of orogeny separated by long periods of quiescence. However, orogeny may be more continuous than many geologists have thought in the past.[17]

Many geologists now accept the theory of *continental drift*, especially because of the results of paleomagnetic studies in the last two decades. The rigid continents are considered to be sufficiently mobile to move over the surface of the earth and crumple sedimentary rocks in front of them. Various hypotheses have been advanced to explain continental drift, including convection currents.

The mantle lies beneath the crust, which ranges from 20 to 50 kilometers in thickness. Although the upper part of the mantle may be relatively rigid, most of it is considered to be sufficiently mobile to permit *subcrustal* convection currents. The mantle extends to a depth of 2900 kilometers. Convection

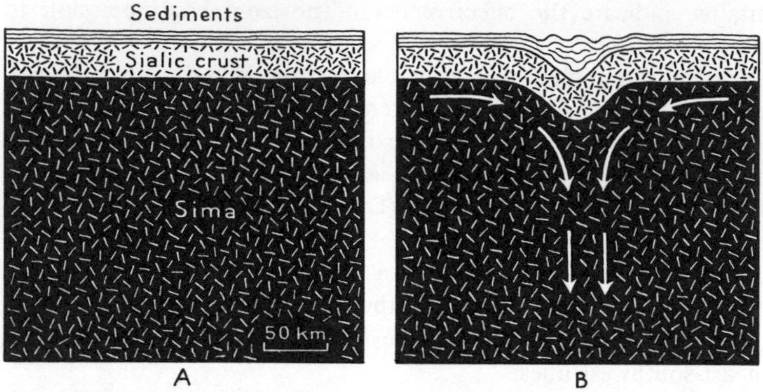

Sediments

Fig. 6-10. Orogeny caused by subcrustal convection currents.
(A) Before orogeny; sialic crust, about 25 km thick, overlain by
sediments and underlain by sima. (B) Convection currents (only
small part of whole convection cell is shown) cause downward
bending of crust and folding of sediments.

currents develop when the lower parts are sufficiently heated. These currents
drag along the base of the crust (Fig. 6-10). The crust is downfolded to form
a *root* and the overlying sediments, forced to occupy a smaller area than
formerly, are thrown into folds.

In recent years the theory of plate tectonics has been developed to explain
sea-floor spreading, continental drift, folding, and thrusting. The Mid-Atlantic
Ridge is believed to be a belt where new basaltic crust is constantly being
formed. The older basaltic crust on either side of this belt is being pushed
away from the ridge at the rate of several centimeters a year (Fig. 23-14).
This is sea-floor spreading. The type of evidence is largely geophysical and
further discussion is reserved for Chap. 23. A continent may ride along on top
of the spreading sea floor. Sediments that have accumulated in front of the
moving continent would be crumpled.[18]

Vertical movements are genetically related to folding in many ways.
Some folds are the direct result of vertical movements, others are an indirect
product.

Large open folds that are tens or hundreds of miles across with limbs that
dip less than a degree or at most only a few degrees, are produced by vertical
movements. Such folds are so large and the dips so low that they can be
deduced only from maps covering large areas. The geologic map of the United
States shows many such broad folds, among them the Cincinnati and Nash-
ville domes and the Michigan and Allegheny basins. Some of these structural
features took tens of millions of years to form. For example, the fact that the
Silurian strata thicken toward the center of the Michigan basin demonstrates
that it was actively subsiding for a long time.

Broad anticlines, such as those in the Southern and Middle Rockies,

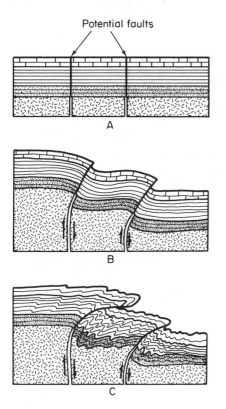

Fig. 6-11. Folding by vertical uplift. (A), (B), and (C) are successive stages.

with cores of Precambrian crystalline rocks, may be the result of vertical movements rather than the product of horizontal compression[19] and the associated folds and thrusts may also be incidental to vertical uplift.

Some Russian geologists believe that the folds in the Caucasus Mountains[20] are due to vertical movements (Fig. 6-11). As the blocks are differentially uplifted, the sediments drape over the edges of the blocks to form folds. Some sliding of the sedimentary rocks may also take place to form overturned and recumbent folds. A great deal of experimental work has produced folds of this type. Such an hypothesis means that the opposite sides of the mountain range are no closer together than prior to the folding. There has been no regional shortening. The extra distance as measured along the folded strata would represent stretching rather than shortening.

In recent years the concept of *gravity sliding* has been greatly emphasized.[21] As a result of vertical movements the upper part of an uplift becomes mechanically unstable and slides into the adjacent trough. Gravity sliding in its simplest form is illustrated in Fig. 6-12. The process may result not only in

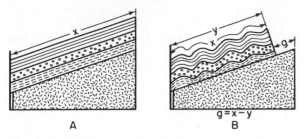

Fig. 6-12. Folding by gravity sliding. (A) Before sliding. (B) After sliding. *x*, width of strata before folding. *y*, width of folded belt after sliding. *g*, gap left at upper end after sliding of block.

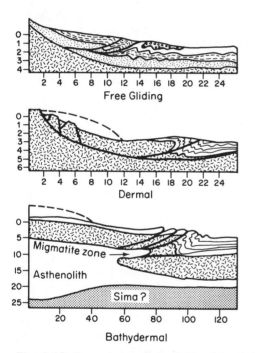

Fig. 6-13. Some types of gravitational tectonics. Distances in kilometers. Black with white hachures represents basement.

folding but, as will be discussed in Chapter 11, in detachment faults. *Free gliding*[22] involves only the sedimentary cover overlying a crystalline basement (Fig. 6-13A). *Dermal gliding* involves the upper part of the basement as well as the overlying sediments (Fig. 6-13B). In *bathydermal gliding* much of the crust is believed to become liquid or partially liquid (Fig. 6-13C).

It has been suggested that the strata of the Jura Mountains were folded (Fig. 3-18) when the whole sheet slid northward on a gently inclined surface.[23]

Many of the recumbent folds in the northern part of the Swiss Alps (Fig. 3-17) are considered by most geologists to be the result of gravity sliding. Unusually complex geometry has developed in the Apennine Mountains of Italy because of gravity sliding.[24,25] The highly incompetent allochthonous formation covering extensive areas of the Apennines cannot have advanced as a nappe, and is best accounted by the successive landslip theory; that is, by a succession of orogenic landslips down the outer slopes of the individual Apennine Ranges, each landslip spreading sufficiently outwards to be picked up and moved forward, together with any freshly deposited unconsolidated deposits by similar landslips on the outer slope of the next range to be uplifted.[26]

NONTECTONIC PROCESSES

Processes operating at or near the surface of the earth, especially in relation to erosion, may cause deformation. The resulting structural features are not only of interest in themselves, but may be confused by the unwary with tectonic features.

In many areas *hillside creep* may greatly modify the attitude of strata, so that incompetent rocks, such as shale, dip into the hillside. Careful study will show, however, that the dips may be very deceiving. In a zone several feet thick at the surface the dip may be unlike the true dip at depth. Figure 6-14A represents a case where the true dip is 60°E., but because of downhill creep the dip (Fig. 6-14B) is 60°W. Similarly, where the true dip is 60°W (Fig. 6-14C), near the surface the dip is only 20°W. (Fig. 6-14D). In such areas it may be necessary to have exposures five or ten feet deep because data obtained at the surface may be very misleading.

Collapse structures[27] are a type of gravity sliding (Fig. 6-15). In Fig. 6-15A a block of limestone has slid downhill. In Fig. 6-15B a block of limestone has folded back on itself to simulate a recumbent fold formed under true tectonic conditions.

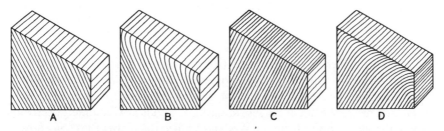

Fig. 6-14. Hillside creep, looking north. (A) True attitude of strata before hillside creep. (B) Same area as (A) after hillside creep. (C) True attitude of strata before hillside creep. (D) Same area as (C) after hillside creep.

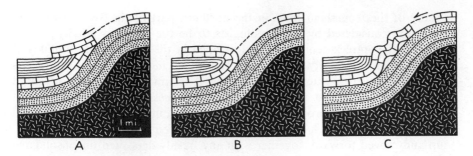

Fig. 6-15. Collapse structures. (A) Landslide simulating klippe. (B) Recumbent fold. (C) Folds caused by gravity sliding. (After J. V. Harrison and N. L. Falcon.[27])

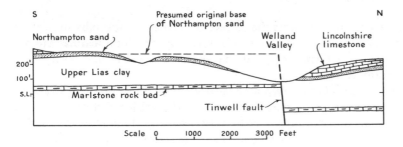

Fig. 6-16. Cambers, Welland Valley, England. (After S. E. Hollingsworth et al.[28])

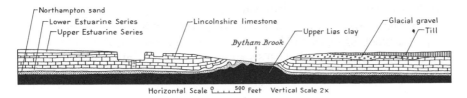

Fig. 6-17. Bulges. Bytham Brook, England. (After S. E. Hollingsworth et al.[28])

Cambers[28] have been described from Great Britain. As shown in Fig. 6-16 the Northampton Sand dips northward toward the Welland Valley. It will be noted, however, that the "Marlstone rock bed" is essentially horizontal, and hence the dip of the Northampton Sand is not tectonic. As erosion cut into the sediments to form the Welland Valley, the Upper Lias Clay flowed toward the valley to be carried away. The Northampton Sand was gradually let down.

Bulges[28] have also been described from Great Britain (Fig. 6-17). After erosion cut the valley occupied by Bytham Brook, the weak Upper Lias Clay bulged upward.

Glacial ice, pushing against the steep slope of a cuesta, may throw the strata, if they are poorly consolidated, into folds. Moreover, ice may override weak sediments and cause drag folds. Folds due to glacial action are well developed near the southern limits of the Pleistocene ice caps on Martha's Vineyard, an island off the southeastern coast of Massachusetts.

Solution of a chemically vulnerable formation may produce large structural features. Figure 6-18 is a structure contour map on the top of the Rustler formation of Permian age.[29] The salt in the underlying Salado and Castile formations, although in several separate beds, originally had a total thickness of about 2500 feet. In general, the domes shown by the top of the Rustler formation coincide with those places where the salt has been least dissolved, whereas the basins coincide with those places where the salt has been completely removed.

Calcium sulfate is most commonly precipitated from evaporating water as anhydrite. Water is subsequently added to convert the anhydrite into gypsum, and the increase in volume is approximately 40 percent. If the beds are flat-lying and if all the expansion takes place upward, the beds thicken, but no folds develop. If, however, much of the expansion is horizontal, compressive forces are set up and folding ensues. The resulting folds are small, with a height of only a fraction of an inch, or, at the most, of a few feet. Moreover, gypsum is not an abundant rock.

Differential compaction of sediments produces broad open folds that in many instances are of economic importance in petroleum geology. The general principles are illustrated by Fig. 6-19. Diagram A represents a land surface carved on solid rocks; the hill is 100 feet high. Later, as shown in diagram B, sediments are deposited on this surface; these sediments are only 100 feet thick on the top of the hill, but are 200 feet thick over the surrounding lowlands. For purposes of illustration, these sediments have been shown as absolutely horizontal. Actually, of course, there would be some minor irregularities, and particularly on the flanks of the hill there would be some outward *initial dips*. If at some subsequent time the mud compacts 20 percent over the top of the hill, the highest sedimentary bed will sink 20 feet—that is, 20 percent of 100 feet. Beyond the limits of the hill, the highest sedimentary bed sinks 20 percent of 200 feet—that is, 40 feet. Consequently, the highest beds will dip away from the center of the buried hill (Fig. 6-19C).

Contemporaneous deformation takes place as sediments are being deposited. Small folds and faults may form in soft sediments due to sliding down gentle slopes (Fig. 6-20). The hinges of the folds and the strike of the thrusts will be at right angles to the direction in which the sediments slide. The axial planes of the folds and the thrust faults dip in the direction from which the slide comes. Movements of this type may occur on slopes as low as two and one-half degrees. In some cases it is probable that the disturbed layer was

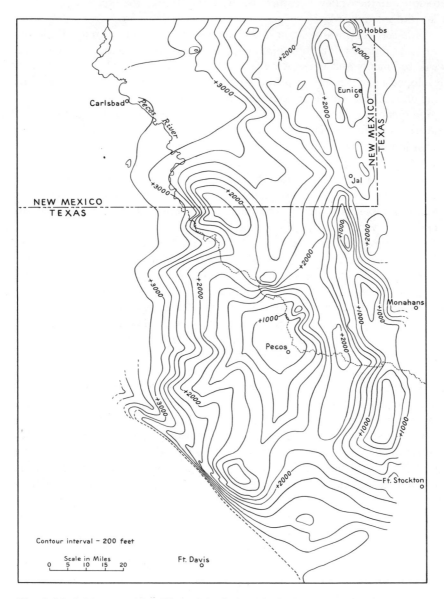

Fig. 6-18. Folds caused by differential solution of salt. Structure contours on top of Rustler formation of Permian age. Folds result from differential solution of underlying salt. (After V. C. Maley and R. M. Huffington.[29])

136

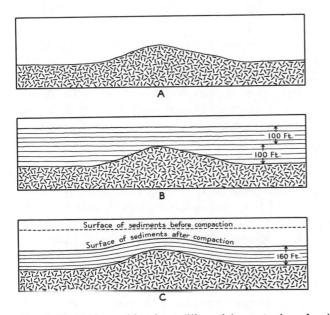

Fig. 6-19. Folds resulting from differential compaction of sediments. (A) Ridge left by erosion. (B) Same ridge covered by mud, but before compaction of mud. (C) Same ridge, but after compaction of mud.

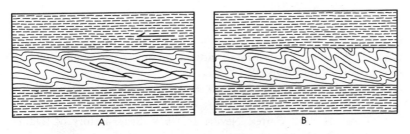

Fig. 6-20. Folds formed by contemporaneous deformation. (A) The uppermost layer slid from right to left over the middle layer. (B) The middle layer slid from right to left before the uppermost layer was deposited upon it unconformably.

covered by younger sediments at the time of the deformation (Fig. 6-20A). Although all the beds were unconsolidated at the time, some were more competent than others. The overlying sediments slide without deforming, hence the folds in the incompetent bed are analogous to drag folds (Fig. 3-14). In other cases it is clear that the beds slid when they were the uppermost strata, covered only by water, because the overlying sediments rest on them unconformably (Fig. 6-20B). After the sediments were deformed, mild subaqueous erosion truncated the folds.

Although in some areas the deformed layer may be only a foot to 15 feet thick, elsewhere it may be hundreds of feet thick.

Beds of soft mud, sand, and ooze may slide for a number of reasons. The slope on which deposition takes place may be inclined and, although the rocks are relatively stable for long periods of time, any disturbing factor may set them in motion. Slight tilting, excessive local deposition, or an earthquake may set masses of mud and ooze into motion. Local, subaqueous erosion may leave the beds with inadequate support. Large slides may even be related to contemporaneous faulting.

It may be difficult to distinguish between features resulting from contemporaneous deformation and those due to later forces. If, as is sometimes observed, the folds are truncated by younger beds of the same sedimentary series, as in Fig. 6-20B, it is evident that the deformation was contemporaneous. The folds resulting from contemporaneous deformation are similar to drag folds. If the major structure is known, the orientation of the minor folds may be useful in deciding whether they are contemporaneous or secondary. If they are true drag folds, they should be related to the major structure in the manner indicated on p. 90; otherwise they are probably the result of contemporaneous deformation. Obviously, however, this criterion is unreliable, and it may lead to erroneous interpretations.

References

[1] Carey, S. Warren, 1962, Folding, *J. Alberta Soc. Petrol. Geol.* 10: 95–144.

[2] O'Driscoll, E. S., 1962, Experimental patterns in superposed similar folding, *J. Alberta Soc. Petrol. Geol.* 10: 145–67.

[3] Hansen, E., 1971, *Strain facies (minerals, rocks, and inorganic materials)*, 220 pp., New York: Springer Verlag.

[4] Dieterich, James H., and Carter, Neville L., 1969, Stress-history of folding, *Amer. J. Sci.* 267: 129–54.

[5] Wilson, J. Robert, 1970, *Geology of the Ossipee Lake Quadrangle, New Hampshire*, New Hampshire Department of Research and Economic Development, 116 pp.

[6] Petrascheck, W. E., Jr., 1953, Die absolute Grösse des Faltungsdruckes, *Comptes rendus de la dix-neuvieme session*, Congr. Geol. Int., section III, Fasc. III, pp. 197–209.

[7] Elkins, W. E., 1942, Test of a quantitative mountain building theory by Appalachian structural dimensions, *Geophysics*, VII: 45–60.

[8] Kienow, S., 1942, Grundzüge einer Theorie der Faltungs–und Schieferungsvorgänge, *Fortschr. der Geol. und Pal.* Berlin.

[9] Hubbert, M. K., 1937, Theory of scale models as applied to the study of geologic structures, *Geol. Soc. Amer. Bull.* 48: 1459–1520.

¹⁰ Currie, John B., et al., 1962, Development of folds in sedimentary strata, *Geol. Soc. Amer. Bull.* 73: 655–74.

¹¹ Biot, M. A., 1961, Theory of folding of stratified viscoelastic media and its implication in tectonics and orogenesis, *Geol. Soc. Amer. Bull.* 72: 1595–1620.

¹² Biot, M. A., et al., 1961, Experimental verification of the theory of folding of stratified viscoelastic media, *Geol. Soc. Amer. Bull.* 92: 1621–32.

¹³ Ramberg, Hans, 1967, *Gravity, deformation, and the earth's crust, as studied by centrifuged models*, 214 pp., London and New York: Academic Press.

¹⁴ Chapple, William M., 1968, A mathematical theory of finite-amplitude rock folding, *Geol. Soc. Amer., Bull.* 79: 47–68.

¹⁵ Willis, Bailey, 1893, The mechanics of Appalachian structure, U.S. Geol. Surv., *13th Annual Report*, part 2, pp. 211–81.

¹⁶ Landes, K. K., 1952, Our shrinking globe, *Geol. Soc. Amer. Bull.* 63: 225–40.

¹⁷ Gilluly, James, 1949, Distribution of mountain building in geologic time, *Geol. Soc. Amer., Bull.* 60: 561–90.

¹⁸ Bird, John M., and Dewey, John F., 1970, Lithosphere plate-continental margin tectonics and the evolution of the Appalachian orogen, *Geol. Soc. Amer. Bull.* 81: 1031–60.

¹⁹ Eardley, A. J., 1963, *Relation of uplifts to thrusts in Rocky Mountains*, Amer. Assoc. Petrol. Geol., Memoir 2, pp. 209–19.

²⁰ Beloussov, V. V., 1962, *Basic problems in geotectonics*, 809 pp., New York: McGraw-Hill Book Company, especially pp. 516–28.

²¹ Bucher, Walter, 1956, Role of gravity in orogenesis, *Geol. Soc. Amer. Bull.* 67: 1295–1318.

²² Van Bemmelen, R. W., 1954, *Mountain building*, 177 pp., The Hague: Martinus Nijhoff.

²³ Lugeon, M., 1941, *Une hypothèse sur l'origine du Jura*, Bull. No. 73. Laboratoires de Géologie, Minéralogic, Géophysique et de la Musée Géologique de l'Université de Lausanne.

²⁴ Maxwell, John C., 1959, Turbidite, tectonic and gravity transport, northern Apennine Mountains, *Amer. Assoc. Petrol. Geol. Bull.* 43: 2701–19.

²⁵ Page, Ben M., 1963, Gravity tectonics near Passa Della Cisa, northern Apennines, Italy, *Geol. Soc. Amer. Bull.* 74: 655–72.

²⁶ Migliorini, C. I., 1952, *Composite wedges and orogenic landslips in the Appenines*, Eighteenth Int. Geol. Congr. (1948), part XIII, pp. 186–98.

²⁷ Harrison, J. V., and Falcon, N. L., 1934, Collapse structures, *Geol. Mag.* LXXI: 529–39.

²⁸ Hollingsworth, S. E., et al., 1944, Large-scale superficial structures in the Northampton Ironstone Field, *Quart. J. Geol. Soc. London*, C: 1–44.

²⁹ Maley, V. C., and Huffington, R. M., 1953, Cenozoic fill and evaporite solution in the Delaware Basin, Texas and New Mexico, *Geol. Soc. Amer. Bull.* 64: 539–46.

7

JOINTS

Observational Data

GEOMETRY

Most rocks are broken by relatively smooth fractures known as *joints* (Plates 15, 16, and 17A and B). The length of such fractures is measured in feet, tens of feet, or even hundreds of feet; the distance between them is likewise to be measured in feet or tens of feet. A special term is often used for fractures that are very closely spaced, where the interval between them is measured in inches or fractions of an inch (Chap. 18, p. 388). Although most joints are planes, some are curved surfaces. There has been no visible movement parallel to the surface of the joint, otherwise it would be classified as a fault. In practice, a precise distinction may not be possible or significant, because within a single set some fractures may show evidence of slight movement, others may not. Slight movement at right angles to the joint may produce an open fissure.

Plate 15. *Vertical joints.* At dam site at Lees Ferry, Coconimo
County, Arizona. Photo: E. C. LaRue, U. S. Geological Survey.

Plate 16. *Joints*. Vertical joint planes in Portage Formation. Tompkins County, New York. Photo: E. M. Kindle, U. S. Geological Survey.

Most joints, at least initially, are tight fractures. But because of weathering the joint may be enlarged into an open fissure; this is especially common in limestone regions. Most joints are smooth, but some display plumose mark-

Plate 17A. *Columnar jointing.* Columbia River Basalt near Race Creek, Riggins Quadrangle, Idaho. Photo: W. B. Hamilton, U. S. Geological Survey.

Plate 17B. *Columnar jointing.* Basalt flow. Devil Postpile National Monument. Madera County, California. Photo: H. R. Cornwall, U. S. Geological Survey.

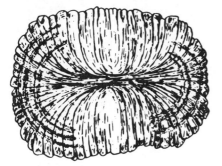

Fig. 7-1. Plumose and conchoidal structure on a joint several feet across. Comb Ridge–Navajo Mountain Area, Arizona and Utah. (After R. A. Hodgson.[2]) Permission American Association Petroleum Geologists.

ings (Fig. 7-1) that are ridges and depressions a millimeter or so in height.[1, 2] Slickensides (p. 201) indicate some movement parallel to the joint.

A knowledge of joints is important in many kinds of geological studies. Quarry operations, especially those involved in obtaining blocks of certain dimensions and sizes, are obviously greatly influenced by the joints. The orientation and concentration of joints is very significant in engineering projects.[3] Closely spaced horizontal joints are obviously of great concern in tunneling. A large joint dipping into a highway cut is the site of a potential landslide. Wells drilled in granites for water supply will be more productive in highly jointed rocks than in less jointed rocks. Many studies of joints have been made in order to deduce the orientation of the stresses to which the rocks have been subjected.

Joints may have any attitude; some joints are vertical (Plates 15 and 16), others are horizontal, and many are inclined at various angles. The strike and dip of joints are measured in the same way as for bedding. The strike is the direction of a horizontal line on the surface of the joint; the dip, measured in a vertical plane at right angles to the strike of the joint, is the angle between a horizontal plane and the joint. In Fig. 7-2 the geographic directions are shown. The front of the block (plane $ABCD$) is a joint that strikes east and has a vertical dip. The right-hand side of the block (plane $BEDF$) is a joint that strikes north and has a vertical dip. The plane $GHIJ$ is a joint that strikes north and dips 50 degrees east.

One type of symbol to show joints on maps is shown in Fig. 7-3; another type is used on Fig. 7-8.

The term joint is said to have originated in the British coal fields because the miners thought that the rocks were "joined" along the fractures, just as bricks are put together in a wall.

Joints may be classified either geometrically or genetically. A geometrical

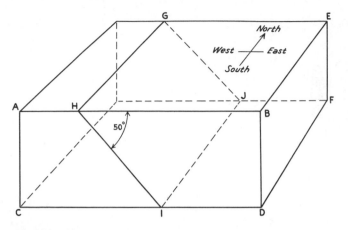

Fig. 7-2. Attitude of joints. Plane *ABCD* represents a vertical joint that strikes east-west; plane *BDEF* represents a vertical joint that strikes north-south; plane *GHIJ* represents a joint that strikes north-south and dips 50° east.

Fig. 7-3. Map symbols for joints. (A) Strike and dip of inclined joint. (B) Strike of vertical joint. (C) Horizontal joint.

classification is strictly descriptive and comparatively easy to apply, but does not indicate the origin of the joints. A genetic classification is more significant, but, as will be seen, is not readily applied in many cases.

In a geometrical classification, the joints may be classified on the basis of their attitude relative to the bedding or some similar structure in the rocks that they cut. *Strike joints* are those that strike parallel or essentially parallel to the strike of the bedding of a sedimentary rock, the schistosity of a schist, or the gneissic structure of a gneiss. In Fig. 7-4, in which the bedding is shown in solid black, *BDEF* and *MNO* are strike joints. *Dip joints* are those that strike parallel or essentially parallel to the direction in which the bedding, schistosity, or gneissic structure dips. *ABCD* and *GHI* are dip joints. *Oblique* or *diagonal* joints are those striking in a direction that lies between the strike and direction of dip of the associated rocks. In Fig. 7-4 *PQR* and *STU* are oblique joints. *Bedding joints* are parallel to the bedding of the associated sedimentary rocks. In Fig. 7-4 *JKL* is a bedding joint.

Characteristically a large number of joints are parallel. A *joint set* consists of a group of more or less parallel joints. A *joint system* consists of two

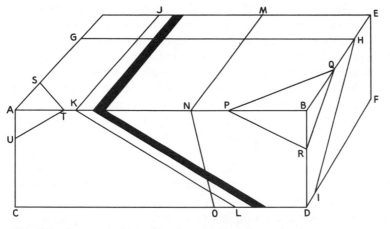

Fig. 7-4. Geometrical classification of joints. Heavy black layer is a bed. *ABCD* and *GHI* are dip joints; *BDEF* and *MNO* are strike joints. *JKL* is a bedding joint. *PQR* and *STU* are diagonal joints.

or more joint sets or of any group of joints with a characteristic pattern.

Joints may be classified according to their strike. It is thus possible to speak of the north-south set, the north-east set, or the east-west set. In some regions there may be several north-south sets; one set may be vertical, a second set may dip 40 degrees east, and a third set may dip 60 degrees west.

FIELD STUDIES

Joints may be readily observed in the field. But it is necessary to gather significant statistical data in order to convey the facts to others and to make pertinent analyses.[4] For many reasons this is not easy, partly because joints are so numerous and varied in attitude, but also because such detailed information cannot readily be portrayed on small scale maps.

In engineering projects, such as dam sites, it is usually possible to make large-scale maps, such as one inch to ten or twenty feet, after the unconsolidated surficial material has been removed and the bedrock exposed. Every joint, fault, and fold may be shown.

In much geological work, however, the maps are on a smaller scale—an inch to 2000 to 5000 feet. It is thus necessary to generalize the data obtained at individual outcrops; this may be difficult.

The size of joints in three dimensions is rarely known. For example, in an area of low relief, even with 100 percent exposure, the vertical dimension of the joints is unknown. Similarly, the size of a horizontal joint exposed on a cliff is known only in a direction parallel to the face of the cliff, but not perpendicular to it. In some places, of course, the face of a cliff or road cut may be parallel to a joint, and give some idea of the minimum dimensions. But in general the size of joints is impossible to analyze statistically.

Most emphasis has been placed on the orientation of the joints. For the area being studied, the attitude of at least 100 joints, and preferably several hundred, should be recorded. The data are then plotted on the lower hemisphere of an equal-area net. The pole of the perpendicular to each joint is represented by a dot, just as in the case of pi diagrams based on bedding (p. 100). Figure 7-5 is a plot of 311 joints in the Adirondack Mountains of New York state.[5] It is apparent that most of the joints dip very steeply and more strike northeasterly than in any other direction. Figure 7-6 is a contour diagram (page 104) of the same 311 joints and brings out the preferred orientation more vividly than the point diagram.

If the investigation covers a region with hundreds of thousands or millions of joints, it is obvious that some statistical approach is necessary. Dozens

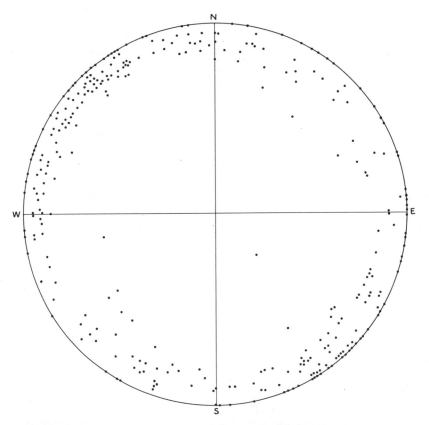

Fig. 7-5. Point diagram of 311 joints in Adirondack Mountains. Plotted on lower hemisphere. (After Balk.[5])

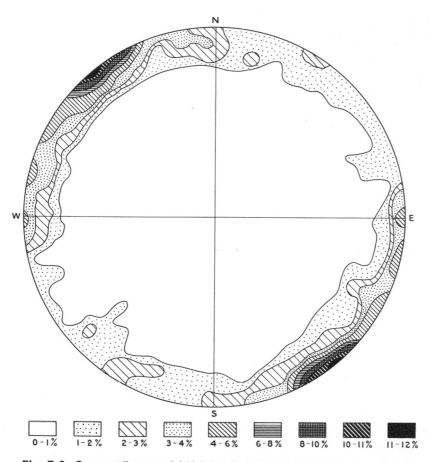

Fig. 7-6. Contour diagram of 311 joints in Adirondack Mountains. Plotted on lower hemisphere. (After Balk.[5])

of separate areas are chosen. In each of these areas the attitudes of *all* the joints within a prescribed radius are measured; the prescribed area is chosen so as to include 100 to 300 joints. A point diagram on an equal-area net, generally expressed as a contour map, is prepared for each of the small areas.[6]

In many areas another sampling problem arises. As in all geological field work, observations are necessarily restricted to outcrops or artificial exposures. The areas without outcrops may have a greater concentration of joints than the outcrops. Another problem is illustrated in Fig. 7-7. This is a very diagrammatic representation of a cubic mass of granite 100 feet on a side. Three sets of joints are present: (1) north-south strike, vertical; (2) east-west strike, vertical; and (3) horizontal. In all three sets the joints are spaced at five foot intervals. If only the surface of the block is exposed—that is, the region is one of low relief—the observer would record nine vertical

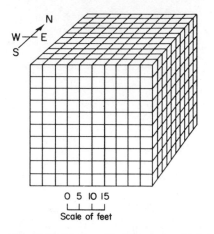

Fig. 7-7. Block with three sets of joints. Demonstrates importance of attitude of surface on which sets are seen.

north-south joints, nine vertical east-west joints, and possibly one horizontal joint. Similarly, if the only exposure were a cliff on the south side of the block, only one vertical east-west joint would be observed. On a north-south vertical cliff most of the north-south vertical joints would not be observed. It is apparent that many supposedly "statistical" analyses on the orientation of joints, because the data are gathered in areas of low relief, tend to de-emphasize gently dipping joints. Moreover, horizontal or gently dipping joints tend to be tighter and hence less readily seen than the steeper joints, partly because of the load of the rocks, partly because of weathering. One or more correction factors can be introduced, but so many variables are involved that they must be prepared separately for each area.

The concentration of joints may also be recorded. The data may be obtained in several ways, depending on the closeness of the joints. If the joints are fairly close, the number in a given set per square yard may be recorded. Or the number of a given set crossed in a traverse of unit length (10 feet or 100 feet, etc.).[7]

In areas of relatively simple stratigraphy and structure, the concentration of joints is partly a function of the thickness of beds of the same lithology. For example, in the Goose Egg Dome Area, Natrona County, Wyoming, there are 10 joints per square yard in a dolomite bed one foot thick, whereas there is only one joint per square yard in a dolomite bed 10 feet thick. Lithology is also a factor. In this same area a sandstone bed one foot thick has 15 joints per square yard, in contrast to 10 per square yard in the dolomite bed one foot thick.[7]

An excellent geologic map of jointing is given in Fig. 7-8; it takes into account the orientation, spacing, and size of the joints.[8] Many other fine field studies have been made.[9,10] Aerial photographs are often used in analyses of fracture patterns.[11]

Relatively few data have been obtained on the age of the jointing, but such information is vital in any analysis of the genesis.

It is axiomatic that joints and other fractures are no older than the rocks in which they occur. They may be essentially contemporaneous with the rocks in which they are present or they may be hundreds of millions of years younger.

A few criteria of age may be mentioned.

(1) If the joints beneath an unconformity have been opened up by weathering and filled by rocks above the unconformity, the joints are older than the overlying rocks.

(2) Joints are older than dikes or veins that have utilized them for emplacement.

(3) Short joints that abut against longer joints are probably the younger ones.

Many joints are symmetrically disposed about such structural features as folds and faults. This suggests that the joints are genetically related to folds and faults. This may be true, but it does not necessarily mean they are contemporaneous. Directional weaknesses due to folding or faulting may be utilized for jointing millions of years later.

Principles of Failure by Rupture

INTRODUCTION

A consideration of the genesis of joints must be preceded by a discussion of failure by rupture.[12,13,14,15,16,17] In Chap. 2 it is pointed out that rocks, when subjected to stress, deform elastically. Under atmospheric pressure and room temperature most rocks are brittle and fail by rupture at the elastic limit. But under high confining pressure and high temperatures most rocks undergo plastic deformation above the elastic limit. But under sufficiently high stress they may eventually fail by rupture.

Several factors are of concern: (1) the nature of the deformation preceding rupture; (2) the physical conditions at the time of rupture; (3) the stresses necessary to cause rupture; and (4) the orientation of the fractures relative to the causative stresses.

In this chapter we are concerned only with joints. But the fracture becomes a fault if displacement takes place parallel to the walls; this subject is discussed in Chaps. 8 through 12.

All ruptures may be classified as *tension fractures and shear fractures.* *Tension fractures* result from stresses that tend to pull the specimen apart. When the specimen finally breaks, the two walls may move away from each

Principal joint sets, with strike and dip
(Bedding joints, sheeting, and occasional joints lacking
systematic attitude are not shown on this map)

Group I.
(Most prominent and best developed; smooth, plane
surfaces; regular spacing, 6 in. to 8 ft.; large and
continuous; numerous joints per set)

Group II.
(Prominent and well developed; smooth to rough
surfaces; irregular spacing, 6 in. to 8 ft.; sometimes
discontinuous; fair number of joints per set)

Group III.
(Least prominent, but well developed; rough, curving
surfaces; irregular, wide spacing, more than 8 ft.;
large and continuous; few joints per set)

Vertical joint sets

Dense jointing, spacing less than 6 inches;
often slickensided

Slickensided joints, with bearing and pitch of
striae; arrow indicates relative displacement of hanging wall where known

Syncline Anticline
Tectonic axes, with bearing and plunge

Faults, observed and inferred

Contracts of Storm King granite,
observed and inferred

Granite lies *within* area bounded by contact lines.

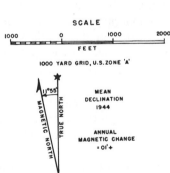

Fig. 7-8. Joints of Bear Mountain area,
New York. (After K. E. Lowe.[8])

152

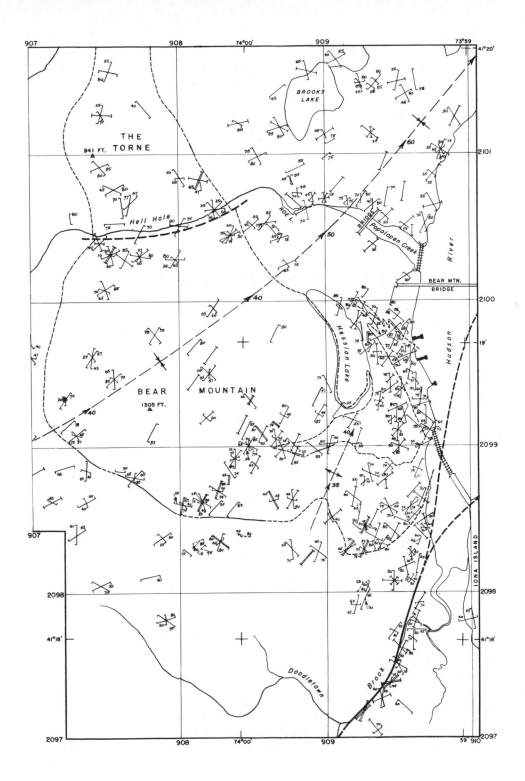

other. *Shear fractures* result from stresses that tend to slide one part of the specimen past the adjacent part. When the specimen finally breaks, the two walls may slide past one another.

It is of the utmost importance to distinguish between the character of the external force and the type of fracture. Tension fractures may result not only from tension, but also from couples and even from compression; as will be shown later, however, a special name is given to tension fractures formed by compression. Shear fractures may develop not only under compression, but also from couples and from tension.

EXPERIMENTAL DATA

In the simplest type of *tension*, the opposite ends of a rod are pulled apart. After elastic and plastic deformation, the specimen fails by rupture. The nature of the rupture depends upon the brittleness of the material. In brittle substances, such as wrought iron or a piece of blackboard chalk, a single tension fracture forms at right angles to the axis of the rod (Fig. 7-9A).

In more ductile substances, rupture may be preceded by "*necking*"; that is, the central part of the rod thins more than the ends (Fig. 7-9B). A conical fracture develops and, when failure ultimately occurs, a conical protuberance withdraws from a conical depression. In this case, the specimen has failed along shear fractures. In some material the rupture is a combination of a shear fracture and a tension fracture (Fig. 7-9C).

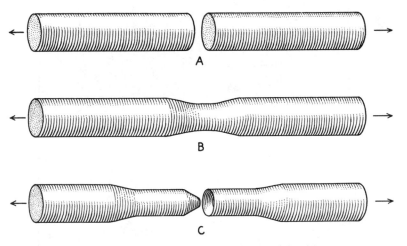

Fig. 7-9. Rod subjected to tension. (A) Brittle material, with a tension fracture at right angles to the axis of the rod. (B) Ductile material that has "necked," but not ruptured. (C) Ductile material that has ruptured; the conical surface is a shear fracture; the blunt end of the cone is a tension fracture.

Consolidated rocks near the surface of the earth are brittle substances and, when they are subjected to tension, we should generally expect them to fail by the formation of tension cracks. In other words, fractures should form at right angles to the tensional stresses.

— Rocks have a much lower tensile strength than compressive strength. Sandstone, for example, has an average compressive strength of 740 kilograms per square centimeter, but the same rock possesses an average tensile strength of only 20 kilograms per square centimeter.

In the simplest type of *compression*, the test specimen, usually a cylinder or a square prism, is subjected to a compressive force at two opposite ends, and the sides are free to expand outward. In other experiments, a square prism is compressed at the two ends, but it is confined on two opposite sides; the other two sides are free to move.

If the block is a square prism, unconfined on the sides, four sets of shear fractures develop. The four planes parallel to which the fractures form are illustrated in Fig. 7-10A by the planes *ABCD, EFG, HIJ*, and *KLMN*. Ordinarily, many fractures develop parallel to each of these four planes. As the com-

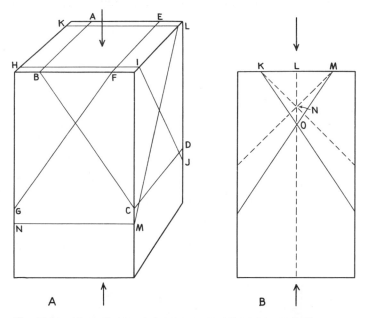

Fig. 7-10. Shear fractures due to compression. Arrows indicate compressive force. (A) In a square prism subjected to simple compression, four sets of shear fractures develop; they are parallel to the planes *ABCD, EFG, HIJ,* and *KLMN*. (B) *KN* and *MN* represent planes of maximum shearing stress deduced mathematically; *KO* and *MO* represent approximate position of shear fractures that form in experiments.

pressive force is increased, the fractures increase in number and size until one fracture eventually cuts all the way across the specimen and the block collapses. Some sets may be much more extensively developed than others, especially if the specimen lacks homogeneity.

The angle that is bisected by the compressive force—angle *KOM* of Fig. 7-10B—is always less than 90°, generally about 60°. That is, the angle between the compressive force and the shear fractures is about 30°.

If the square prism is confined on two opposite sides, two sets of fractures dip toward the unconfined sides of the specimen. If the front and back of the block shown in Fig. 7-10A were confined, only two sets of fractures, represented by *ABCD* and *EFG*, would form.

If the test specimen is cylindrical, the surfaces of rupture tend to assume a conical form; this is similar to the shear fractures that form in ductile rods under tension.

In many cases, however, specimens under compression fail along fractures parallel to the sides of the prism, especially if a lubricant is placed along the contact of the specimen and the piston exerting the compressive force (Fig. 7-11A). From one point of view these are tension fractures, on the principle that active compression in one direction sets up tensional forces at right

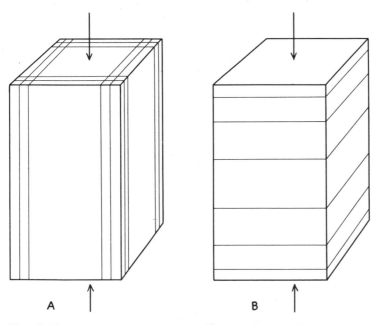

A   B

Fig. 7-11. Extension fractures and release fractures due to compression. Arrows indicate compressive force. (A) Extension fractures form parallel to sides of the prism. (B) Release fractures form parallel to top of prism.

angles. There are, however, theoretical objections to such an analysis, and ruptures of this type are preferably called *extension fractures*.

Yet another type of fracture results indirectly from compression. The specimen, while immersed in a fluid and under high confining pressure, is subjected to compression. After the load is released and the specimen removed, there are numerous fractures at right angles to the axis of compression (Fig. 7-11B). Such fractures are, in a sense, tension fractures caused by expansion of the specimen upon the release of load, but because there is no active tension they may be called *release fractures*.

The relation of ruptures to couples is illustrated by Fig. 7-12. A sheet of rubber is placed across a square iron frame, with hinges at each of the four corners. The rubber is then coated with a thin layer of paraffin. If the frame is then sufficiently deformed by a couple, the paraffin is broken by numerous cracks. The first ruptures are vertical tension fractures (*t* of Fig. 7-12B) that strike parallel to the short diagonal of the parallelogram.

This is not unexpected, for obviously the paraffin is being stretched parallel to the long diagonal of the parallelogram. After further deformation, vertical shear fractures (*s* of Fig. 7-12B) develop parallel to the sides of the wooden frame. These fractures are also not unexpected for they are analogous to the shear fractures that develop under compression. Their orientation, however, is controlled by the sides of the frame. Small thrust faults (*th* of Fig. 7-13B), which develop in the last step of the deformation, strike parallel to the long diagonal of the parallelogram.

Torsion results if opposite ends of a body are rotated in opposite directions. If the two ends of a piece of blackboard chalk are pulled apart, a tension fracture forms at right angles to the long axis of the specimen, as in Fig.

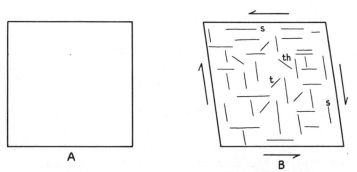

A   **B**

Fig. 7-12. Ruptures due to a couple. (A) Square frame that is covered by a sheet of rubber, on which there is a layer of paraffin. (B) Fractures that develop because of a couple: *t*, tension fractures (perpendicular to plane of paper); *s*, shear fractures (perpendicular to plane of paper); *th*, thrust faults (inclined to plane of paper).

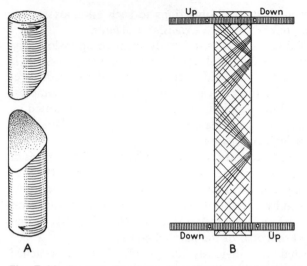

Fig. 7-13. Rupture due to torsion. (A) Helical fracture due to twisting of a piece of blackboard chalk. (B) Tension fractures on a piece of glass. Fractures that extend from upper left-hand to lower right-hand corner are on upper side of sheet of glass; those that extend from upper right-hand to lower left-hand corner are on underside of the sheet of glass.

7-9A. If the same material is twisted, a helical fracture develops, as is shown in Fig. 7-13A. Although chalk illustrates the difference between the ruptures produced by tension and torsion, the rock masses with which the structural geologist deals can scarcely be compared to rods. The twisting of sheetlike bodies, involving either a single bed, a formation, or the whole outer shell of the earth, is far more significant than the twisting of a rod.

The stresses developed in a twisted sheet can be resolved into the simpler stresses, tension or compression, and the ruptures obey the generalizations already set forth. Many years ago rather simple experiments were performed by twisting a sheet of glass. As shown in Fig. 7-13B, the upper right and lower left-hand corners were moved down, whereas the upper left and lower right-hand corners were moved up. Two sets of fractures developed. On the upper surface the fractures were diagonal, and they extended from the upper left-hand corner to the lower right-hand corner. On the underside the fractures were also diagonal, but they extended from the upper right-hand corner to the lower left-hand corner. A brief consideration explains why the glass fractures in this manner. If a sheet is bent, as in Fig. 6-2A, the upper part is subjected to tension, the lower part to compression. Between the two parts is a surface of no strain. In the sheet of glass that was twisted, the upper side was subjected to tensional forces acting from the upper right to the lower left-hand corner. Conversely, the tensional forces on the lower side acted from the upper left- to the lower right-hand corner. The tension fractures formed because the tensile strength is less than the compressive strength.

It is difficult to evaluate the importance of torsion as a cause of rupture in the rocks of the earth. It seems probable that torsion is an important type of deformation, but in any particular case it may be hard to decide whether local tension or local compression was due to regional torsion.

RELATION OF RUPTURE TO STRESS

Experimental data on the orientation of fractures under various conditions of deformation have been discussed in the preceding section. The same problem may now be treated from a more general and theoretical point of view using the *stress ellipsoid*.

The stresses acting at a point may be referred to three mutually perpendicular axes known as the principal stress axes.[14] In Fig. 7-14, σ_1 is the greatest principal stress axis, σ_2 is the intermediate stress axis, and σ_3 is the least principal stress axis. All three principal stresses can be compressive, all three can be tensile or various combinations are possible. In geology, a compression (pressure) is treated as positive, whereas a tension is considered as negative. In physics and engineering the reverse convention is generally followed. The *stress difference* is the algebraic difference between the greatest and least principal stresses.

A few specific examples will illustrate these concepts. If a compressive stress of 1000 kg/cm² is applied to the ends of a solid cylinder, the sides of which are confined only by air, the greatest principal stress, σ_1, is +1000 kg/cm², whereas the intermediate and least principal stresses are one atmos-

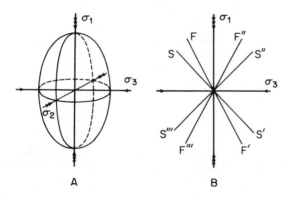

Fig. 7-14. Stress ellipsoid and rupture. (A) Stress ellipsoid. (B) Planes of maximum shearing stress (*SS'* and *S''S'''*) and planes of rupture (*FF'* and *F''F'''*).

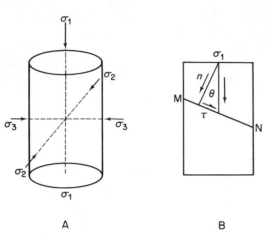

Fig. 7-15. Stresses on imaginary plane. (A) Solid cylinder under compression; σ_1, greatest principal stress; σ_2, σ_3, intermediate and least principal stresses. In this case σ_2 and σ_3 are equal. (B) Section through cylinder; τ, shearing stress; n, normal stress; θ, angle between σ_1 and plane.

phere, that is, $+1$ kg/cm². The stress difference is $+999$ kg/cm². If the sides of the cylinder were subjected to a hydrostatic pressure of 500 kg/cm², whereas the ends were subjected to a compressive stress of 2000 kg/cm², the greatest principal stress is $+2000$ kg/cm² and the intermediate and least principal stresses are $+500$ kg/cm². The stress difference is $+1500$ kg/cm². If a tensile stress of 1000 kg/cm² is applied to the ends of the cylinder, the sides of which are confined by air, the least principal stress is -1000 kg/cm², whereas the intermediate and greatest principal stresses are 1 atmosphere, that is, $+1$ kg/cm². The stress difference is $+1001$ kg/cm². If the sides of the cylinder are subjected to a hydrostatic pressure of 500 kg/cm², whereas the ends are subjected to a tensile stress of 2000 kg/cm², the least principal stress is -2000 kg/cm², whereas the intermediate and greatest stresses are $+500$ kg/cm². The stress difference is $+2500$ kg/cm², because the greatest principal stress is compressive, the least is tensile.

Figure 7-15B is an imaginary section through the cylinder shown in Fig. 7-15A and containing σ_1 and σ_3; σ_2 is perpendicular to the page and is neglected in the analysis. The principal stresses are compressive. We wish to calculate the stresses in a plane MN that makes an angle of θ with the greatest principal stress axis. The shearing stress, τ, and the normal stress, n, along the plane are:

$$\tau = \frac{\sigma_1 - \sigma_3}{2} \sin 2\theta \tag{1}$$

$$n = \frac{\sigma_1 + \sigma_3}{2} - \frac{\sigma_1 - \sigma_3}{2} \cos 2\theta \tag{2}$$

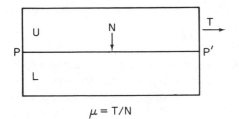

$$\mu = T/N$$

Fig. 7-16. Coefficient of friction. N, weight of upper block; T, force to move block; μ, coefficient of friction.

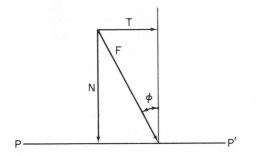

Fig. 7-17. Angle of internal friction. PP', plane; N, force perpendicular to plane, T, force parallel to plane, F, resultant of T and N; ϕ, angle of internal friction.

Thus τ is at a maximum when $\theta = 45°$; whereas n is at a minimum when $\theta = 0°$ and at a maximum when $\theta = 90°$ (when $\theta = 90°$, cos $2\theta = -1$).

Since actual shear fractures make an angle of less than 45° with the greatest principal stress axis, it is apparent that the normal stress is important. Shear fractures develop when n and τ combine to make the shear stress most effective. This problem is discussed further in a later section.

Before analyzing this problem it is necessary to consider two additional concepts. One is the theory of *internal friction*. The second is the concept of *cohesive strength* or *adhesive strength*.

First let us review the concept of the *coefficient of friction*. Figure 7-16 plane represents two blocks in contact along a horizontal plane PP'. N is the load exerted by block U on block L. The force to pull block U over block L is T. The coefficient of sliding friction, μ_s, is

$$\mu_s = \frac{T}{N} \tag{3}$$

The forces T and N can be thought of as two components of F making an angle ϕ with the perpendicular to PP' (Fig. 7-17). This is the *angle of friction*.

In like manner, along any potential plane of rupture within a rock,

$$\mu_i = \frac{\tau}{n} = \tan \phi \tag{4}$$

where μ_i is the *coefficient of internal friction*, $\tau =$ shearing stress along plane, $n =$ normal stress across plane, and ϕ is the *angle of internal friction*.

The total shearing resistance given by an isotopic substance to shear fracturing is the sum of the internal friction and the *cohesive strength*. The method to determine the value of these two parameters follows.

The relations between stress and rupture may be determined graphically by Mohr's stress circles. The accompanying diagrams are based on *triaxial compression tests*. The ends of a cylinder under confining pressure are compressed. Thus the greatest principal stress axis, σ_1 is parallel to the axis of the cylinder, whereas σ_2 and σ_3 are equal. Figure 7-18 is a plot of a single run. The origin is at 0. The vertical axis is shearing stress. The values of the confining pressure, σ_3, and compressive stress, σ_1, are plotted on the horizontal axis. The stress difference is $\sigma_1 - \sigma_3$. A circle is drawn through σ_3 and σ_1 with the center on the horizontal axis; the center of the circle is obviously $(\sigma_1 + \sigma_3)/2$ and the radius is $(\sigma_1 - \sigma_3)/2$.

Assume a plane making an angle θ with the greatest principal stress axis. The line cl is plotted to make an angle 2θ with the horizontal axis; 2θ is plotted clockwise from the horizontal axis. This line cuts the circle at l. The coordinates of l, which are τ' and n', give the shearing and normal stresses on the plane.

On a plane parallel to the greatest principal stress axis ($2\theta = 0$) the normal stress across the plane is σ_3 and the shearing stress is 0. If the plane makes an angle of 45° with the greatest principal stress axis ($2\theta = 90°$), the shearing stress is at a maximum and the normal stress is $(\sigma_1 + \sigma_3)/2$. If the plane makes an angle of 90° with the greatest principal stress axis ($2\sigma = 180°$), the shearing stress is 0 and the normal stress is σ_1.

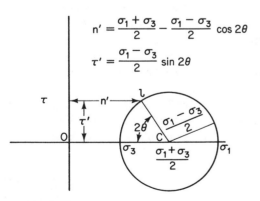

$$n' = \frac{\sigma_1 + \sigma_3}{2} - \frac{\sigma_1 - \sigma_3}{2} \cos 2\theta$$

$$\tau' = \frac{\sigma_1 - \sigma_3}{2} \sin 2\theta$$

Fig. 7-18. Mohr stress circle.

Experiments of the type outlined could be continued until the rock fails by rupture.

To determine the cohesive strength and angle of internal friction a series of experiments must be run with differing values of the confining pressure. Each experiment is run until rupture occurs. Figure 7-19 is the plot for five runs with different stresses. In the first experiment the confining pressure was atmospheric pressure. Each circle cuts the horizontal axis in two places. The left-hand intersection is the confining pressure for each experiment. The right-hand intersection is the compressive stress causing rupture in each experiment. The circles show that as the confining pressure is increased, the stress as well as the stress difference must be increased to produce rupture.

A line is now drawn tangent to these circles; this is the Mohr envelope. The angle that this line makes with the horizontal axis of the diagram is the angle of internal friction, ϕ.

The intercept on the vertical axis, τ_0, is the cohesive strength of the rock. The curve for the Mohr envelope is

$$\tau = \tau_0 + \sigma_n \tan \phi \tag{5}$$

The angle that fractures should theoretically make with the greatest principal stress axis is

$$\theta = \pm 45 \pm \frac{\phi}{2} \tag{6}$$

Thus, if the angle of internal friction is 32°, the fractures would make an angle of 29° with the greatest principal stress axis.

In experimental work it is difficult to measure the fracture angles with

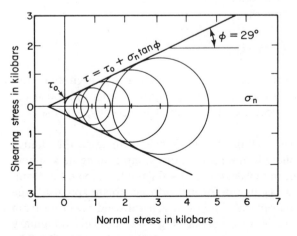

Fig. 7-19. Mohr stress envelope.

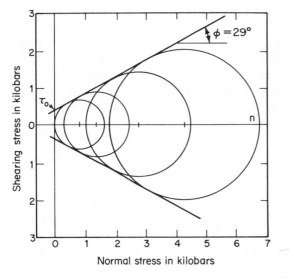

Fig. 7-20. Mohr envelope, Berea sandstone, at 24°C, dry-jacketed specimens under compression; four experiments with confining pressures of 0, .05, 1, and 2 kilobars. (After Handin et al.[12]) Permission American Association Petroleum Geologists.

great precision. Nevertheless, observations tend to confirm the fracture angles indicated by Mohr's envelopes.

A specific example of a Mohr's envelope is given in Fig. 7-20.[12] The rock is Berea sandstone tested at 25°C. Four tests were run, at confining pressures of 0, 0.5, 1, and 2 kilobars. The angle of internal friction is 29°, and the cohesive strength about 0.35 kilobars. Shear fractures should theoretically form at 31° (equation 6).

RELATION OF RUPTURE TO STRAIN

In the preceding section the relation of rupture to stress has been analyzed. Another method of analysis is to relate the rupture to strain, and the *strain ellipsoid*. This approach has been extensively used in geology, especially in North America.

A convenient way of visualizing deformation is to imagine the change in shape of an *imaginary* sphere in the rocks. For example, imagine a sphere in a body of granite. If the granite were compressed from the top and bottom, the imaginary sphere would become deformed into an oblate spheroid, the short axis of which would be vertical. The most general solid resulting from the deformation of a sphere is an ellipsoid (Fig. 7-21A). This imaginary figure may be called the *strain ellipsoid* or the *deformation ellipsoid*. The

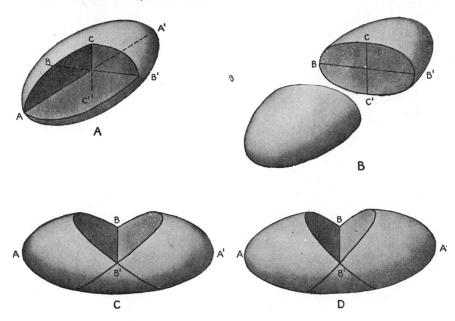

Fig. 7-21. Strain ellipsoid. (A) *AA'* is the greatest strain axis, *BB'* is the intermediate strain axis, and *CC'* is the least strain axis. (B) Tension fractures form perpendicular to the greatest strain axis. (C) Every ellipsoid has two circular sections that intersect at the intermediate axis *BB'*. Acute angle between circular sections is bisected by *AA'*. (D) Obtuse angle between shear fractures is bisected by *AA'*.

largest axis of the ellipsoid may be called the *greatest strain axis* (Fig. 7-21A); the intermediate axis is the *intermediate strain axis*, and the shortest axis is the *least strain axis*.

If tension fractures form, they are parallel to the plane that contains the least and intermediate strain axes (Fig. 7-21A); that is, tension fractures form at right angles to the greatest strain axis (Fig. 7-21B). If the attitude of the strain ellipsoid is known, the position of the tension fractures may be predicted. Conversely, if fractures can be identified as of tensional origin, the greatest strain axis is readily determined; the plane containing the least and intermediate strain axes is also defined, but the position of these axes within this plane can be determined only if additional data are available.

Most sections through ellipsoids are ellipses (Fig. 7-21A). Two of the sections, however, are circles. These *circular sections* pass through the intermediate axis *BB'* (Fig. 7-21C). Experiments show that shear fractures are closer to the least strain axis than the circular sections. In Fig. 7-22, a two-dimensional representation of the strain ellipsoid, the intermediate axis is perpendicular to the page. The planes *SS'* and *S"S'''* are the traces of the

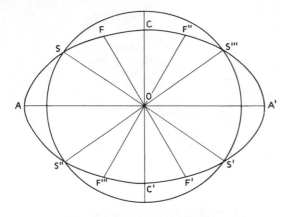

Fig. 7-22. Strain ellipse. *AA'* is the greatest strain axis. *CC'* is the least strain axis. *SS'* and *S''S'''* are the traces of the circular sections of the ellipsoid. *FF'* and *F''F'''* are the traces of the planes parallel to which shear fractures form.

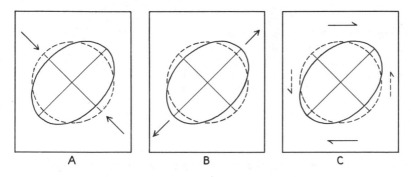

Fig. 7-23. Deformation of circle into an ellipse. (A) Compression. (B) Tension. (C) Couple.

circular sections. The angle between them and *CC'* is always greater than 45°. If this particular ellipse were the result of compression parallel to *CC'*, we know from experimental evidence that shear fractures would form parallel to *FF'* and *F''F'''*; that is, the angle between shear fractures and *CC'* is approximately 30°. (See also Fig. 7.21D.)

We may utilize this concept as follows. If two sets of shear fractures are present and are the product of the same deformation, the line formed by their intersection is parallel to the intermediate axis of the strain ellipsoid. Moreover, the least strain axis *CC'* bisects the *acute* angle between the shear fractures.

A point of fundamental importance is that the strain gives us no direct evidence of the external forces that caused the deformation. An ellipsoid may be formed from a sphere by simple compression, by tension, or by a couple. This fact is illustrated in two dimensions by Fig. 7-23, where the intermediate axis is perpendicular to the page. The three ellipses are identical. The ellipse in Fig. 7-23A, however, is the result of compression; that in Fig. 7-23B is the result of tension; and that in Fig. 7-23C is the result of a couple, as is shown by the solid arrows. The dotted arrows would give the same result. Thus even though the field geologist may accurately describe the strain, he cannot directly deduce the external forces without some additional evidence.

A simple example may serve to illustrate the use of the concept of the strain ellipsoid in relation to ruptures. Figure 7-24A is a cross section through a fault—that is, a fracture along which the blocks on opposite sides have been displaced relative to each other. Scratches on the surface of the fault indicate that the movement was parallel to the dip of the fault. The problem is to decide whether the eastern block moved up or down relative to the western block—that is, which arrows, those at a or those at b, represent the movement.

On the east side of the fault are short, open tension fractures, arranged *en échelon* as shown in Fig. 7-24A. The long axis of the strain ellipse AA' is therefore vertical, as is shown in Fig. 7-24B, and the ellipse is oriented as shown. If the couple along the fault acted as shown by the arrows at b, the long axis of the ellipse would be horizontal. If the couple acted as shown by the arrows at a, the long axis of the ellipse would be vertical. Inasmuch as the latter corresponds to the orientation deduced from the tension cracks, it is

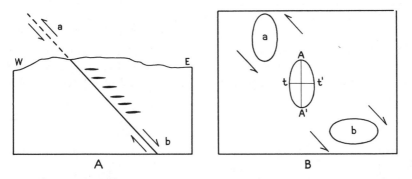

A	B

Fig. 7-24. Use of strain ellipse in a structural problem. (A) Cross section; heavy diagonal line is a fault. The problem is to decide whether movement has been of type represented by arrows at a or by arrows at b. Small gashes are open tension cracks. (B) The tension cracks of diagram A are represented by tt'; therefore the greatest strain axis lies in direction AA', which is the long axis of the strain ellipse. The movement represented by the arrows at a would give such an orientation of the strain ellipse; the movement represented by the arrows at b would not.

concluded that the east wall moved up relative to the west wall. This problem has been analyzed by omitting the third dimension, but in this case the intermediate axis of the strain ellipsoid is perpendicular to the plane of the paper and may be neglected.

Genetic Classification of Joints

Joints may be classified genetically as either shear fractures or tension (including extension) fractures. The ultimate causes may be several: (1) tectonic stresses, causing fracturing essentially contemporaneously with the tectonic activity; (2) residual stresses, due to events that happened long before the fracturing; (3) contraction due to shrinkage because of cooling or dessication; and (4) surficial movements, such as downhill movements of rocks or mountain glaciers.

Many joints are systematically disposed about folds and it has generally been assumed that they have resulted from the same compressive forces as those that produced the folds.

Joints perpendicular to the axes of folds are common in orogenic belts (Fig. 7-25, plane *ABCD*). Such joints may be *extension joints*, resulting from slight elongation parallel to the axes of the folds. They would be analogous to the ruptures that form parallel to the sides of specimens under compression (Fig. 7-11A).

Joints parallel to the axial planes of folds (Fig. 7-25, plane *EFGH*) may be *release joints* similar to those that form at right angles to the axis of compression when the load is released (Fig. 7-11B). Other joints with this attitude may be due to tension on the convex side of a bent stratum (Fig. 6-2).

Shear joints are difficult to recognize. If a joint is slickensided (p. 201), the opposite walls have obviously slipped past one another. But this is not proof that the fracture originated under shearing stress. The stress may have been tensional, and the sliding of walls past each other could be a later phenom-

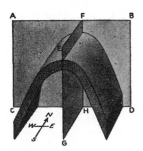

Fig. 7-25. Fold with vertical dip joint and vertical strike joint. *ABCD*, vertical dip joint. *EFGH*, vertical strike joint.

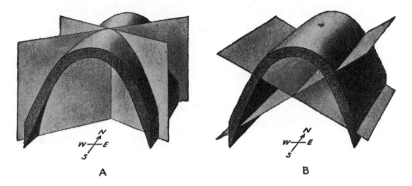

Fig. 7-26. Folds with conjugate joint systems. (A) Fold with vertical diagonal joints. (B) Fold with strike joints dipping about 30°.

enon. It is often supposed that tension fractures break around the pebbles of conglomerates and that only shear fractures cut indiscriminately across pebbles and matrix. Although this may be true for loosely consolidated conglomerates, apparently it is not a reliable criterion in well-cemented conglomerates.

Two sets of joints that intersect at a high angle to form a conjugate system are often considered shear fractures, especially if they are symmetrically disposed about the strain axes. Figure 7-26A is a block diagram of an area in which horizontal fold axes trend north-south. There are two sets of vertical joints, one of which strikes west-northwest, and the other of which strikes east-northeast. The attitude of the folds indicates that the compressive force was acting along east-west lines. In the light of experimental observations, the joints could be interpreted as shear fractures that developed due to a compressive force acting in an east-west direction, with the easiest relief in a north-south direction.

Figure 7-26B is a block diagram of an area in which the horizontal fold axes trend north-south and in which the axial planes are vertical. Two sets of joints strike north-south, but one set dips about 30 degrees east, whereas the other set dips about 30 degrees west. Such a conjugate system can also be interpreted as shear fractures due to an east-west compressive force, but under such conditions that the easiest relief was upwards.

Residual stresses may cause jointing. Some may be tectonic stresses that were not fully relieved. But another type of residual stress may be of even greater importance, especially in massive granitic rocks. Such rocks consolidate at depths of several miles in the crust under high lithostatic pressures. The stresses are initially essentially hydrostatic. Such rocks gradually come nearer to the surface of the earth due to erosion and tend to expand. The vertical expansion is readily accomplished, as only air is in the way. But the lateral stresses remain and become increasingly greater as erosion progresses.

Sheeting, a form of rupture similar to jointing, is best exposed in artificial openings such as quarries. The sheeting surfaces are somewhat curved and are essentially parallel to the topographic surface, except in regions where there has recently been rapid erosion. The fractures are close together at the surface of the earth, and in many places the interval between them is measured in inches. The interval increases with depth, and a few tens of feet beneath the surface the visible sheeting disappears. At greater depths, however, invisible planes of weakness parallel to the sheeting are utilized by quarrymen in their work. Although sheeting is best developed in granitoid rocks, it is also observed in sandstone.

Compressional forces parallel to the surface of the earth increase as erosion goes on. Eventually the strength of the rocks is exceeded. Shear fractures would be inclined at angles of about 30 degrees to the surface of the earth. But extension fractures (p. 157) would develop parallel to the surface.

Exfoliation domes are formed in some granitic areas. These are mountains composed of granite that is essentially unjointed except for the fractures that are parallel to the surface of the dome, dipping gently on the top but steeply on the cliffs. The rock in the surrounding lowlands is more easily eroded, either because of its lithology or because of its abundant joints. The least compressive stress is perpendicular to the surface of the dome, whatever its attitude, and the greatest compressive stress is parallel to the surface of the dome. Extension fractures would form parallel to the surface of the dome.

Many steeply dipping joints elsewhere may also be due to residual stresses. The greatest principal compressive stress would be horizontal but if the least compressive were horizontal, extension fractures or shear fractures would dip steeply.

Tension fractures due to a couple are represented by some of the crevasses in glaciers. Figure 7-27A is a diagrammatic map of vertical crevasses that are diagonal to the contact between a glacier and the rock walls. The direction in which the ice flows is shown by the large arrow. The friction with the wall sets up couples, which are indicated by the smaller arrows. The greatest strain axis is *AA'*, perpendicular to which tension fractures form.

Some joints are tension fractures resulting from the shrinkage of the rock. The shrinkage may be due to cooling or desiccation. Igneous rocks contract on cooling. Mud and silt contract because of desiccation.

But one of the best analyses of tension due to shrinkage has come from a study of permanently frozen ground in Alaska.[18] Irregular polygons, generally four-sided, from 10 to more than 100 feet in diameter, are especially spectacular when observed from the air. The polygons are separated by shallow trenches, several feet wide, each underlain by a vertical tabular mass of clear ice that extends downward for 10 to 20 feet. It is deduced that in the middle of the winter, whenever there is a rapid drop in temperature with-

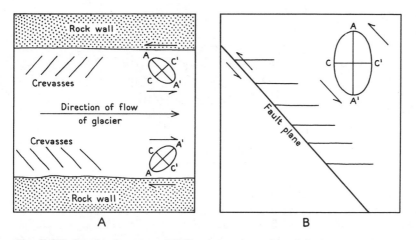

Fig. 7-27. Tension fractures. (A) Crevasses along side of glacier. Couples caused by friction are shown by smaller arrows. Ellipses represent orientation of strain ellipsoid. (B) Cross section to show feather joints, which are represented by horizontal lines to right of fault plane. Arrows near fault show relative movement along it. Orientation of ellipse resulting from this movement shown in upper right-hand corner.

in a few hours, a narrow vertical tension crack forms. During the ensuing spring thaw, which affects only the top few feet of soil, water filters down into this crack and eventually freezes. The process is repeated year after year so that the ice wedge gradually widens. These tabular bodies of ice are veins, or at least analogous to veins.

The deformation is viscoelastic. The temperature in the upper few feet of the frozen soil in midwinter ranges from $-20°$ to $-25°C$. At such low temperatures the "viscosity"—strictly the viscoelastic parameter—is large (1.25×10^{28} dynes^3 cm^{-6} sec). A sudden drop in temperature, at a rate of $5°C$ a day for a day or two, results in a thermal tension of 10 kg cm^{-2}. But this is approximately the tensile strength of the frozen ground. Consequently a tension fracture several feet deep would develop.

Such an analysis explains a simple crack, but does not explain a series of cracks 10 to 100 feet apart. The formation of a crack relieves 95 percent of the stress within a few feet of the crack. But if the cracks were 10 feet deep, only 10 percent of the stress would be relieved at a distance of 50 feet from the crack. Future cooling could easily build up stresses that result in renewed fracturing. Of course, many variables are involved, but fractures 10 to 100 feet apart may be expected.

The pattern of the cracks and the resulting polygons must also be explained. Once a crack has formed, tension at right angles to the crack is

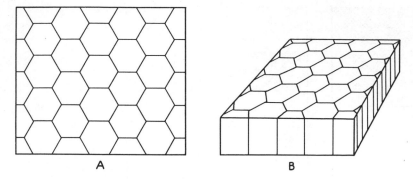

Fig. 7-28. Hexagonal fractures in a sheet. The tension results from loss of volume either because of cooling of lava or desiccation of mud. (A) Upper surface of sheet. (B) Block diagram of sheet.

reduced to zero in the immediate vicinity of the crack. But stress relief parallel to the crack is at a minimum. Consequently a new crack will form at right angles to the first crack. The resulting polygons would tend to be four-sided. This is the most common pattern in frozen ground. According to this theory the fractures develop successively and not simultaneously.

Mud cracks present a similar problem and will not be discussed further.[19]

The columnar jointing (Plate 17A) found in igneous rocks,[20] especially basaltic flows, sills, and dikes, is a contraction phenomenon (Fig. 7-28). The columns are generally a few feet to many feet in diameter, but are many feet long. Cross joints, perpendicular to the axis of the column, are commonly present. Locally the cross fractures have a "cup and socket" form similar to the conical fracture found experimentally in rods subjected to tension. The axes of the columns are generally perpendicular to the contact of the igneous body. Some of the columns are hexagonal, others are five-sided or four-sided. Joints formed at the surface of a cooling lava are very irregular.[21]

Such joints are undoubtedly the result of contraction due to cooling.[22] The basalt solidifies at about 1000°C and slowly cools. There is a general opinion that hexagonal columns are more common than hexagonal polygons in frozen ground or mud cracks. But no statistical data are available. However, the conditions may not be strictly analogous. In cooling lavas, dikes, and sills, a uniform tension may develop in the plane parallel to the contacts, and thus result in hexagonal fracture.

References

[1] Hodgson, R. A., 1961, Classification of structures on joint surfaces, *Amer. J. Sci.* 259: 493–502.

2 Hodgson, R. A., 1961, Regional study of jointing in Comb Ridge–Navajo Mountain area, Arizona and Utah, *Amer. Assoc. Petrol. Geol. Bull.* 45: 1–38.

3 Trollope, D. H., 1966, The stability of trapezoidal openings in rock masses, *Felsmechanik und Ingenieurgeologie* IV/I: 232–41.

4 Kirchmayer, M., and Denis, J. G., 1966, Information from joint diagrams, *Felsmechanik und Ingenieurgeologie* IV/I: 37–40.

5 Balk, Robert, 1931, Structural geology of the Adirondack anorthosite, *Mineralogische und Petrographische Mitteilungen*, Leipzig, 41: 308–433.

6 Wise, Donald U., 1964, Microjointing in basement, middle Rocky Mountains of Montana and Wyoming, *Geol. Soc. Amer. Bull.* 75: 287–306.

7 Harris, John F., et al., 1960, Relation of deformational fractures in sedimentary rock to regional and local structure, *Amer. Assoc. Petrol. Geol. Bull.* 44: 1853–73.

8 Lowe, Kurt E., 1950, Storm King granite at Bear Mountain, New York, *Geol. Soc. Amer. Bull.* 61: 137–90.

9 Spencer, E. W., 1959, Geologic evolution of the Beartooth Mountains, Montana and Wyoming, part 2, Fracture patterns, *Geol. Soc. Amer. Bull.* 70: 467–508.

10 Nickelson, R. P., and Hough, V. D., 1967, Jointing in the Appalachian Plateau of Pennsylvania, *Geol. Soc. Amer. Bull.* 78: 609–29.

11 Kelley, V. C., and Clinton, N. J., 1960, *Fracture systems and tectonic elements of the Colorado Plateau*, Pub. Univ. New Mexico Geol., No. 6, 104 pp.

12 Handin, John, et al., 1963, Experimental deformation of sedimentary rocks under confining pressure: Pore pressure tests, *Amer. Assoc. Petrol. Geol. Bull.* 47: 717–55.

13 Price, Neville J., 1966, *Fault and joint development in brittle and semi-brittle rock*, 176 pp., Oxford: Pergamon Press.

14 Anderson, E. M., 1951, *The dynamics of faulting and dyke formation, with applications to Britain*, 2d ed., Edinburgh: Oliver and Boyd.

15 Jaeger, J. C., 1956, *Elasticity, fracture, and flow*, 152 pp., New York: John Wiley & Sons; London: Methuen & Co. Ltd.

16 Mogi, Keyoo, 1966, Some precise measurements of fracture strength of rocks under uniform compressive stress, *Felsmechanik und Ingenieurgeologie* IV/I: 41–55.

17 Scholz, C. H., 1968, Microfracture and the inelastic deformation of rock in compression, *J. Geophys. Res.* 73: 1417–32.

18 Lachenbruch, Arthur H., 1962, *Mechanics of thermal contraction cracks and ice-wedge polygons in permafrost*, Geol. Soc. Amer., Spec. Paper 70, 65 pp.

19 Neal, James T., 1968, Giant dessication polygons of Great Basin playas, *Geol. Soc. Amer. Bull.* 79: 69–90.

20 Huber, N. King, and Rinehart, C. Dean, 1967, *Cenozoic volcanic rocks of the Devils Postpile Quadrangle, Eastern Sierra Nevada, California*, U.S. Geol. Surv., Prof. Paper 554–D, pp. D1–D19.

21 Peck, Dallas L., and Minakim, Takeshi, 1968, Formation of columnar joints in the upper part of the Kilauea Lava Lake, *Geol. Soc. Amer. Bull.* 79: 1151–69.

22 Jaeger, J. C., 1961, The cooling of irregularly shaped igneous bodies, *Amer. J. Sci.* 259: 721–34.

8

DESCRIPTION AND CLASSIFICATION OF FAULTS

General Characteristics

Faults (Plates 18 and 19) are ruptures along which the opposite walls have moved past each other. The essential feature is differential movement parallel to the surface of the fracture. Some faults are only a few inches long, and the total displacement is measured in fractions of an inch. At the other extreme, there are faults that are hundreds of miles long with a displacement measured in miles and even tens of miles.

The *strike* and *dip* of a fault are measured in the same way as they are for bedding or jointing. The strike is the trend of a horizontal line in the plane of the fault. The dip is the angle between a horizontal surface and the plane of the fault; it is measured in a vertical plane that strikes at right angles to the fault. The *hade* is the complement of the dip; that is, the hade equals 90 degrees less the angle of dip. The hade may also be defined as the angle between the fault plane and a vertical plane that strikes parallel to the fault.

Plate 18. *Fault.* Small high-angle fault in sandy shale one mile southwest of Little River Gap in Chilhowee Mountains. Blount County, Tennessee. Photo: A. Keith, U. S. Geological Survey.

Plate 19. *Thrust fault.* Lower Ordovician limestone, Gusta, southeast of Östersund, Norrland, Sweden. Cliff about 10 feet high. Separation along fault about 2½ feet. Photo: J. Haller.

Hade is a relatively obsolete term, but is common in the older literature. In Fig. 8-1 the front of the block is a plane that strikes east-west and dips vertically. The right-hand side of the block strikes north-south and has a vertical dip. The fault is an inclined plane that strikes north-south, dips 35 degrees east, and has a hade of 55 degrees east.

The block above the fault is called the *hanging wall* (Fig. 8-1); the block below the fault is the *footwall*. A person standing upright in a tunnel along a fault would have his feet on the footwall, and the hanging wall would hang over him. It is obvious that vertical faults have neither a footwall nor a hanging wall.

Although many faults are clean-cut, in many instances the displacement is not confined to a single fracture, but is distributed through a *fault zone*, which may be hundreds, even thousands, of feet wide. The fault zone may consist of numerous interweaving small faults, or it may be a confused

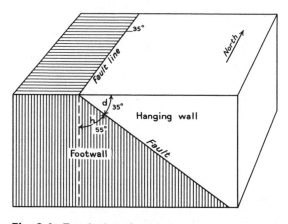

Fig. 8-1. Terminology for a fault plane. *d*, dip; *h*, hade.

zone of breccia or mylonite (p.202). *Distributive faulting* occurs if the differential movement takes place by systematic small displacements along a large number of closely spaced fractures.

The intersection of the fault with the surface of the earth is known as the *fault line, fault trace,* or *fault outcrop* (Fig. 8-1). In most instances, the fault line, as it appears on a map, is reasonably straight or somewhat sinuous. If, however, the dip of the fault is low and the topographic relief high, the fault line may be exceedingly irregular.

Nature of Movement Along Faults

TRANSLATIONAL AND ROTATIONAL MOVEMENTS

The movement along faults may be translational or rotational. In Fig. 8-2, diagrams *A* and *B* illustrate translational movement, whereas diagrams *C* and *D* illustrate rotational movement.

In *translational movement* there has been no rotation of the blocks relative to each other; all straight lines on opposite sides of the fault and outside the dislocated zone that were parallel before the displacement are parallel afterwards.

In Fig. 8-2A, two points, *a* and *a'*, contiguous before faulting, have been separated by the faulting. The right-hand block has moved directly down the dip of the fault relative to the left-hand block. The lines *bc* and *c'd*, which were parallel before faulting, are also parallel after faulting. In Fig. 8-2B, the right-hand block has moved diagonally down the fault; the lines *bc* and *c'd*, parallel to each other before faulting, are also parallel after faulting.

Rotational movements are those in which some straight lines on opposite

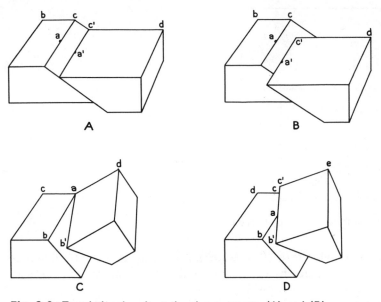

Fig. 8-2. Translational and rotational movements. (A) and (B), translational movements, (C) and (D), rotational movements. Lowercase letters are referred to in the text.

sides of the fault and outside the dislocated zone, parallel before the displacement, are no longer parallel afterwards. In Fig. 8-2C the right-hand block has gone down relative to the left-hand block, but the displacement increases toward the front; at point *a* there has been no displacement, but *b* and *b'* were contiguous before faulting. Lines *ca* and *ad*, parallel before faulting, are not parallel after faulting. In Fig. 8-2D, the back part of the right-hand block has gone up relative to the left-hand block, but the forward part of the right-hand block has gone down. The lines *dc* and *c'e*, parallel before faulting, are no longer parallel after faulting. The center of rotation is a point in the fault plane and the axis of rotation is a line that is perpendicular to the fault plane at this point. The center of rotation may shift with time, thus complicating the analysis of any movement.

In a sense, all faults have a certain amount of rotational movement. The displacement increases or decreases along the strike of all faults, and the blocks must rotate somewhat relative to one another. But if the rotation is not too great, the movements at any one locality may be treated as if the fault were a translational one.

RELATIVE MOVEMENTS

A rather elaborate terminology has, of necessity, been devised to describe the movement along faults and the effects on disrupted strata. The terminology has been devised primarily for translational movements, but it may be used with modifications for rotational movements. Throughout this book most of the terminology advocated by a committee of the Geological Society of America in 1913 has been followed.[1] A few additional terms proposed since the committee made its report have been incorporated and some suggested modifications have been accepted[2,3,4].

Faults in themselves never offer any direct evidence as to which block actually moved. Thus in Fig. 8-2A, the right-hand block may have gone down and the left-hand block may have remained stationary, or the left-hand block may have gone up and the right-hand block may have moved down; both blocks may have gone down, but the right-hand block may have gone down more than the left-hand block, or both blocks may have gone up, but the left-hand block may have gone up more than the right-hand block. Because in most cases no direct evidence is available concerning the absolute movements, the terminology is based chiefly on relative movements.

Figure 8-3 illustrates some of the various kinds of relative movements that may take place along a translational fault. In diagram A the hanging wall has moved directly down the dip relative to the footwall; in diagram B the hanging wall has moved parallel to the strike; and in diagram C the hanging wall has moved diagonally down the fault plane. In diagram D the hanging wall has moved directly up the dip of the fault, and in diagram E the hanging wall has moved diagonally up the fault plane.

The term *slip* is used to indicate the relative displacement of formerly adjacent points on opposite sides of the fault, and it is measured in the fault surface. The *net slip* (*ab* of Fig. 8-3) is the total displacement; it is the distance measured on the fault surface between two formerly adjacent points situated on opposite walls of the fault. It is defined in terms of the distance and the angle it makes with some line in the fault plane, such as a horizontal line or a line directly down the dip. In Fig. 8-3C, the net slip *cb* makes an angle of 35° with a horizontal line in the fault plane; the distance depends, of course, on the scale. It is also necessary to state the relative movement; in this case the hanging wall went down relative to the footwall. It is equally correct to say that the footwall went up relative to the hanging wall. The *strike slip* is the component of net slip parallel to the strike of the fault; it is indicated by *ac* in Fig. 8-3C. The *dip slip* is the component of the net slip measured parallel to the dip of the fault plane; it is *bc* of Fig. 8-3C. In Figs. 8-3A and 8-3D, the dip slip equals the net slip, and the strike slip is zero. In Fig. 8-3B, the strike slip equals the net slip *ab* and the dip slip is zero. In Fig. 8-3E, inasmuch as the movement is diagonal, there is both a dip slip component *ac* and a strike slip component *bc* to the net slip *ab*.

The *rake* is the angle that a line in a plane makes with a horizontal line in that plane. Thus in Fig. 8-4, the plane *ABGH* dips to the right and contains the line *BH*. The angle *ABH* is the rake of *BH*. Plunge was defined on p. 58; the only vertical plane in Fig. 8-4 that contains *BH* is the plane

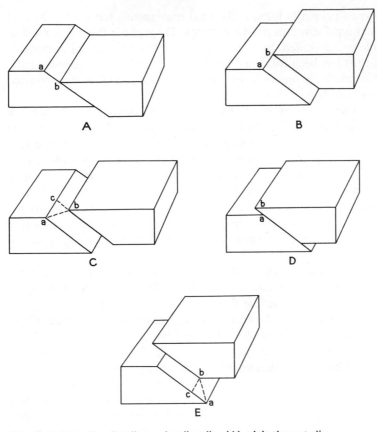

Fig. 8-3. Net slip, dip slip, and strike slip. (A) *ab* is the net slip, which in this case equals the dip slip; the strike slip is zero. (B) *ab* is the net slip, which in this case equals the strike slip; the dip slip is zero. (C) *ab* is the net slip; *cb* is the dip slip; *ac* is the strike slip. (D) *ab* is the net slip, which in this case equals the dip slip; the strike slip is zero. (E) *ab* is the net slip; *bc* is the strike slip; *ac* is the dip slip.

BCFH. The angle *CBH* is the plunge. The rake, the angle *ABH*, as measured in the plane *ABGH*, is about 57°; the plunge, the angle *CBH*, as measured in the plane *BCFH*, is about 45°.

In Fig. 8-3C, the rake of the net slip is the angle *bac*; in Fig. 8-3E, the rake is the angle *cba*.

In Fig. 8-5A, the fault intersects a bed or vein, which is shown in solid black; the fault dips toward the reader. The *trace slip* is that component of the net slip parallel to the trace of the bed on the fault. In Fig. 8-5A, the net slip is *ab*, the strike slip is *ac*, and the dip slip is *bc*. The trace slip is *db*. The *perpendicular slip* is that component of the net slip measured perpendicularly to the trace of the bed on the fault; it is *ad* of Fig. 8-5A.

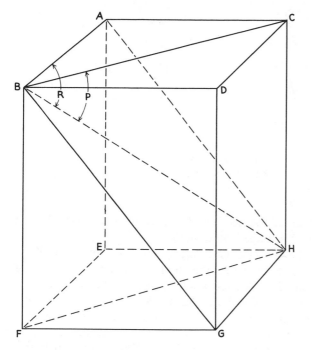

Fig. 8-4. Rake and plunge. The line *BH* lies in the plane *ABGH*. Angle *ABH* is the rake; angle *CBH* is the plunge.

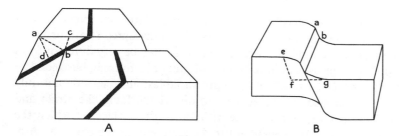

Fig. 8-5. Slip and shift. (A) Fault dips toward the reader; the black is a stratum of rock. *ab* is the net slip, *ac* is the strike slip, *cb* is the dip slip, *db* is the trace slip, *ad* is the perpendicular slip. (B) *ab* is the net slip, which in this case equals the dip slip; the strike slip is zero. *ef* is the net shift, which in this case equals the dip shift; the strike shift is zero.

A vertical plane perpendicular to the strike of the fault contains the dip slip. The front of the block shown in Fig. 8-6 represents such a plane, and *ad* is the dip slip. The *vertical slip, ae,* is the vertical component of the net slip and dip slip. The *horizontal slip* is the horizontal component of the net slip. The *horizontal dip slip, ed,* is the horizontal component of the dip slip.

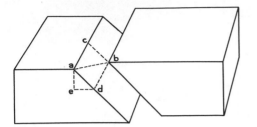

Fig. 8-6. Vertical slip and horizontal dip slip. *ab* is the net slip; *ac* is the strike slip; *cb* and *ad* are the dip slip; *ae* is the vertical slip, called *throw* by many geologists. *ed* is the horizontal dip slip, called *heave* by many geologists.

By some geologists *ed* is called the heave and *ae* is called the throw. As will be shown later, however, this usage is undesirable.

The slip refers to the displacements along the fault plane itself. If there is drag along the fault, however, the total displacement may be of more significance than the slip. The term *shift* is used to refer to the displacement on opposite sides of the fault and outside the dislocated zone. Figure 8-5B illustrates a fault along which there has been drag; the movement has been directly down the dip. The dip slip *ab* equals the net slip in this case. The net shift is *ef*; the dip shift is the same. The strike shift in this case is zero.

EFFECTS ON DISRUPTED STRATA

The preceding discussion has been confined to the relative movements along faults, and it has not considered the effects on the disrupted strata or veins. The apparent movement of the disrupted stratum may be very different from the net slip. This point is very important and cannot be overemphasized. The apparent movement is a function of many variables, and depends not only on the net slip, but also on the strike and dip of the fault, the strike and dip of the disrupted stratum, and the attitude of the surface on which the observations are made. It is possible for the apparent movement to be zero, although the net slip may be great.

Figures 8-7 to 8-16 show the relationship between the net slip and the apparent movement under different conditions. In Fig. 8-7, the beds are horizontal, and the net slip *ab* is directly down the dip. Figure 8-7A illustrates relations before erosion, and Fig. 8-7B illustrates the relations after the left-hand block has been eroded down to the level of the right-hand block. On the map—the upper surface of Fig. 8-7B—different beds outcrop on opposite sides of the fault. On the front of the blocks, the apparent movement equals the net slip. A deep valley or an artificial opening, such as a quarry or mine, might produce an exposure of this sort.

In Fig. 8-8, the net slip *ab* is parallel to the strike of the fault. In Fig.

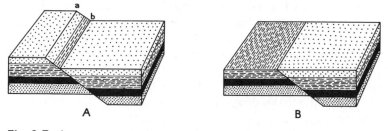

Fig. 8-7. Apparent movement in a vertical section equals the net slip. (A) Before erosion; *ab* is the net slip, which in this case equals the dip slip. (B) After erosion of top of footwall block.

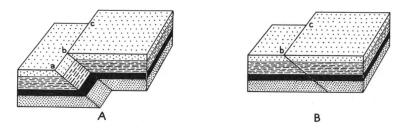

Fig. 8-8. Apparent movement in a vertical section is zero. (A) *ab* is the net slip, which in this case equals the strike slip. (B) After removal of front of footwall block.

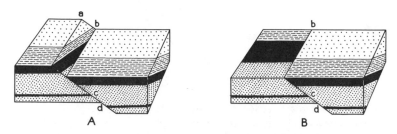

Fig. 8-9. Apparent movement in vertical section equals the net slip. (A) *ab* is the net slip, which in this case equals the dip slip. (B) After erosion of top of footwall block. A right separation.

8-8B, the front of the left-hand block has been eroded back to coincide with the front of the right-hand block. Such an exposure might be found on the side of a valley or in an artificial opening. The apparent movement in such a section is zero, although the net slip may be considerable. If the net slip were diagonally down the fault plane, a vertical section at right angles to the strike of the fault would show an apparent movement, but the value would be less than the net slip.

Figures 8-9 to 8-12 illustrate faults that strike at right angles to the strike of the bedding. In Fig. 8-9, the net slip *ab* is directly down the dip. Figure 8-9A shows the relations before erosion; Figure 8-9B shows the relations after the left-hand block has been eroded down to the level of the right-hand block. On the map—the upper surface of Fig. 8-9B—the apparent movement is such as to suggest that the left-hand block moved back a considerable distance parallel to the strike of the fault. If the beds have a low dip, a comparatively small net slip down the dip can give a large apparent displacement on the map. The apparent movement on the front of the blocks, Fig. 8-9, equals the net slip.

In Fig. 8-10, the net slip is parallel to the strike of the fault; Fig. 8-10A depicts the relations before erosion; Fig. 8-10B indicates the relations after the front of the left-band block has been eroded back to coincide with the front of the right-hand block. On the map the apparent movement equals the net slip. But the apparent movement on the front of the block, Fig. 8-10B, gives the false impression that the hanging wall has moved up.

In Fig. 8-11 the net slip *n* has been diagonally down the dip. After the surface and front of the left-hand block have been eroded to the surface and front of the right-hand block, respectively, the relations are those illustrated in Fig. 8-11B. On the map, the left-hand block has apparently moved back; in the structure section the right-hand block has apparently moved down.

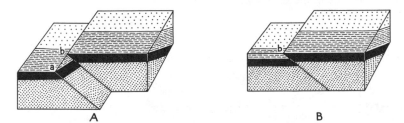

Fig. 8-10. Apparent movement in a vertical section gives the erroneous impression that the hanging wall has gone up. (A) *ab* is the net slip, which in this case equals the strike slip. (B) After removal of front of footwall block. A left separation.

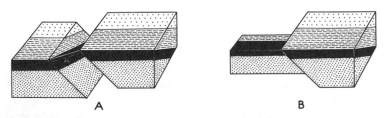

Fig. 8-11. Apparent movement in vertical section is less than net slip. (A) *n* is the net slip. (B) After removal of top of footwall block.

Figure 8-12 illustrates the special case in which the net slip is parallel to the trace of the bedding on the fault plane. Fig. 8-12B shows the relations after erosion of the top and front of the left-hand block. There is no apparent movement either on the map or on the front of the block. In fact, the generalization may be made that wherever the net slip is parallel to the trace of the disrupted stratum on the fault plane, there is no apparent movement on the map or cross sections. Figure 8-8 also illustrates this principle.

Figure 8-13 is an example of a fault that strikes parallel to the strike of the disrupted strata. The hanging wall has gone down relative to the footwall. The apparent movement in the front of the block in Fig. 8-13A equals the net slip. On the map of Fig. 8-13B, some of the beds are repeated because of the faulting. If the net slip were parallel to the strike of such a fault, there would be no apparent movement because the net slip would be parallel to the trace of the beds on the fault. If the hanging wall were to move diagonally down the dip, some of the beds would be repeated, but the apparent movement on a cross section at right angles to the strike of the fault would be less than the net slip.

Figure 8-14 illustrates a fault that strikes parallel to the strike of the strata, but the hanging wall has moved up relative to the footwall. The net slip is the same as the apparent movement on the front of the block in Fig.

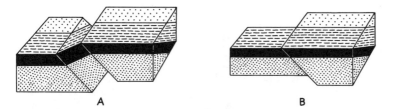

Fig. 8-12. Net slip parallel to trace of bedding on fault. Apparent movement in a vertical section and in map is zero. (A) Immediately after faulting. (B) After removal of top and front of footwall block.

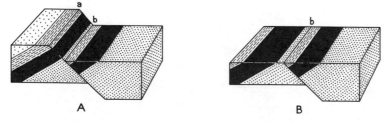

Fig. 8-13. Apparent movement in vertical section equals net slip. (A) *ab* is the net slip, which in this case equals the dip slip. (B) After removal of top of footwall block.

8-14A. If the right-hand block is eroded to the level of the left-hand block, the bed shown in solid black does not crop out at the surface.

Figure 8-15 is the special case in which the fault and the strata have not only the same strike, but have also the same dip. It is obvious that, in such a case, the apparent movement on the map and in the cross section is zero, regardless of the value of the net slip.

Figure 8-16 represents the case where the fault strikes diagonally to the strata and where the hanging wall has moved directly down the dip of the

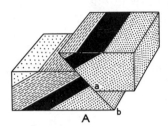
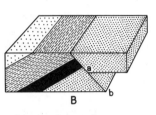

Fig. 8-14. Apparent movement in vertical section equals net slip. (A) *ab* is the net slip, which in this case equals the dip slip. (B) After removal of top of hanging wall block.

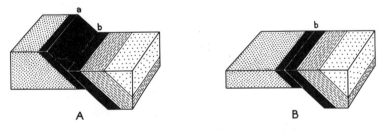

Fig. 8-15. Fault is parallel to bedding, and hence there is no apparent movement. (A) *ab* is the net slip, which in this case equals the dip slip. (B) After removal of top of footwall block.

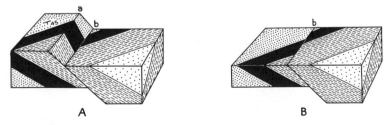

Fig. 8-16. Apparent movement on map does not equal the net slip. (A) *ab* is the net slip, which in this case equals the dip slip. (B) After removal of top of footwall block.

fault. After erosion, as shown in Fig. 8-16B, the apparent movement on the map suggests that the hanging wall has moved back relative to the footwall.

These numerous examples have been cited to emphasize that the apparent movement may be very deceiving. Moreover, it is disconcerting to realize that even if we know the dip and strike of the fault, the dip and strike of the disrupted strata, and the apparent movement, it is nevertheless impossible to determine the net slip. Suppose, for example, that the geologist mapped an area such as that shown on the top of the block diagram in Fig. 8-9B. The attitude of the bedding and of the fault are known; moreover, the apparent movement along the fault is given by the map. Actually, the net slip was directly down the dip. But the observed, apparent movement on the map could have resulted equally well from a horizontal movement parallel to the strike of the fault—that is, if the left-hand block moved backward relative to the right-hand block. Moreover, diagonal movement down the dip of the fault would have produced the same effect on the map.

CALCULATION OF NET SLIP

The amount and nature of the movement can be determined, however, if the strike and dip of the fault and the strike and dip of two or more planes with different attitudes are broken by the fault. Figure 8-17A is an example of two veins, $aa'a''a'''$ and $bb'b''b'''$, that have different attitudes and are displaced along a fault. This problem can be solved by graphical methods; the rake of the net slip is 85 degrees toward the northeast, and the hanging wall (north block) has moved up relative to the footwall; the value of the net slip is 175 feet. The disrupted bands may be dikes, veins, bedding planes, or older faults. (See also pages 559 to 568.)

Even if data are available for only one disrupted band, the problem can be solved if the direction of movement is known. The striations on a slicken-

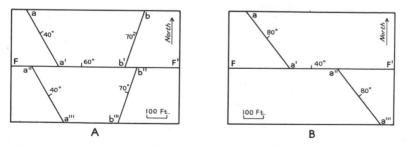

Fig. 8-17. Calculation of net slip. (A) $aa'a''a'''$ and $bb'b''b'''$ are two veins displaced along the fault FF'. With these data the net slip can be calculated. (B) $aa'a''a'''$ is a vein displaced along the fault FF'. The net slip cannot be calculated unless some additional data are available.

sided surface may show the direction of movement, but this method is
dangerous, for as is stated on p. 201, slickensides may record only the last
movements along the fault plane. Figure 8-17B shows how this method may
be used. In addition to the data indicated on the map, it is known that the
striations rake 45 degrees toward the northeast. The net slip may be calculated by graphical methods to be 260 feet.

SEPARATION

Separation indicates the distance between two parts of the disrupted horizon
measured in any indicated direction. The *horizontal separation* is the separation measured in any horizontal direction. Figure 8-18A is a map of a fault
and a disrupted horizon. The line *we* is the horizontal separation in an east-
west direction. The line *ns* is the horizontal separation in a north-south
direction. The strike separation, *hi*, is the horizontal separation parallel to the
strike of the fault. The bed in Fig. 8-18A shows a *right separation*[2] because an
observer following the bed along the strike must turn to the right to find the
same bed across the fault. The terms *right-handed* and *right-lateral* have also
been used, but the word "separation" should be used to avoid confusion with
similar terms used to refer to slip (p. 179). The bed in Fig. 8-18B shows a
left separation; the terms *left-handed* and *left-lateral* have also been used, but
here also the word separation should be used. It should be emphasized that
these terms apply to the disrupted stratum, and not to the fault, as two strata
may show different separations, as shown in Fig. 8-17A. The *offset* or *normal
separation* is measured perpendicular to the disrupted horizon; *ji* in Fig.
8-18A is the offset. The overlap is *hj*. In Fig. 8-18B, *mn* is the offset and *ln*
is the *gap*.

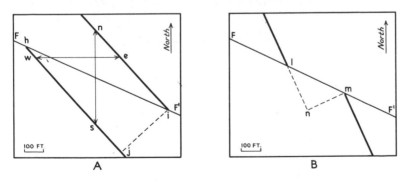

Fig. 8-18. Separation as shown on geological map. Heavy black
band is disrupted bed. *FF'*, fault trace. (A) *ns*, north-south
horizontal separation; *ew*, east-west horizontal separation; *hi*,
strike separation; *ji*, offset; *hj*, overlap. (B) Geological map. *lm*,
strike separation; *mn*, offset; *ln*, gap.

The *vertical separation* is the separation measured along a vertical line. In Fig. 8-19B, which is a vertical section perpendicular to the strike of the fault, *eg* is the vertical separation. The *dip separation* is the separation measured directly down the dip of the fault. In Fig. 8-19B, *ej* is the dip separation.

Faults bring beds into contact that are normally separated by intervening strata with a definite thickness. The thickness of these intervening beds is the *stratigraphic separation* or *stratigraphic throw* along the fault. Figure 8-20 shows how the stratigraphic separation may be determined. Along the fault, bed *m* in the right-hand block is brought into contact with bed *n* in the left-hand block. From the top of the right-hand block it is possible to calculate the thickness of the beds between *m* and *n* according to the equation

$$t = gf \times \sin \delta$$

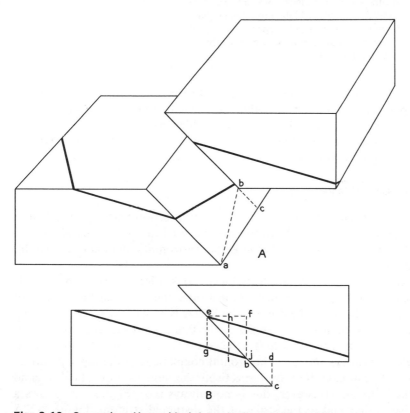

Fig. 8-19. Separation. Heavy black layer is disrupted bed. (A) Block diagram; *ab*, net slip; *be*, dip slip; *ac*, strike slip. (B) Vertical cross section perpendicular to strike of fault. *eg* and *hi* are the vertical separation; *eb*, apparent movement, which is the dip separation; *fj*, throw; *ef*, heave.

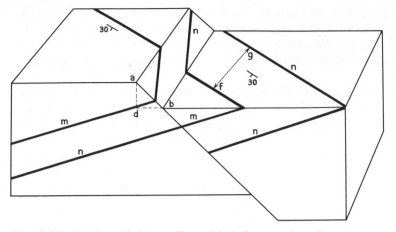

Fig. 8-20. Stratigraphic throw. Heavy black lines are two disrupted beds, *m* and *n*. *ab* is the net slip, which in this case equals the dip slip; *ad* and *db* are the vertical and horizontal slips. *gf* is the breadth of outcrop between beds *m* and *n*. Stratigraphic throw can be calculated as shown in the text.

In this equation t = thickness of beds between *m* and *n*; gf = breadth of outcrop between *m* and *n* measured perpendicularly to the strike of the strata; and δ = angle of dip of beds. The stratigraphic separation is one of the most important measurements of all, because in sedimentary rocks it can be determined with precision if the stratigraphy is known.

THROW AND HEAVE

The *throw* and *heave* are measured in a vertical section that is perpendicular to the strike of the fault. The throw is the vertical component of the dip separation in such a section; heave is the horizontal component of the dip separation.

Figures 8-21 and 8-22 illustrate these terms. Figure 8-21 is an example in which the fault strikes parallel to the strata, and in which the movement has been directly down the dip. Figure 8-21A is a block diagram, and Fig. 8-21B, a cross section at right angles to the fault, is the same as the front of the block diagram. The throw (*ad*) is the vertical component of the dip separation (*ab*). The heave, *db*, is the horizontal component of the dip separation. In this particular case, they are respectively the same as the vertical slip and horizontal dip slip. However, this is not always true. Figure 8-22 shows a fault that is diagonal to the strike of the disrupted stratum. Figure 8-22A is a block diagram, and Fig. 8-22B is a vertical section perpendicular to the strike of the fault; the throw is *df*, and the heave is *fe*. The vertical slip is *ib*, and the horizontal dip slip is *ci*. They are obviously different from the throw and heave.

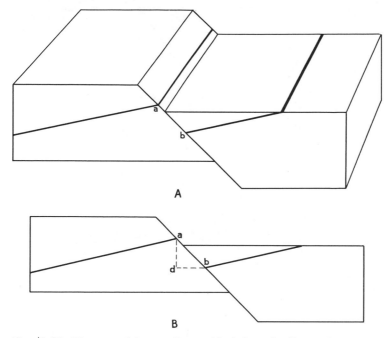

Fig. 8-21. Throw and heave. Heavy black layer is disrupted stratum. (A) Block diagram; *ab* is the net slip, which in this case equals the dip slip. (B) Vertical cross section perpendicular to strike of fault. Same as front of block diagram above. *ab* is the net slip, which in this case equals the dip slip; *ad*, throw, *db*, heave.

It is clear that throw and heave refer to the effects on the disrupted band. If two bands with different attitudes, such as two dikes, are broken by a fault, each has a different value for the throw and heave. Many geologists feel that the terms should refer only to the movement along the fault and that they should be independent of the effect on the disrupted bands. These geologists use throw for the vertical slip (*ib* of Fig. 8-22B) and they use heave for the horizontal dip slip (*ci* of Fig. 8-22B).

Classifications

GEOMETRICAL CLASSIFICATIONS

Bases of classifications

Faults, like joints, may be classified on the basis of their geometry or their genesis. Because no interpretation is involved, geometrical classifications are obviously less hazardous than genetic classifications. It is partly for this reason that the geometrical classifications will be considered first.

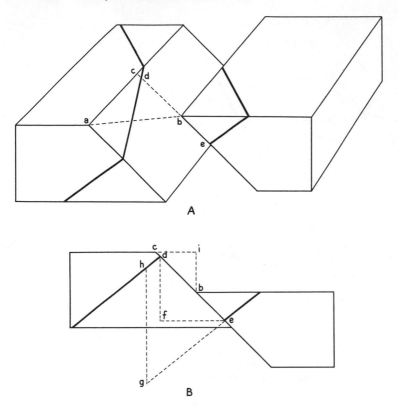

Fig. 8-22. Throw and heave. Heavy black layer is disrupted stratum. (A) Block diagram; *ab* is the net slip, *ac* is the strike slip, *cb* is the dip slip. (B) Vertical cross section perpendicular to the strike of the fault; along same plane as front of hanging wall block of block diagram. *cb* is the dip slip; *ib* is the vertical slip, *throw* to some geologists; *ci* is the horizontal dip slip, *heave* to some geologists; *de* is the apparent movement, which equals the dip separation; *df* is the throw; *fe* is the heave; *hg* is the vertical separation.

The bases of five different geometrical classifications are: (1) the rake of the net slip; (2) the attitude of the fault relative to the attitude of the adjacent rocks; (3) the pattern of the faults; (4) the angle at which the faults dip; and (5) the apparent movement on the fault.

Classification based on rake of net slip

A *strike-slip fault* is one in which the net slip is parallel to the strike of the fault (Fig. 8-10A); that is, the strike equals the net slip and there is no dip-slip component. The rake of the net slip is therefore zero.

A *dip-slip fault* is one in which the net slip is up or down the dip of the fault (Fig. 8-9A); that is, the dip slip equals the net slip and there is no strike-slip component. The rake of the net slip is therefore 90 degrees.

A *diagonal-slip fault* is one in which the net slip is diagonally up or down the fault plane (Fig. 8-11A). There is both a strike-slip and dip-slip component; the rake of the net slip is greater than zero but less than 90 degrees

Classification based on attitude of fault relative to attitude of adjacent beds

The second of these geometrical classifications, which is based on the attitude of the faults relative to the attitude of the adjacent rocks, would be highly involved if all the variables were considered. In general, therefore, the terms refer merely to the relations as observed in plan—that is, on a geological map. A *strike fault* is one that strikes essentially parallel to the strike of the adjacent rocks. Figures 8-13. and 8-14 are examples of strike faults. The strike of the adjacent rocks is ordinarily measured on the bedding, but if the bedding is absent, the strike may be measured on the schistosity of metamorphic rocks or on the flow structure of igneous rocks.

A *bedding fault* is a variety of strike fault that is parallel to the bedding; Figure 8-15 is an example of a bedding fault. A *dip fault* strikes essentially parallel to the direction of dip of the adjacent beds; that is, its strike is perpendicular to the strike of the adjacent beds. Figures 8-10, 8-11, and 8-12 are examples of dip faults. An *oblique* or *diagonal fault* is one that strikes obliquely or diagonally to the strike of the adjacent rocks; Fig. 8-16 is an example of a diagonal fault.

A *longitudinal fault* strikes parallel to the strike of the regional structure; *abcd* of Fig. 8-23 is an example of a longitudinal fault. Along most of its

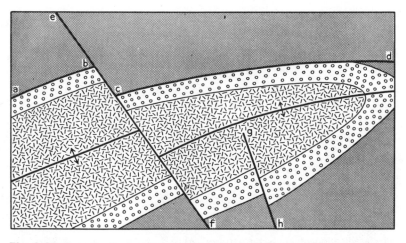

Fig. 8-23. Longitudinal and transverse faults. Anticline plunging toward the east (right) is broken by faults. *abcd* is a longitudinal fault; *ebcf* and *gh* are transverse faults.

course it is a strike fault, but locally the adjacent rocks may strike at a high angle to the fault. A *transverse fault* strikes perpendicularly or diagonally to the strike of the regional structure; *ef* and *gh* of Fig. 8-23 are examples of transverse faults. Along most of its course a transverse fault is a dip or diagonal fault, but locally the adjacent rocks may strike parallel to the fault.

Classification based on fault pattern

A third geometrical classification is based on the pattern shown by the faults; ordinarily the classification is based on the pattern on a map, but it may be based on the pattern in a cross section. The attitude of the adjacent rocks is unimportant. In some localities, the faults have essentially the same dip and strike; they thus belong to a set of *parallel faults* (Fig. 8-24A). If the strikes are the same but the dips differ the faults are assigned to two or more sets of parallel faults. *En échelon faults* are relatively short faults that overlap each other (Fig. 8-24B). *Peripheral faults* are circular or arcuate faults that bound a circular area or part of a circular area (Fig. 8-24C). *Radial faults* belong to a system of faults that radiate out from a point (Fig. 8-24D).

In some areas, the strike and dip of the faults may differ so markedly that the arrangement appears to be quite haphazard. In many such cases however, the faults may be grouped into several sets.

Classification based on value of dip of fault

The fourth geometrical classification is based on the angle of dip of the fault. *High-angle faults* are those that dip greater than 45°; *low-angle faults* are those that dip less than 45°.

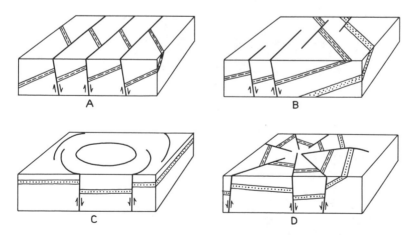

Fig. 8-24. Geometrical classification of faults by pattern. (A) Parallel faults. (B) *En échelon* faults. (C) Peripheral faults. (D) Radial faults.

Classification based upon apparent movement

A fifth geometrical classification is based upon the apparent movement in vertical sections at right angles to the fault. For various reasons the usage proposed by the Committee of the Geological Society of America[1] has not been followed.

An *apparent normal fault* is one in which the hanging wall, in a vertical section at right angles to the strike of the fault, appears to have gone down relative to the footwall. The Committee advocated using the term *normal fault* in such cases. An *apparent thrust fault* is one in which the hanging wall, in a vertical section at right angles to the strike of the fault, appears to have gone up relative to the footwall. The Committee advocated using the term *reverse fault* for such cases. Apparent normal faults are illustrated by Figs. 8-7, 8-9, 8-11, and 8-13. Apparent reverse faults are illustrated by Figs. 8-10 and 8-14. It should be noted that the apparent movement is not necessarily the same as the true movement. In Fig. 8-10 the movement along the fault was parallel to the strike of the fault. In Fig. 8-25 the hanging wall moved

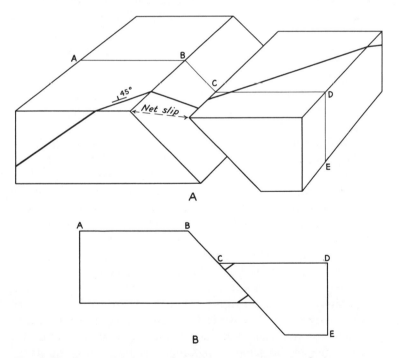

Fig. 8-25. Contrasting true and apparent movement on a fault. (A) Block diagram. Heavy black layer is disrupted bed. (B) Vertical cross section along *ABCDE* of the block diagram. Although the hanging wall has gone down relative to the foot-wall, the disrupted horizon suggests the opposite.

relatively down, but in the vertical section at right angles to the strike of the fault (Fig. 8-25B), the hanging wall appears to have moved relatively up.

Summary. It is obviously hopeless to attempt to establish a single set of terms that will take into consideration all the factors enumerated above. A far better system is to describe faults by using several terms from the various classifications given above. Thus the faults in one locality may be described as high-angle, *en échelon*, dip faults. In another locality the faults may be low-angle, parallel, longitudinal faults.

GENETIC CLASSIFICATION

The most satisfactory genetic classification at present is based on the nature of the relative movement along the fault.

A *thrust fault* or *thrust* is a fault along which the hanging wall has moved up relative to the footwall. Three categories are usually recognized. A *reverse fault* is a thrust that dips more than 45°, the term *thrust (sensu restricto)* (Plate 19) is one that dips less than 45°, but *overthrust* (Plate 22) is used for a fault that dips less than about 10° and has a large net slip. Overthrust has also been used with a different connotation (p. 198). Thrust faults indicate shortening of the rocks involved.

A *normal fault* (Plate 18) is a fault along which the hanging wall has moved relatively downward. The Committee recommended using the term *gravity fault* for such cases, but this proposal has not been generally followed by geologists. A *detachment fault* is a special category of low-angle normal faults due to the downhill sliding of rocks from an uplifted region. They are discussed further on page 226.

Strike-slip faults, also called *wrench faults,* are those along which displacement has been essentially parallel to the strike of the fault—that is, the dip-slip component is small compared to the strike-slip component. If an observer approaches a strike-slip fault from one side, the opposite wall will have moved relatively either to the right or left. A *right-slip fault* is one in which the opposite wall moved relatively to the right. A *left-slip fault* is one in which the opposite wall moved relatively to the left. A clear distinction must be made between the net-slip and the separation. In Fig. 8-17A the bed *aa'a''a'''* shows a right separation, but bed *bb'b''b'''* shows a left separation. In this case the net slip is directly down the dip of the fault. Figure 8-17B shows a left separation, but the net slip is unknown. There has been a lot of confusion in terminology in the literature. The terms "dextral," "right-handed," and "right-lateral" have been used but often on the assumption that the net slip was the same as the separation. Similarly, the terms "sinistral," "left-handed" and "left-lateral" have been used. This is why the terms "separation" and "slip" should be introduced into the terminology.[2]

Because the relative movement along the fault plane is not necessarily parallel to either the strike or the dip, but may be diagonal, the terminology

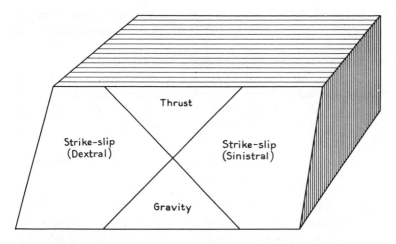

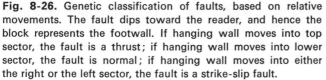

Fig. 8-26. Genetic classification of faults, based on relative movements. The fault dips toward the reader, and hence the block represents the footwall. If hanging wall moves into top sector, the fault is a thrust; if hanging wall moves into lower sector, the fault is normal; if hanging wall moves into either the right or the left sector, the fault is a strike-slip fault.

should make allowance for these possibilities. The most satisfactory solution is to divide the fault plane into four quadrants by two lines making an angle of 45° with a horizontal line on the fault plane. This has been done in Fig. 8-26, which represents the footwall of the fault. If the relative movement of the hanging wall is such that the net slip is toward the upper quadrant, the fault is a thrust fault; if the net slip is toward the lower quadrant, the fault is a normal fault; if the net slip is toward the right or left quadrants, the fault is a strike-slip fault. If the net slip is not directly down the dip or parallel to the strike, a modifying clause should be added to the appropriate term. Thus if the net slip is in the upper quadrant but makes an angle of 60° with a horizontal line in the fault plane, the fault can be called a thrust fault with a large component parallel to the strike of the fault.

CLASSIFICATION BASED ON ABSOLUTE MOVEMENTS

The classification outlined above is based on relative movements. A more elaborate classification would be based on absolute movements relative to some datum plane, such as sea level. Thus five kinds of normal faults might be recognized: (1) those in which the footwall did not move, but in which the hanging wall moved down; (2) those in which the footwall moved up, while the hanging wall remained stationary; (3) those in which the hanging wall moved down and the footwall moved up; (4) those in which both blocks moved down, but in which the hanging wall moved a greater amount; and (5) those in which both blocks moved up, but in which the hanging wall moved less than the footwall.

Similarly, five kinds of thrust faults might be established.

In most instances, however, data are not available to indicate the absolute movement on faults. Many attempts have been made to establish criteria based on the pattern of the faults, the dip of the fault plane, or the comparative intensity of the deformation in the two blocks, but such criteria are unreliable.

In a few instances, where movements along faults near the ocean have occurred within historic times, it is possible to ascertain which block moved[5,6,7]. Moreover, from a consideration of crustal forces, it is sometimes possible to theorize about the absolute movements along faults. Under certain conditions, therefore, terms based on absolute movements may be of value. *Upthrusts* are high-angle faults along which the relatively uplifted block has been the active element. If the hanging wall of a high-angle thrust fault has moved up while the footwall stayed in place, or if the footwall of a high angle gravity fault has moved up while the hanging wall stayed in place, the fault is an upthrust.

Sometimes the term *underthrust* is used for those thrust faults in which the footwall has been the active element, whereas *overthrust* is used for those thrust faults in which the hanging wall has been the active element.

References

[1] Reid, H. F., et al., 1913, "Report of the committee on the nomenclature of faults," *Geol. Soc. Amer. Bull.* 24: 163–86.

[2] Kelley, Vincent C., 1960, Slips and separations, *Geol. Soc. Amer. Bull.* 71: 1545–46.

[3] Hill, M. L., 1959, Dual classification of faults, *Amer. Assoc. Petrol. Geol. Bull.* 43: 217–21.

[4] Crowell, J. C., 1959, Problems of fault nomenclature, *Amer. Assoc. Petrol. Geol. Bull.* 43: 2653–74.

[5] Plafker, George, 1965, Tectonic deformation associated with the 1964 Alaskan earthquake, *Science* 148: 1675–87.

[6] Plafker, George, 1969, *Tectonics of the March, 1964 Alaskan earthquake*, U. S. Geol. Surv., Prof. Paper 543–I, 74 pp.

[7] Witkind, Irving J., et al., 1964, *The Hegben Lake, Montana, earthquake of August 17, 1959*, U. S. Geol. Surv., Prof. Paper 435, 242 pp.

9

CRITERIA
FOR FAULTING

Introduction

Numerous criteria demonstrate faulting, but only a few of these criteria may be available in any specific case. Moreover, useful criteria depend in part on the size of the area involved. For example, small faults may be observed in natural and artificial exposures, such as highway cuts (Fig. 4-1), mines,[1] and tunnels.[2] On the other hand, the presence of large faults is often deduced on stratigraphic and physiographic evidence, and only small segments, if any, of the fault may be exposed. For some faults many criteria may be available, whereas for others very little evidence may be obtained. Moreover, faults may be difficult to distinguish from other structural features, notably unconformities.

The criteria for the recognition of faults may be considered under the following headings: (1) discontinuity of structures; (2) repetition or omission of strata; (3) features characteristic of fault planes; (4) silicification and mineralization; (5) sudden changes in sedimentary facies; and (6) physiographic data.

Discontinuity of Structures

If strata suddenly end against different beds, a fault may be present (Plates 18 and 19). On a map, cliff, or artificial exposure, the discontinuity occurs along a line, but this is merely the trace of a surface of discontinuity. In some instances the disrupted strata may be found in the same outcrops or in nearby exposures, but usually this is not so. The discontinuity of strata along faults is illustrated in Figs. 8-7, 8-9, 8-10, 8-11, 8-13, 8-14 and 8-16. Dikes, veins, or older faults also may end suddenly along some line[3], and the displaced parts may appear elsewhere. In such cases, however, the observer must realize that dikes, veins, or faults may form with a discontinuous pattern. Moreover, discontinuity of structures is, in itself, not proof of faulting; the truncation of structures is also typical of unconformities (p. 283), intrusive contacts (p. 332), and, on a small scale, cross-bedding.

In summary, discontinuity of structures is characteristic of faults, but it is a proof of faulting only if other possible interpretations have been eliminated.

Repetition and Omission of Strata

Figure 9-1 is a geological map of a region of folded and faulted sedimentary rocks. A syncline lies near the center of the map, as is shown by the dips, and the formations are progressively younger from *a* to *e*. In certain places, however, one or more formations are missing, as, for example, along the line *FF*, where formation *b* is absent, and along the line *F'F'*, where *c* and *d* are missing. The lines *FF* and *F'F'* must be the traces of faults, but no data are given to indicate the direction and value of the dip of the faults.

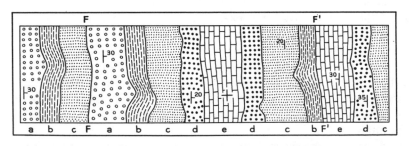

Fig. 9-1. Faults indicated by repetition and omission of strata. Strata, from oldest to youngest, are *a, b, c, d*, and *e. FF* and *F'F'* are faults. Dip-strike symbols indicate attitude of bedding.

The omission of strata, however, may be due to an unconformity (Chap. 13).

Features Characteristic of Fault Planes

Many faults are accompanied by such distinctive features as slickensides, grooving, gouge, breccia, and mylonite; these features are conclusive proof of faulting, but some of them may be confused with phenomena of a different origin.

Slickensides are polished and striated surfaces that result from friction along the fault plane. The scratches or striations are parallel to the direction of movement, but caution is necessary in employing such information because some faults show many slickensided layers, in each of which the striations have different trends. Moreover, a slickensided layer may record only the last movements along the fault, and the earlier displacements may have been in some other direction.

Many slickensided surfaces are accompanied by sharp, low steps that trend at right angles to the striations (Fig. 9-2A). These sharp little rises are commonly only a fraction of an inch high and may be so small that they are difficult or even impossible to see. These rough surfaces can be used to determine the relative movement along the fault plane, in much the same way that *roches moutonnées* indicate the direction in which glacial ice was moving. In Fig. 9-2A, the upper block, which is not shown, moved from left to right relative to the lower block, on which the slickensides are shown. Even if the small irregularities are not visible, however, a person with sensitive fingers may be able to tell the direction of movement. The surface feels smooth if the fingers slide in the direction that the missing block was displaced, whereas in the reverse direction the fault feels rough. Although some experimental evidence suggests that this conclusion is not always reliable,[4] field evidence indicates it is generally correct.

Some faults show large grooves or furrows several feet from crest to

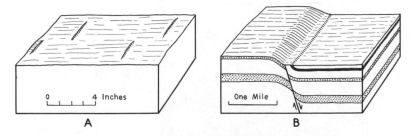

Fig. 9-2. Features associated with faults. (A) Slickensides with small associated steps; lines parallel to front of block are scratches; at right angles to them are small "steps." (B) Drag along a fault. Stippling and solid black represent special beds.

crest and several inches deep; they are parallel to the direction of displacement.

Drag is in some cases an aid in divulging the relative motion along the fault, as is shown in Fig. 9-2B. Because of friction the beds in the hanging wall are dragged up in this particular case, whereas the beds in the footwall are dragged down. This method is subject to the same limitation as are slickensides because the observed drag may be due to the last movements along the fault and may even be the opposite of the major movement.

Some of the rock along a fault may be pulverized to a fine-grained *gouge*, which looks and feels like clay. In fact, gouge differs in no important way from clays of glacial origin because both are pulverized rock.

Breccia (Plate 20) consists of angular to subangular fragments of various sizes, characteristically associated with a more finely crushed matrix. The fragments typically range from an inch to several feet in diameter, but much larger blocks may occur. Fault breccias may be many tens of feet thick.

Large blocks may be caught along faults. Such blocks are separated from the foot and hanging walls by faults that may or may not be accompanied by breccia. Such large blocks are called *horses* or *slices*. The term *horse* usually refers to such a block caught along a normal fault (p. 196), and *slice* refers to a block caught along a thrust fault (p. 196). This distinction is of no great importance, and merely reflects the fact that the terms were first used in two different countries.

A *mylonite* is a microbreccia that maintained its coherence during the deformation. It is characteristically dark and fine-grained and may be difficult to distinguish from sedimentary or volcanic rocks. The brecciated character

Plate 20 *Photomicrograph of tectonic breccia.* About 2 centimeters across. Derived from slate and chert along Melones fault zone. Dutch Flat Quadrangle, California. Photo: L. D. Clark, U. S. Geological Survey.

is generally apparent only from microscopic study. Although usage varies,[5] the term *mylonite* should be restricted to those microbreccias with a streaked or platy structure; they may look like slate. Uncrushed fragments of the parent rock can be recognized under the microscope. An *ultramylonite* forms if the crushing is so complete that no such fragments remain. An ultramylonite may be difficult to recognize unless transitions to mylonite and the parent rock are preserved. *Flinty crush-rock* and *pseudotachylite* are massive microbreccias that lack the platy structure. Flinty crush-rock looks like chert. Pseudotachylite looks like tachylite, which is a variety of basaltic glass. Flinty crush-rock and pseudotachylite that are exceedingly fine-grained —the individual grains are 0.001 millimeter in diameter—may fill irregular fractures near the fault and may simulate dikes of igneous rocks. Although some geologists believe that these rocks were actually molten at one time, there is no unanimity of opinion on this matter.

Although slickensides, gouge, breccia, mylonite, and related phenomena are found along many faults, they are not necessarily present. It is often assumed that the larger the fault, the greater the amount of breccia, gouge, and mylonite. This is by no means true. In general, gouge and breccia form near the surface of the earth, where the confining pressures are comparatively small, and mylonite forms at greater depth, where the confining pressure forces the rocks to retain their coherence. Parts of some of the great overthrusts in the Alps are so devoid of slickensides, gouge, breccia, and mylonite that they passed unnoticed and were for a time mapped as sedimentary contacts. It was only after paleontological evidence was obtained and after areal mapping was extended that the existence of the great faults was recognized.

Silicification and Mineralization

Faults, because they are extensive fractures or branches of large fractures, are often the avenues for moving solutions. The solutions may replace the country rock with fine-grained quartz, causing *silicification*. This phenomenon in itself is not proof of faulting, but in some localities it may be highly suggestive. *Mineralization* along faults is typical of many mining districts.

Differences in Sedimentary Facies

Different *sedimentary facies*[6] of rocks of the same age may be brought into juxtaposition by large horizontal displacements. Figure 9-3 illustrates very diagrammatically a basin of deposition in which sandstones are deposited near shore, shales farther out, and limestones farthest from shore. The transition from sandstone to shale, and from shale to limestone, will be gradual and there will be considerable interfingering of beds. The rocks of this par-

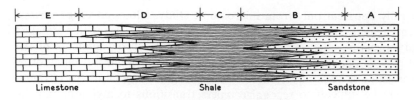

Fig. 9-3. Changes in sedimentary facies across a basin of deposition. *A*, Sandstone facies, *B*, sandstone-shale facies, *C*, shale facies, *D*, shale-limestone facies, *E*, limestone facies.

ticular age are said to be represented by a sandstone facies in region *A*, a sandstone-shale facies in region *B*, a shale facies in region *C*, a shale-limestone facies in region *D*, and a limestone facies in region *E*. Even if the strata are strongly folded and exposed by erosion, the various facies will grade into each other. On the other hand, a large overthrust may bring the sandstone facies of region *A* into contact with the limestone facies of region *E*. Similarly, a strike-slip fault that strikes diagonally to the facies boundaries and along which there has been considerable displacement, may bring different facies together. This is the case along some of the large strike-slip faults in California.

Physiographic Criteria

Resistant sedimentary formations are generally expressed topographically by ridges. A dip fault or diagonal fault will displace the strata, as in Fig. 8-9B, and, consequently, the ridge held up by some resistant bed will be discontinuous, and an *offset ridge* will result.

A scarp is a relatively steep, straight slope that may range in height from a few feet to thousands of feet.[7,8] A scarp is not, of course, proof of the presence of a fault, because scarps may originate in many other ways. But a scarp that truncates topographic features, such as alternating ridges and valleys (Fig. 9-4), is highly suggestive of a fault. A *fault scarp* (Plate 21) owes its relief directly to the movement along the fault (Fig. 9-5). A *fault-line scarp* owes its relief to differential erosion along a fault line. Figure 9-6 shows two ways in which such scarps may form. The downthrown block may be topographically high. A *composite fault scarp* owes its height in part to differential erosion, but also in part to direct movement along the faults.

Piedmont scarps, sometimes called *scarplets* (Fig. 9-7), lie at or near the foot of mountain ranges. They are confined to areas of active faulting, such as the Basin and Range Province of the United States. The height is generally measured in feet or tens of feet. They are generally straight and uneroded or only slightly eroded; that is, the face of the scarp is also a fault plane. Some cut the bedrock, but others are confined to the overlying unconsolidated

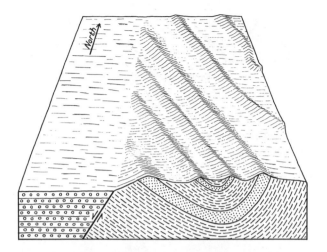

Fig. 9-4. Mountain front truncating internal structure of a range. Resistant sandstones, shown by dots, hold up ridges; main valley is underlain by unconsolidated alluvium, shown by open circles.

Plate 21 *Fault scarp.* Red Canyon fault scarp, east valley wall of Red Canyon, resulting from Hebgen Lake earthquake in 1959. Montana. Photo: I. J. Witkind, U. S. Geological Survey.

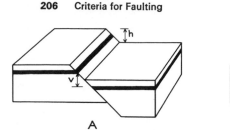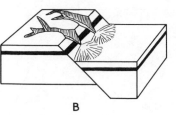

Fig. 9-5. Fault scarp. (A) Before erosion; h, height of scarp, equals v, the vertical slip. (B) After some erosion the material removed from deep valleys on the footwall block has been deposited as alluvial fans on hanging-wall block.

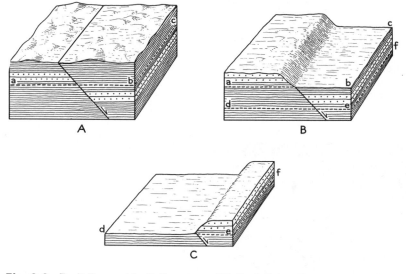

Fig. 9-6. Fault line and fault-line scarp. (A) Fault line without scarp. Dotted formation is resistant to erosion; formations shown by parallel lines are not resistant to erosion; *abc* new baselevel of erosion. (B) Easily eroded rocks on hanging-wall block have been reduced to new baselevel; rocks resistant to erosion on footwall have been only partially eroded; the plane *def* represents a new baselevel. (C) Scarp developed on relatively down-dropped block.

deposits such as alluvial fans, glacial moraines, and lake terraces. Piedmont scarps that cut unconsolidated deposits may be a direct continuation of a large fault in the bedrock, or they may be indirectly related to the master fault and due to tension in the unconsolidated deposits during faulting in the bedrock.

Triangular facets are found on some scarps resulting from normal faulting (Fig. 9-8). But triangular facets may also result from glacial erosion, marine erosion, or even fluviatile erosion.

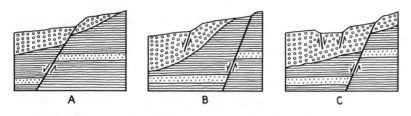

Fig. 9-7. Scarplets or piedmont scarps. Dotted and lined patterns represent bedrock. Open circles represent unconsolidated material. (A) Scarplet that is a direct continuation of a fault in the bedrock. (B) Scarplet on right is direct continuation of fault in the bedrock, but scarplet on left is not. (C) Graben bounded by faults that do not extend into bedrock.

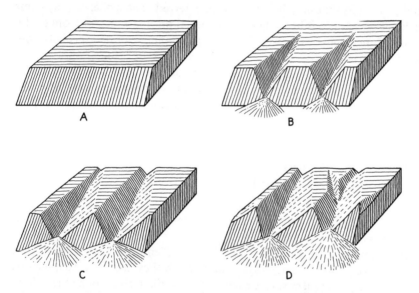

Fig. 9-8. Evolution of triangular facets. (A) Fault scarp prior to erosion. (B) Partially eroded fault scarp. (C) Triangular facets representing remnants of original fault scarp. (D) Triangular facets that represent the original fault scarp driven back somewhat by erosion.

Springs aligned along the foot of a mountain range are highly suggestive of faulting, especially if the water is hot. The alignment suggests the presence of a major plane of weakness, and the hot water indicates a fracture that permits deep penetration of circulating waters.

Offset streams are found along active strike-slip faults. If a southwesterly flowing stream is offset by displacement along a northwesterly trending steep strike-slip fault, along which the northeast wall moved relatively southeast, the stream on the southwest wall will be offset relatively toward the northwest. Such offset streams have been described from California.

A *break in a stream profile* may be present near a fault line. The net slip is the cumulative effect of many small displacements taking place over tens of thousands of years. If the stream cannot erode with sufficient rapidity to maintain grade during this time, the profile of the stream may be unusually steep in the vicinity of the fault. After the 1915 earthquake at Pleasant Valley, Nevada, a stream flowed over a waterfall 10 feet high.[9] Such breaks are not common, however, because the time between successive movements is long enough to permit incision of the streams. Moreover, the deposition of alluvium on the downthrown block tends to smooth out the profile. Broken stream profiles, however, may be due to causes other than faulting.

In the broad sense of the word, a *lineament* is a line, resulting from natural processes, that may be observed or inferred. The evidence may come from direct observations in the field, from aerial photographs, or from geologic maps. Lineaments expressed in the topography may be especially conspicuous in aerial photographs. The lineament may be a long depression or a long ridge. Many of the depressions are undoubtedly due to erosion along lines of weakness, either faults or sets of joints. Lineaments may be easily confused with ridges and valleys resulting from the erosion of steeply dipping interbedded strata of varying hardness.

Distinction Between Fault Scarps, Fault-line Scarps, and Composite Fault Scarps.

It is not always easy to distinguish fault scarps, fault-line scarps, and composite fault scarps from one another.

The following features suggest a fault scarp. Piedmont scarps are evidence of active faulting, but the larger scarp with which they are associated may be a fault-line scarp. Frequent severe earthquakes in the vicinity of a scarp along a fault line indicate a fault scarp. Lakes aligned along a fault line suggest a fault scarp. A lake may form if a fault cuts across a stream, and the block on the downstream side is uplifted (Fig. 9-9A). Depressions also develop if the downdropped block settles different amounts along the strike of the fault, and lakes or swamps may occupy these depressions (Fig. 9-9B). *Sag ponds* occupy depressions along active faults in California. But lakes associated with faults are not likely to last very long because the outlet may be lowered very rapidly, or sediments from the nearby hills and mountains may fill up the lake.

Lakes are not a normal accompaniment of fluviatile erosion, and they are not to be expected along fault-line scarps. On the other hand, lakes at the foot of fault-line scarps may result from local overdeeping by glaciers, damming by landslides and lava flows, and from other causes.

Several features suggest a fault-line scarp. If the downthrown block is

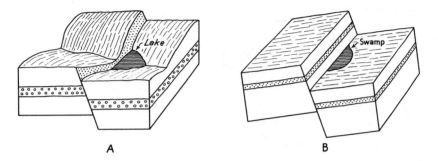

Fig. 9-9. Lakes and swamps along a fault scarp. (A) Stream that flowed toward the left has been dammed by the fault scarp. (B) Swamp occupies depression at foot of fault scarp.

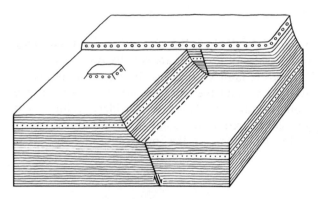

Fig. 9-10. Fault-line scarp. Dots represent sandstone; parallel lines, shale; circles, conglomerate. (See text.)

topographically higher than the upthrown block (Fig. 9-6C), the scarp is an fault-line scarp. The relief along the fault must be caused by differential erosion.

Locally the evidence may be noncommittal, as in the front part of the block diagram shown in Fig. 9-10. The scarp could be either a fault scarp or a fault-line scarp. At the back of the block, however, a younger conglomerate, shown by circles, overlies the older rocks and is unaffected by the fault. It is evident that the original topographic expression of the fault had been destroyed by erosion of the whole region to the surface directly beneath the conglomerate. The conglomerate was then deposited. Finally, in a renewed cycle of erosion, the conglomerate in the foreground, and many older beds in the right foreground, were eroded, and a fault-line scarp developed.

A close correlation between topography and rock resistance is indicative of a fault-line scarp. The scarp is high and abrupt where the contrast in

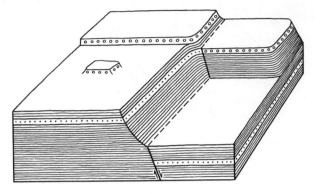

Fig. 9-11. Composite fault scarp. Dots represent sandstone; parallel lines, shale; circles, conglomerate. (See text.)

lithology is greatest, is more subdued where the contrast is slight, and may disappear if rocks on opposite sides of the fault are equally resistant.

The recognition of composite fault scarps must be based upon combinations of the criteria given above, and the local conditions are so variable that a general discussion here is inadvisable. One type of evidence is illustrated by Fig. 9-11, which may be considered a later stage than that illustrated by Fig. 9-10. Part of the height of the scarp in the foreground of Fig. 9-11 is directly due to movement along the fault because the conglomerate has been displaced by the fault.

Map Symbols

A fault is generally shown on a geological map by a line, usually heavier than the line representing contacts between formations. In addition, the map should, if possible, give additional information.

The *attitude* can be shown by symbols similar to those used for bedding and other planar features (Figs. 9-12a and b). The trend is the strike of the fault, the tick gives the direction of dip, and an arabic numeral gives the value of the dip. Obviously such data can be readily obtained and portrayed on large-scale maps of mines and tunnels. On the other hand, in large areas being mapped on a smaller scale, although the strike of the fault may be readily deduced from the distribution of outcrops, the value of the dip may be unknown. Or, if known, it may differ from place to place; of course, the dip may be shown at those localities where it is known. Figures 9-12c, d, and e, as indicated in the caption, may be used in some cases.

The nature of the relative movement may be shown in various ways. Figures 9-12f and g show two alternate symbols for normal faults, whereas

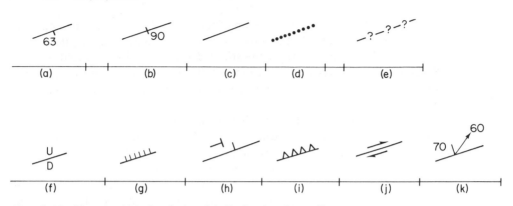

Fig. 9-12. Map symbols for faults. (a) Fault, showing strike and dip. (b) Strike of vertical fault. (c) Strike of fault, dip uncertain. (d) Concealed fault, i.e., fault overlain by younger beds not affected by fault. (e) Possible fault. (f) Normal fault; *U* on upthrown block, *D* on downthrown block. (g) Normal fault, hachure on downthrown block. (h) Thrust; *T* on upthrown block. (i) Thrust, sawteeth on upthrown block. (j) Strike-slip fault, showing relative movement. (k) Fault with dip, also shows bearing and plunge of slickensides.

Figs. 9-12h and i show two alternate symbols for thrust faults. The relative movement along strike-slip faults is shown by arrows, as shown in Fig. 9-12j. For diagonal slip, the symbols can be combined. These symbols are, of course, strictly qualitative. Usually sufficient data are not available to be quantitative, but appropriate figures could be placed on the map. Figure 9-12k shows how a line symbol may be added to show the bearing and plunge of slickensides.

In oil fields where extensive drilling has been done, precise data on faults may be obtained. Figure 9-13 is a map of such an area. Structure contours on a key bed show that it dips about 11°NNW. A fault trending northeast is shown by stippling; as shown by structure section *XY*, this is the area where a vertical drill hold would miss the key bed. The altitude of the fault at any point is the same as the altitude of the structure contour in contact with it. Thus, at *c* the altitude of the fault is −3200 feet. The altitude of the fault at *a* is −3300 feet, and the altitude at *b* is also −3300 feet; hence *ab* is the strike of the fault. Note that the azimuth of the stippled belt is not the strike of the fault. To get the dip of the fault, the line *cd* is drawn perpendicular to the strike. At *c* the altitude of the fault is −3200 feet, whereas at *d*, by interpolation of the structure contours, it is −3550 feet. Thus the fault dips 350 feet in 510 feet, that is, at an angle of 34°. It should be noted that the width of the stippled band representing the fault is not only a function of the angle of dip of the fault, but also of the dip of the key bed, as well as the angle between the strike of the fault and the strike of a key bed. However, in the case of a vertical fault the width of the stippled zone would be reduced to zero. Because of variations in the attitude of the fault and the attitude of the strata, the width of the band represented by stippling will change from place to place.

A little reflection will show that it may not be possible to show on a map all the faults that have been observed. It is obvious that the scale of the map is a prime factor. In many areas every outcrop contains small faults. It would be quite impossible to show all such faults on a map on a scale of one inch to a mile. Thus one can show only the significant faults. But what is significant depends on the purpose of the investigator, a factor that also controls the scale. A fault with a net slip of 10 feet is not very important on an inch-to-a-mile scale, but it may be of great significance at a dam site. Everyone reading geological maps should appreciate that this problem exists. There is no absolute rule to follow. Normally the strike separation should be great enough to affect the map if the fault is to be shown. Thus, a strike separation of 20 feet is 0.05 inch on an inch-to-a-mile scale, but 1 inch on a scale of 1 inch to 20 feet. The stratigraphic throw along a strike fault may be very large —many thousands of feet—but on the map the contacts may be parallel to the fault for many miles. Thus a great deal of judgement must be used in deciding which faults should be shown on the map. Similarly, people using geological maps should realize this problem exists.

References

[1] Meyer, Charles, Shea, Edward P., Goddard, Charles C., Jr., et al., 1968, Ore deposits at Butte, Montana, pp. 1373–1416, in *Ore Deposits of the United States, 1933–1967*, John Ridge, ed., Amer. Inst. Mining, Metal., and Petrol. Eng., 1880 pp., New York.

[2] Billings, Marland P., 1967, The significance of faults in tunnels, *Economic Geology in Massachusetts*, 568 pp. Graduate School, Univ. Mass., 267–72.

[3] Smith, G. I., 1962, Large lateral displacement on the Garlock fault, California, as measured from offset dike swarm, *Amer. Assoc. Petrol. Geol.* 46: 85–104.

[4] Paterson, M. S., 1958, Experimental deformation and faulting in Wombeyan marble, *Bull. Geol. Soc. Amer.* 69: 465–76.

[5] Christie, John M., 1960, Mylonitic rocks of the Moine thrust-zone in the Assynt region, North-west Scotland, *Edinburgh Geol. Soc. Trans.* 18, pt 1, 79–93.

[6] Pettijohn, F. J., 1957, *Sedimentary rocks*, 2d ed., 718 pp., New York: Harper and Row.

[7] Blackwelder, E., 1928, The recognition of fault scarps, *J. Geol.* 36: 289–311.

⁸ Cotton, C. A., 1950, Tectonic scarps and fault valleys, *Bull. Geol. Soc. Amer.*, 61: 717–58.

⁹ Jones, J. C., 1915, The Pleasant Valley, Nevada, earthquake of October 2, 1915, *Bull. Seismol. Soc. Amer.* 5: 190–205.

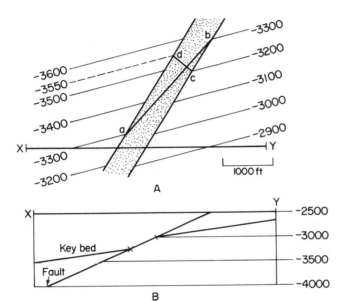

Fig. 9-13. Common method of showing faults in oil fields. (A) Map. Fault plane is stippled; structure contours on top of same key bed. (B) Structure section; apparent dips are less than true dips.

10

REVERSE FAULTS, THRUST FAULTS, AND OVERTHRUSTS

Introduction

Reverse faults and *thrust faults* are similar in that the hanging wall has moved upward relative to the footwall. Reverse faults dip more than 45°, whereas thrust faults dip less than 45°. The term *overthrust* has been used for low-angle thrust faults along which the net slip is large, generally in miles. *Detachment faults* are a variety of low-angle normal faults. But they are considered in this chapter because they have many features in common with overthrusts.

Some flexibility and additional qualifications may be necessary in using these terms. The attitude of a fault may differ considerably in different places. This may be due either to initial differences in dip or to later folding.

Thrusts and Reverse Faults

Some thrusts and reverse faults may be unrelated to folding. They may form in crystalline rocks, such as granite. Moreover, they may develop in sedimen-

Plate 22 Champlain overthrust. Lake Champlain, Vermont. Lower Cambrian dolomites (white) are thrust over Ordovician shales (dark gray). Stratigraphic throw is about 6000 feet. (A) Shows gentle easterly dip of thrust plane. (B) Looking east. At left side of photograph, smeared-out blocks of Ordovician limestone lie along the overthrust plane. Photo: Vermont Geological Survey, Burlington, Vermont.

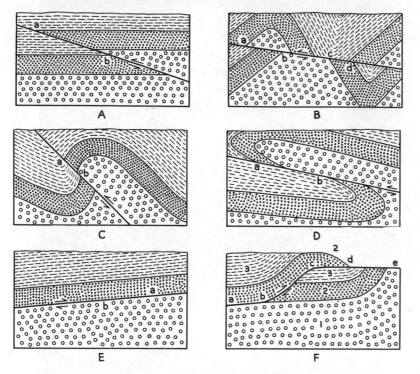

Fig. 10-1. Thrust faults. Sections assumed to be at right angles to strike of fault and beds, and to be parallel to net slip. (A) Thrust that forms without prior folding; *ab* is the net slip. (B) Thrust that forms later than folding; *ab* is the net slip; at *c* younger rocks are thrust over older; at *d* the rocks above and below the thrust plane are the same age; between *a* and *b* older rocks are thrust over younger. (C) Thrust plane diagonal to limb of anticline; *ab* is the net slip. (D) Thrust formed by stretching of overturned limb of fold; *ab* is the net slip. (E) Bedding thrust. (F) Erosion thrust; *1*, oldest beds; *2*, beds of intermediate age; *3*, youngest beds. Although between *a* and *b* the fault is a bedding thrust, between *b* and *c* it transgresses the bedding in the footwall; between *c* and *d* it is an erosion thrust. If thrust block moves to *e*, younger rocks would be thrust over older.

tary rocks before they are folded (Fig. 10-1A) or after they are folded (Fig. 10-1B).

A second category of thrusts and reverse faults are related to folding. Where rocks are relatively brittle, a sharp break may develop on one limb of an anticline, as a result of which the older rocks are thrust over the younger rocks (Fig. 10-1C). Where the rocks are relatively ductile, one limb of a fold may be greatly stretched before fracture actually occurs and the older rocks slide over the younger (Fig. 10-1D).

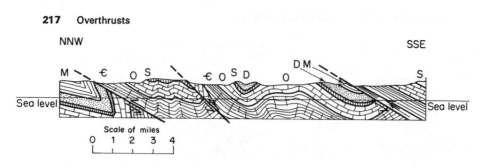

Fig. 10-2. Folding and thrusting in Valley and Ridge Province of Appalachian Highlands in Virginia. Є, Cambrian; O, Ordovician; S, Silurian; D, Devonian; M, Mississippian. (After B. N. Cooper.[1])

Thrust faults are generally readily recognized because older rocks rest on younger rocks (Fig. 10-1A, C, and D). It is apparent, however, that under certain conditions younger rocks may be thrust over older rocks. In some thrusts, such as *c* in Fig. 10-1B, younger rocks rest on older rocks; elsewhere, as at *a*, older rocks are on younger; locally, as at *d*, the rocks above the fault are the same stratigraphic unit as those beneath it.

Some faults may follow bedding planes for long distances. In some instances the upper plate is pushed up the fault plane (Fig. 10-1E); these are bedding thrusts. In other instances the upper plate may move down the fault plane; these are now called detachment faults (page 226). Bedding thrusts and detachment faults may be difficult to recognize, inasmuch as there may be little or no stratigraphic throw.

Thrusts may emerge at the surface of the earth and the upper plate may ride out over an erosion surface (Fig. 10-1F). Such faults are known as erosion thrusts.

Thrusts may be folded, as at the southeast end of Fig. 3-18. The data here are especially good because a tunnel gave a continuous section and crossed the fault in three places.

The term *imbricate structure* is used in geology in several ways. It may be used for slabs of stream boulders that overlap one another. It is also used for those cases in which several adjacent thrust planes dip in the same direction, as a result of which the intervening thrust sheets overlap one another. Excellent examples are found in the Canadian Rockies and in the southern Appalachian Mountains (Fig. 10-2).[1]

Overthrusts

Overthrusts are spectacular structural features that have intrigued geologists for a century (Plates 22A and B). Overthrusts are low-angle faults along which

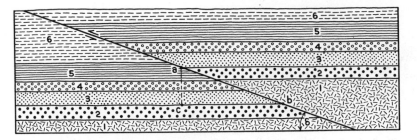

Fig. 10-3. Calculation of net slip from a cross section. Assumes no important strike-slip component. Section assumed to be at right angles to strike of fault. Formations in order of decreasing age are *1, 2, 3, 4, 5,* and *6. ab,* net slip; *ac,* stratigraphic throw; δ, dip of thrust plane.

there has been great horizontal displacement, essentially at right angles to the regional trend of the fault. They were called overthrusts because it was assumed by many that initially, at least, the hanging wall had moved upward and over the footwall (Fig. 10-3); it was assumed that later folding explained those instances where the hanging wall appeared to have moved down relative to the footwall (as at *d* in Fig. 10-4). But, as some geologists recognized long ago, this is not necessarily so: the hanging wall may have been pushed down the dip or have slid down the dip under its own weight.

The term *slide* is used extensively in the Central and Southern Highlands of Scotland and in northern Ireland[2]. A slide is a low-angle fault with large displacement, but the following qualifications must be made: (1) later folding may greatly modify the dip of the fault; (2) either wall may do the moving, but determining this is largely subjective; and (3) a *gravity slide* is one in which the hanging wall moved down relative to the footwall—it is the same as a detachment fault (page 226). An overthrust is one type of slide. The term slide is probably preferable to overthrust in many cases, as it lacks genetic implications, but the term overthrust is so firmly ingrained in the American literature it will be used here.

Our knowledge of overthrusts results largely from the fact that they are folded and eroded. Figure 10-5 is a map and structure section of a folded overthrust in a region of low relief. The attitude of the thrust plane is given by dip-and-strike symbols. The rocks above are shown by a dotted pattern, whereas those beneath the thrust plane are shown by diversely oriented short dashes. The overthrust sheet originally extended as far west as the line *xy,* but has been subsequently eroded from the central part of the area. The broken line in the upper part of the structure section shows the original top of the overthrust sheet. The symbol *K* is a remnant of the overthrust sheet, now isolated by erosion from the main thrust sheet. This is one way in which a *klippe* may originate. Klippe is a German word for cliff, and refers to the

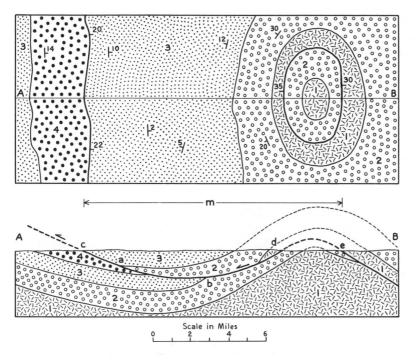

Fig. 10-4. Measuring net slip on an overthrust. Assumes no important strike-slip component. Map above, structure section below. Formations in order of decreasing age are *1, 2, 3,* and *4.* Thrust is shown in heavier black line, with figures to indicate dip. *ab*, net slip. *m*, minimum breadth of overthrust block is horizontal component of *cabde.*

fact that many klippe in the Alps have steep cliffs. Erosion has broken through the upper sheet at *F*, exposing the rocks beneath the fault. This area is a *fenster* or *window* because it is possible to look through the upper sheet to the lower. In Fig. 10-5, in which the surface of the earth is flat, the klippe is preserved in a doubly plunging syncline, and the fenster is due to a doubly plunging anticline. Erosion alone may be chiefly responsible for a klippe or fenster in regions of high relief. The term *window* is frequently misused in American geological literature to refer to *inliers*, which occur in a normal stratigraphic sequence wherever erosion has broken through the younger strata to expose a circular, elliptical, or irregular area of older rocks.

The *root zone* of an overthrust is the exposure of the overthrust nearest · its source. Thus in Fig. 10-5 the root zone is at *h.*

The stratigraphic throw along overthrusts may be many thousands of feet, and in many instances it may be determined with considerable precision.

The calculation of the net slip on an overthrust may be difficult. It is apparent, however, that if the strata were essentially flat when thrusting began and that if the dip of the fault plane were low, the net slip would be many times the stratigraphic throw. In fact, a simple trigonometric relation exists, and, as shown in Fig. 10-3:

$$ab = \frac{ac}{\sin \delta}$$

In this equation ab = net slip, ac = stratigraphic throw, and δ = dip of the fault plane. This assumes that the movement is directly up the dip of the fault; that is, there is no strike-slip component.

Even if the fault and the beds have been subsequently folded, δ may be taken as the angle between the bedding and the fault. In practice this method can seldom be applied because measurement of this critical angle may be impossible due to the drag and brecciation along the fault.

If sufficient data are available to prepare an adequate structure section, the net slip may be measured by noting the position of some key horizon both above and below the plane of the overthrust. In the structure section of Fig. 10-4, the net slip ab is five miles. The bottom of formation 3 below the thrust plane is truncated by the fault at point b. The same horizon above the thrust is truncated by the fault at a.

The erroneous assumption is often made that a minimum value for the net slip may be obtained by measuring m (Figs. 10-4 and 10-5); this is the distance between the most advanced exposure of the fault in a klippe and the most recessive exposure in a window, and it is measured at right angles to the strike of the main outcrop of the overthrust. Actually, m, as measured on the map, is the horizontal component of a somewhat greater distance measured along the fault plane. In Fig. 10-5, for example, m of the map is the horizontal component of $abcdefgh$ of the structure section; in the structure section of Fig. 10-4, m is the horizontal component of $cabde$. The distance m may be called the *minimum breadth* of the overthrust. The true breadth of the overthrust is a much larger value and would be measured along the plane of the fault from the original most advanced position of the thrust block to that unknown place within the crust where the fault originated. Obviously, it is impossible to measure the true breadth of overthrusts. In Fig. 10-4, the distance m is 13.7 miles, but the net slip ab is only 5 miles.

In many regions there may be more than one overthrust. In Fig. 10-6, three overthrusts, ab, cd, and ef, dip east, and along each of them the net slip is measured in miles. Each block between two overthrusts is an over-thrust sheet; there are three overthrust sheets in Fig. 10-6.

The region in front of the overthrusts is often called the *foreland* (Fig. 10-6). Small thrusts, with a net slip of hundreds or even thousands of feet, may occur in the foreland, but there are no overthrusts.

The rocks of the foreland are essentially where they were deposited and are said to be *autochthonous*—that is, developed where found; these rocks

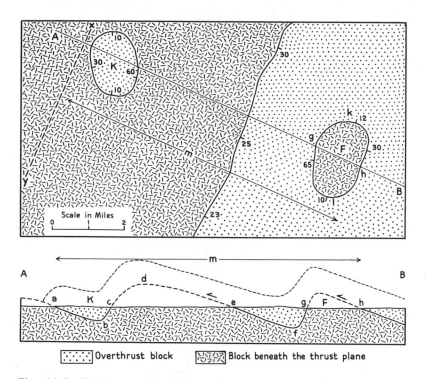

Overthrust block   Block beneath the thrust plane

Fig. 10-5. Folded overthrust. Map above, structure section below. Dip of thrust plane shown by figures. *K*, Klippe, *F*, Fenster (window). *AB*, line of structure section. *m*, minimum breadth of overthrust block. *xy*, original western limit of overthrust block. Other letters are referred to in text.

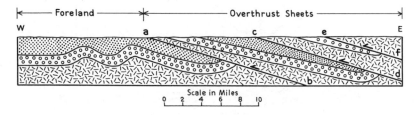

Fig. 10-6. Foreland of an overthrust belt. *ab, cd,* and *ef* are thrust planes.

are sometimes called the *autochthon*. Although *foreland* and *autochthon* are somewhat similar terms, the former refers to the place, the latter to the rocks in that place.

The rocks in the overthrust sheets have traveled many miles from their original place of deposition and are said to be *allochthonous*—that is, formed somewhere else; these rocks are sometimes called the *allochthon*.

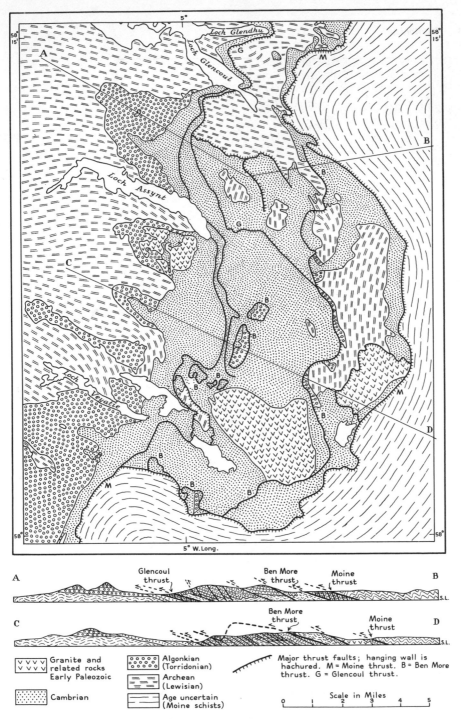

Fig. 10-7. Overthrusts of the Assynt District in the North-West Highlands of Scotland. (Simplified from the Geological Survey of Scotland, Assynt district, 1923.)

A *nappe* is a large thrust sheet or recumbent fold. It is best described as a large body of solid rock that has moved over the underlying rocks a long distance by either recumbent folding or overthrusting. The words "long distance" are not precise, but are generally understood to mean a mile or more. A large recumbent fold is a nappe, but a small recumbent fold is not. Likewise, a low-angle thrust with large net slip is a nappe, but a low-angle thrust with a net slip of a few hundred feet is not a nappe.

The North-West Highlands of Scotland, a classic region of overthrusting[2,3] are illustrated by a map and structure sections of the Assynt district in Fig. 10-7. To the west is the foreland, composed of the Precambrian and Cambrian rocks. The western edge of the overthrust belt trends somewhat east of north across the center of the map. The Moine overthrust, designated by the letter *M*, is the major overthrust, and along it the Moine Schists have been thrust westward. It is apparent that the Moine Schists at one time extended at least as far west as Loch Assynt, but they have been removed from the central part of the area by erosion. Evidence to the north of the Assynt district indicates that the net slip along the Moine overthrust is at least 15 miles.

The Ben More thrust, designated by the letter *B*, outcrops west of the Moine thrust. Several klippe of the Ben More thrust lie in the vicinity of section line *CD*.

The lowest thrust, separating the rocks that are essentially in place from those that have been moved, is known as a *sole*. Many minor thrusts, all dipping steeply to the east, produce a well-developed imbricate structure.

Some of the large thrusts of the Glarus District of Switzerland are shown in Fig. 10-8. The autochthonous rocks in the lower part of the section are separated from the overlying allochthonous rocks in the upper part of the section by a great overthrust. The plane of overthrusting extends for many tens of kilometers east-west and north-south. The net slip is at least 25 kilometers, the allochthon having moved relatively northward. Somewhat south of the center of the section, Permian rocks rest on Eocene rocks. A thin layer of limestone is present in many places along the thrust plane.[4] This unit, called the Lochenseitkalk, is generally less than one meter thick, but ranges

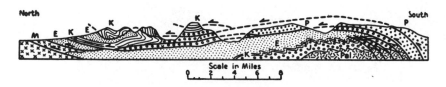

Fig. 10-8. Overthrusts of the Glarus District, Switzerland. Thrusts shown in heavy black. Pal, Paleozoic; P, Permian; J, Triassic and Jurassic; K, Cretaceous; E, Eocene; M, Miocene. (After J. Oberholzer.)

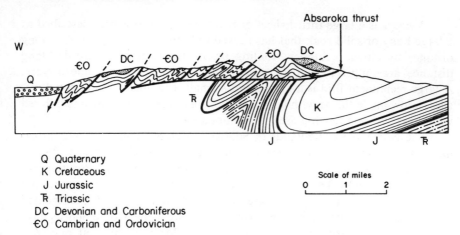

Q Quaternary
K Cretaceous
J Jurassic
Ṟ Triassic
DC Devonian and Carboniferous
€O Cambrian and Ordovician

Scale of miles
0   1   2

Fig. 10-9. Absaroka overthrust in Wyoming. (After W. W. Rubey)

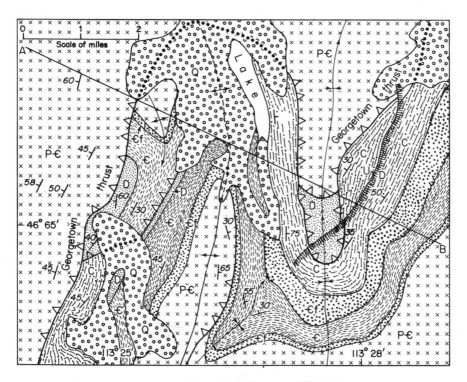

Fig. 10-10. Folded overthrust. Map of Georgetown Thrust, Montana, P€, Precambrian; €f, Cambrian Flathead sandstone; €, Cambrian above Flathead sandstone; D, Devonian; C, Carboniferous; Q, Quaternary. (After Poulter.[6])

Georgetown Thrust, Montana

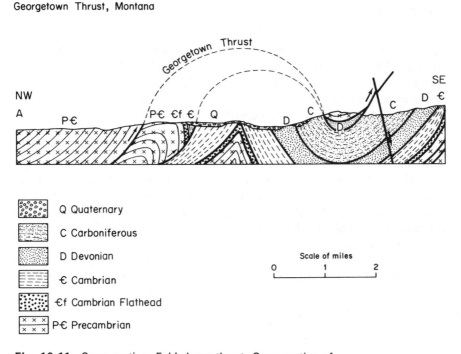

Fig. 10-11. Cross section. Folded overthrust. Cross section of Georgetown Thrust, Montana. P€, Precambrian; €F, Cambrian Flathead sandstone; €, Cambrian above Flathead sandstone; D, Devonian; C, Carboniferous; Q, Quaternary. (After Poulter.[6])

from zero to 20 meters. It is a Jurassic and Cretaceous unit that has been smeared out along the fault. Although precise data are obviously not available, the displacement may have been accomplished in an interval of time lasting from 100,000 to 1,000,000 years.

Figure 10-9 is a geologic cross section of the Absaroka overthrust in southwestern Wyoming.[5] In the area covered by this figure Paleozoic strata, deformed into open and isoclinal folds, and cut by minor thrust faults, have been thrust over Mesozoic strata in the overturned limb of a syncline. This fault is younger than the folding. In places Upper Cambrian strata have been thrust onto Cretaceous strata, indicating a stratigraphic throw of about 12,500 feet. The minimum value of the net slip is 4 miles, and is probably much greater.

Figure 10-10 is a map and Fig. 10-11 a cross section of a folded overthrust,[6] the Georgetown thrust in Montana. Along the line of the cross section, Precambrian sedimentary strata have been thrust over Devonian strata; the stratigraphic throw is at least 10,000 feet, and the net slip is much greater.

Detachment Faults

The Heart Mountain detachment fault in northwest Wyoming is illustrated in Figs. 10-12 and 10-13. The rocks above the Heart Mountain fault are preserved as isolated blocks, a few hundred feet to 5 miles across, in a triangular area 30 by 60 miles. Heart Mountain (Fig. 10-12), the largest block, is a mile long and a quarter of a mile wide. At Heart Mountain, Paleozoic rocks are above the fault, whereas Eocene rocks are below it; here the stratigraphic throw is 15,000 feet. But in the western part of the area the fault is a bedding fault and the normal stratigraphic section occurs across the fault. In places here the fault is marked by gouge and breccia up to 50 feet thick. Figure 10-13 shows in diagrammatic fashion the isolated blocks above the fault. For many years it was assumed that the isolated blocks were erosional remnants of a once-continuous sheet. The upper plate was presumed to have been pushed from the southwest.

The structure is now interpreted as a detachment fault (Fig. 10-13). As a result of an uplift to the northwest, the Paleozoic rocks above the Cambrian slid southeasterly.[7,8,9] The *breakaway* is the surface bounding the rocks that remained in place. For several miles downdip the fault was a bedding fault. Further southeast it became a transgressive fault, cutting diagonally upward across the strata. Still further east the upper plate moved across

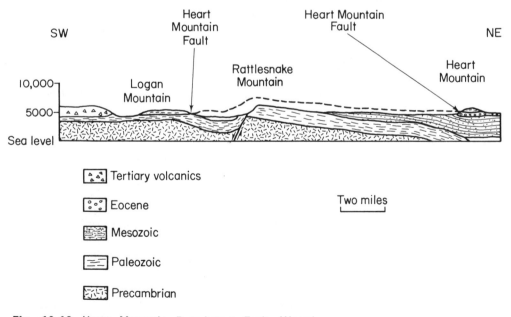

Fig. 10-12. Heart Mountain Detachment Fault, Wyoming. (After Pierce.[7,8])

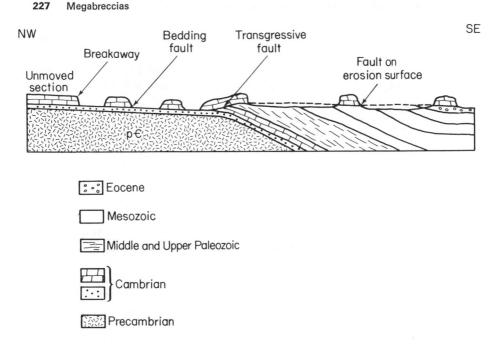

Fig. 10-13. Origin of Heart Mountain Detachment Fault, Wyoming. (After Pierce.[9])

an erosion surface. Since the net slip must be several times greater than the stratigraphic throw, it is apparent that the net slip at Heart Mountain is probably 10 miles or more. It is difficult to determine the extent to which the blocks above the fault moved as isolated blocks. But some klippe, rather than being remnants of a once-continuous sheet, may be isolated blocks that moved independently.

But it is rare indeed that one can see the breakaway of a detachment fault. Since this is in an uplifted area it is almost invariably destroyed by erosion. But in the northeast corner of Yellowstone Park thick volcanic deposits buried the breakaway before it could be eroded.

Detachment faults, such as Heart Mountain, pose several problems. How high was the breakaway above Heart Mountain? What was the dip of the fault? What was the coefficient of friction along the fault? The nature of these problems is considered more fully in a later section.

Megabreccias

Megabreccias will be discussed here, although they are not exclusively associated with thrust faults.

Breccias are consolidated rocks that consist of angular rock fragments set in a finer-grained matrix. In most cases the fragments are inches, feet, or tens of feet in diameter, with the matrix of relatively finer grain. There may be a great range in the ratio of fragments to matrix. Such breccias are of many origins. Fault breccias have already been discussed (p. 202). Volcanic and plutonic breccias are described on pages 329 and 381. Other breccias may form from talus slopes, landslides, the collapse of caverns, or the collapse of walls of open fractures.

But in this chapter we are concerned with megabreccias, which may be defined as very coarse breccias in which the larger fragments are hundreds or thousands of feet long—or even miles; moreover, it is obvious that the areas covered by such megabreccias must be large, measured in square miles. Various terms have been used for these coarse breccias, but most of them have genetic connotations that may be difficult to prove. The term mega-breccia is a descriptive term without genetic implications. On the other hand, the field geologist should be fully cognizant of the various ways in which megabreccias may form.

The term *megabreccia* was used[10] in the Lake Mead area of Arizona and Nevada for extensive coarse unsorted debris, in which the individual fragments may be as much as tens or hundreds of feet long. These megabreccias were believed to be the result of landslides from rising fault blocks.

Chaos has been applied to a coarse breccia in the Death Valley region of California. The so-called Amargosa Chaos has been described as follows:[11]

(1) The arrangement of the blocks is confused and disordered-chaotic. (2) The blocks, though mostly too small to map, are vastly larger than those in anything that could be called a breccia; most of them are more than 200 feet in length, some are as much as quarter of a mile, and a few are more than half a mile in length. (3) They are tightly packed together, not separated by much finer-grained material. (4) Each block is bounded by surfaces of movement—in other words, each is a fault block. (5) Each block is minutely fractured throughout, yet the original bedding in each block of sedimentary rock is clearly discernible and is sharply truncated at the boundary of the block. Commonly the bedding, even of incompetent beds, is not greatly distorted.

This chaos is probably formed under a large thrust sheet by the imbrication of small thrust plates; many of the plates broke up into large lenses and blocks.[12]

The term *mélange* has been used to describe widespread coarse breccias in Wales[13] and Turkey.[14] In the example from Turkey, the Ankara Mélange consists of huge blocks of limestone, some of them kilometers in length, and some small blocks of graywacke, set in a matrix of graywacke. This mélange is believed to be a gigantic breccia under an overthrust sheet. It has also been suggested that the Franciscan series in the California Coast Range is a mélange.[15,16] But various suggestions have been made to explain its origin: fragmentation due to compression, gravity sliding during sedimentation, or brecciation due to thrust faulting.

Olistostromes have been described from Newfoundland[17] and Italy. Those found in the Appenine Mountains are composed of highly deformed black shales containing blocks of various kinds of rocks, that range in size from a few centimeters to many kilometers and range in age from Triassic to Miocene.[18] They are considered to be the product of submarine landsliding of blocks derived from uplifted areas. Highly folded rocks associated with this chaos are shown in Plate 23.

These various kinds of megabreccias have much in common. The method of investigation depends on the map-scale relative to the size of the blocks. If a large-scale base map is available, individual blocks may be mapped separately. Individual blocks may be dated by their fossil content—if they have any fossils—or by comparison with an undisturbed section in the vicinity. The megabreccia is as young as or younger than the youngest fossils contained in it.

Mechanics of Reverse Faulting, Thrust Faulting, and Overthrusts

INTRODUCTION

In Chap. 7 the relation of rupture to stress was considered. It was shown that fractures are of two general types, tension fractures and shear fractures. Along tension fractures the walls move apart. Along shear fractures the displacement is parallel to the walls, and there is no movement perpendicular to the fracture. Tension fractures are not faults, at least when they first form. It is entirely possible, of course, for later displacement to be parallel to the walls. The tension fracture could thus become a fault. Moreover, the displacement along shear fractures may be later than and independent of the stresses that produced the initial fracture.

Most of the present section will deal with those faults that were originally shear fractures. Moreover, it is assumed that the displacements are caused by the same stresses as those that produced the initial rupture. But it should be clearly understood that the displacement may be the result of stresses other than those that caused the original rupture. A discussion of this phase of the problem is reserved to the end of the chapter.

In any attempt to analyze faulting on the basis of mechanical principles, it is necessary to make some simplifying assumptions. For the present it is sufficient to say that the analysis treats the rocks as if they were isotropic, that is, as if their properties were the same in all directions. A more detailed discussion of the nature of these assumptions and their possible significance is also discussed near the end of the chapter.

Plate 23 *Folds in chaos.* Highly deformed strata in chaos. Looking southeast. Shale, sandstone, marly limestone of Cretaceous-Eocene age. Gulf of Genoa, southwest of Camogli, Italy. Photo: Kurt Lowe.

RELATION OF THRUSTS TO STRESSES

In Chap. 7 it was shown that shear fractures and tension fractures are systematically related to the principal stress axes. We are concerned here with the shear fractures. The intersection of the two sets of shear fractures is a line parallel to the intermediate principal stress axis; moreover, the angle between the fractures and the greatest principal stress axis is approximately 30°.

As a first approach, it may be assumed that during deformation one of the principal stress axes is vertical, whereas the other two are horizontal. The three possible combinations are illustrated in the upper part of Fig. 10-14. If continued pressure leads to movement along the fractures they become faults. If the greatest principal stress axis is horizontal and the least principal stress axis is vertical (Fig. 10-14A), the hanging wall moves relatively over the footwall (Fig. 10-14B); the resulting faults are thrusts that dip 30° ±. The strike-slip faults and normal faults illustrated in Fig. 10-14D and F are discussed on pages 273 and 257.

Figure 10-15 shows that under the conditions postulated in Fig. 10-14A, two sets of fractures develop. In Fig. 10-15B a fracture belonging to one of

West   East

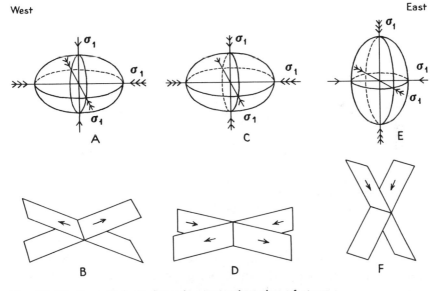

Fig. 10-14. Type of fault depends upon orientation of stress ellipsoid. σ_1, greatest principal stress axis; σ_2, intermediate principal stress axis; σ_3, least principal stress axis. (A) Least principal stress axis is vertical. (B) Thrusts form under conditions postulated in (A). (C) Intermediate principal stress axis is vertical. (D) Strike-slip faults form under conditions postulated in (C). (E) Greatest principal stress axis is vertical. (F) Normal faults form under conditions postulated in (E).

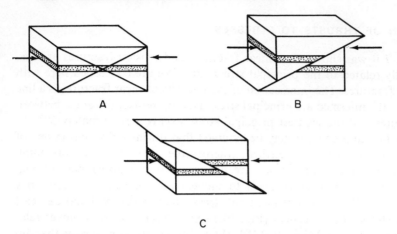

Fig. 10-15. Thrust faults dipping 30° ± form when greatest and intermediate principal stress axes are horizontal.

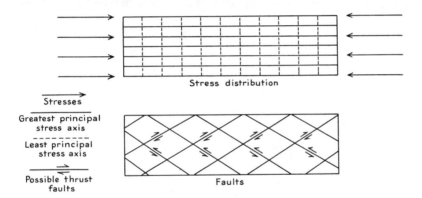

Fig. 10-16. Faults if superimposed horizontal stress is same at all depths. Upper diagram shows trajectories of stress axes; lower diagram shows attitude of possible thrust faults. (Based on description by W. Hafner.[19])

the sets has become a thrust fault. In Fig. 10-15C a fracture belonging to the alternate net has become a thrust. Both sets of faults may develop. But in many areas one set is favored, as discussed below.

The least principal stress axis is not necessarily vertical, and hence thrusts may have an infinite variety of dips. But in analyzing this problem certain simplifying assumptions are necessary. We assume that rocks are brittle and isotropic. Moreover, the analysis is made in two dimensions. In the illustrations given in Figs. 10-16, 10-17, and 10-18 this means that the intermediate stress axis is assumed to be horizontal and perpendicular to the paper. We

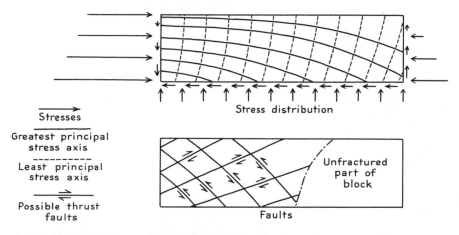

Stress distribution

Stresses

Greatest principal
stress axis

Least principal
stress axis

Possible thrust
faults

Unfractured
part of
block

Faults

Fig. 10-17. Faults if superimposed horizontal stress increases downward. Rupture strength assumed to be 3000 kg/cm². Upper diagram shows trajectories of stress axes; lower diagram shows attitude of possible thrust faults; details depend on several variables, including strength of rocks. (After W. Hafner.[19])

assume that the additional stresses are imposed on the *standard state*; that is, the lithostatic pressure, which increases downward at a rate of about 1 lb/in² per foot (about 5000 lb/in² per mile), is uniform on all sides of any imaginary sphere in the crust.

In the case illustrated in Fig. 10-16 an additional horizontal pressure is superimposed upon the standard state.[19] This superimposed horizontal pressure is assumed to have the same value at all depths.

The upper diagram shows the distribution of the stresses. The solid horizontal lines represent the *trajectories*—that is, the paths—of the greatest principal stress axes. They are everywhere horizontal, parallel to the plane of the paper. The broken lines are the trajectories of the least principal stress axes. They are everywhere vertical. The intermediate principal stress axes are everywhere perpendicular to the plane of the paper.

The rocks rupture when the superimposed pressure becomes great enough, and eventually the fractures may develop into thrust faults. Of course, the regular spacing of the faults is entirely diagrammatic.

In the case illustrated in Fig. 10-17 an additional horizontal pressure is superimposed upon the standard state. It is assumed that this additional pressure increases downward; this is indicated at the left end of the upper diagram by the increased length of the arrows toward the bottom. It is also assumed that the effectiveness of the superimposed pressure decreases in intensity from left to right; this fact is indicated by the greater length of the horizontal arrows on the left side of the diagram compared to those on the right. The additional stresses, which are shown in the diagram and which are necessary to preserve equilibrium, need not concern us here.

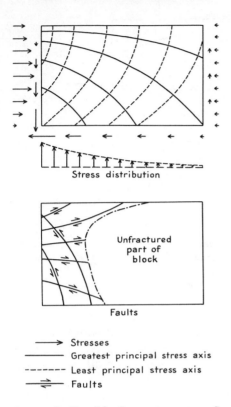

Stress distribution

Unfractured part of block

Faults

⟶ Stresses
—————— Greatest principal stress axis
-------- Least principal stress axis
⇌ Faults

Fig. 10-18. Faults if supplementary stress decreases exponentially from left to right. Rupture strength assumed to be 3000 kg/cm². Upper diagram shows trajectories of stress axes. Lower diagram shows attitude of possible faults; one set consists of high-angle thrusts; second set consists of low-angle faults, which are thrusts above, normal faults below. (After W. Hafner.[19])

The solid lines in the upper diagram show the orientation of the trajectories of the greatest principal stress axes. The broken lines are the trajectories of the least principal stress axes. The intermediate principal stress axes are everywhere parallel and are perpendicular to the plane of the paper.

If the supplementary stress becomes great enough, the resulting shear fractures will be oriented as shown in the lower part of Fig. 10-17. Inasmuch as the plunges of the greatest principal stress axes differ from place to place, the dips of the faults will also differ. As shown at the left end of the lower diagram, one set of faults dips 30° to the left, the other set 30° to the right, because the greatest principal stress axes are here horizontal. At the right-hand end of the diagram, because the greatest principal stress axes plunge to the right, one set of faults dips about 20° west, the other set about 40° east. All the faults are thrusts. Because the supplementary pressure was applied to the left end of the block, and its effectiveness is assumed to decrease in intensity to the right, the right-hand end of the block does not fracture. The boundary between the fractured and unfractured parts of the blocks depends upon a number of factors.

In Fig. 10-18 the superimposed horizontal stress decreases exponentially in a horizontal direction from left to right. In this case the trajectories of the greatest principal stress axes plunge to the right. As before, the fractures will be inclined 30° to the greatest principal stress axes. There will be one set of high-angle thrusts and one set of low-angle thrusts. In fact, progressing downward in the diagram the low-angle thrusts become horizontal faults and at the bottom of the diagram become low-angle normal faults. The right-hand part of the block is not fractured. The location of the boundary between the fractured and unfractured part of the block depends upon a number of factors.

In Fig. 10-19 it is assumed that the supplementary pressure system consists of two parts. One is a variable vertical stress represented in the upper diagram by the vertical arrows under the block. A variable shearing stress is shown by the horizontal arrows at the bottom of the block. In order to give the diagram some geological significance, Hafner[19] assumed that the block was 10 miles thick. In the upper diagram the trajectories of the greatest principal stress axes are shown by solid lines. They plunge toward the center of the diagram at relatively low angles at the margins and at progressively higher angles toward the center. Conversely, the trajectories of the least principal stress axis form a gigantic arc. The resulting faults are shown in the lower diagram. In the upper central part of the diagram they are high-angle normal faults. Toward the margins one set of these normal faults dips at progressively lower angles. The second set becomes progressively steeper, eventually passes through the vertical, and nearer the margins becomes high-angle thrust faults. The curved faults at the very bottom of the diagram change from low-angle thrusts to low-angle normal faults.

This two-dimensional analysis shows that both thrusts and normal faults may dip at angles ranging from 0° to 90°. We thus find a rational theoretical basis for field observations. Strike-slip faults have not been mentioned in this section because, in order to restrict the discussion to reasonable lengths, it has been assumed that the intermediate principal stress axis was always horizontal. If, however, the intermediate principal stress axis were vertical, then the diagrams in Figs. 10-16, 10-17, 10-18, and 10-19 could be looked upon as if they were maps rather than sections. Moreover, the principal stress axes are not necessarily horizontal and vertical. If all the axes are inclined, diagonal-slip faults may form.

MODIFYING FACTORS

It has been necessary in the preceding sections to make a number of simplifying assumptions. Consequently the patterns shown by faults in the field will undoubtedly be more complex and variable than those shown in the diagrams. Again it should be emphasized that the regular distribution of the faults is entirely diagrammatic. Some of the other factors that make the problem more complex are discussed below.

The forces along the margins of the blocks in the preceding figures have a symmetry and regularity that is very unlikely within the crust of the earth. Thus the distribution of stress within the blocks would be more complex.

It has already been emphasized that the crust of the earth lacks homogeneity. This would have an important effect on the distribution of the stresses even if the applied forces varied in some uniform and systematic way. Moreover, inhomogeneities influence the orientation of the shear fractures. For example, if the theoretical shear fractures at some place are horizontal, but the bedding dips 10 to 20 degrees, the actual fracture would probably follow the bedding. Foliation planes, joints, and older faults influence the orientation of new fractures. Once the rocks have ruptured, the stress distribution may be profoundly modified.[20] A few small fractures in a large block would not be significant, but obviously a few large ruptures would exert a great influence on the behavior of the rocks. If large blocks become completely isolated from one another by fractures, the stress distribution might differ considerably from what it was originally.

In the preceding analyses it was tacitly assumed that the rocks were relatively brittle, with little or no plastic deformation preceding the rupture. At depths of 5 or 10 miles many rocks are relatively plastic. If they are sufficiently plastic they might never fail by rupture during the period of deformation; they would continue to flow indefinitely. Such cases need not concern us here. But we are concerned with rocks that yield plastically a considerable amount before they rupture. Even in such cases the principles enunciated above are pertinent because they deal with the stresses just before and at the time of rupture. Nevertheless, it must be emphasized that a considerable plastic deformation may precede the rupture.

As already indicated at the beginning of this section, it has been tacitly assumed that all faults are initially shear fractures. It is clear, however, that tension fractures may also become faults.

Moreover, the attitude of a fault plane may be modified by later deformation. A vertical fault might cut horizontal sediments; if the sediments are later folded to a vertical position, the fault will be horizontal. Low-angle thrust planes may be subsequently folded, in some cases to assume very high dips (see pages 218 to 224).

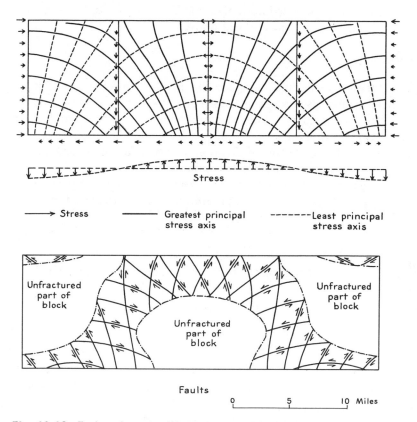

Fig. 10-19. Faults when supplementary stress consists of two parts. Rupture strength assumed to be 3000 kg/cm². Upper diagram shows trajectories of stress axes. Supplementary stresses are: (a) a variable vertical stress represented by vertical arrows; (b) a variable shearing stress represented by arrows at bottom of block. Lower diagram shows attitude of possible faults. (After W. Hafner.[19])

A shear fracture may form, with or without faulting, under one condition of stress. Subsequently, under very different conditions of stress, displacement may take place. For example, a vertical strike-slip fault, along which there had been relatively little displacement, might be utilized later for large differential vertical movements.

It is apparent from the preceding discussion that no simple rules can be established to determine the nature and magnitude of the forces involved in the formation of a fault. The geological history of the region must be known. It may be necessary to extend the study to a relatively large area. Moreover, several alternate hypotheses may explain the observed facts. Nevertheless, any conclusions must be based not only on field observations, but also on a mechanical basis such as that outlined above.

OVERTHRUSTS

Large overthrusts, where the upper plate has traveled tens or scores of miles, have puzzled geologists for years. Elementary mechanics suggest that the compressive force necessary to move such large masses of rock would have to be so great that the sheet would rupture before it would move.

An analysis of a rather simple situation may serve as an introduction to a more complete analysis given in later paragraphs.

Obviously the coefficient of friction between the block being pushed and the underlying block is an important parameter. If the fracture is horizontal (Fig. 10-20):

$$\mu = \frac{F}{W} \tag{1}$$

where μ is coefficient of friction, W is weight of upper block, and F is force necessary to push it.

$$W = bcd\rho g \tag{2}$$

where b, c, and d are dimensions of block, ρ is specific gravity of rock in block, g is acceleration of gravity.

$$F = scd \tag{3}$$

where s is force per unit area, and c and d are dimensions of face on which compression is being exerted.

Substituting equations (2) and (3) in equation (1),

$$\mu = \frac{scd}{bcd\rho g} \tag{4}$$

or

$$s = \mu b\rho g \tag{5}$$

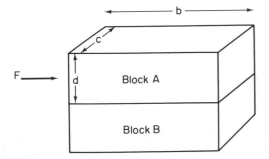

Fig. 10-20. Horizontal fault. Block *A* pushed over block *B* when subjected to compressive force *F*.

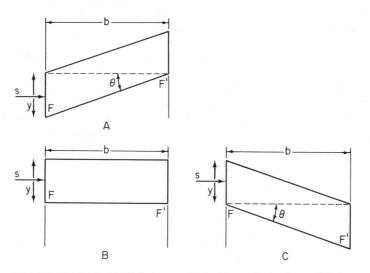

Fig. 10-21. Fault block being pushed along a fault plane. (B) Horizontal fault plane. (C) Fault block pushed downslope. (A) Fault block pushed upslope.

For example, if b is 20 km ($= 2 \times 10^6$ cm), if μ is 0.5, and ρ is 2.3 g/cm³,

$$s = 2.25 \times 10^9 \text{ dynes} \tag{6}$$

But since s exceeds the crushing strength of the average rock (7×10^8 dynes/cm²), the block would break up before it could move.

Equation (5) can be recast to the form

$$B = s/\mu\rho g \tag{7}$$

where B is maximum possible size of thrust sheet, s is crushing strength of a rock (7×10^8 dynes/cm²), and other parameters are as before. Substituting in equation (7)

$$B = 6.2 \times 10^5 \text{ cm} = 6.2 \text{ km} \tag{8}$$

But most fault planes are inclined. If the fault plane dips toward the source of the active pressure the thrust block is pushed uphill. The compressive force must be great enough not only to overcome the friction but also to lift the block.*

If the thrust block is shaped as in Fig. 10-21, the horizontally directed stress to move it is

$$s = \frac{b\rho g(\mu + \tan \theta)}{(1 - \mu \tan \theta)} \tag{9}$$

where θ is the dip of the fault plane, and other parameters are as indicated above.

*The ensuing equations were prepared by Mr. Claude Dean.

The maximum possible size of the thrust block is

$$B = \frac{s(1 - \mu \tan \theta)}{\rho g(\mu + \tan \theta)} \tag{10}$$

If $\mu \tan \theta > 1$, $s = \infty$ and $B = 0$.

For example, if $\mu = 0.5$, $\mu \tan \theta = 1$ when $\theta = 63°.5$. In other words, under postulated conditions, a block can not be thrust up along a fault dipping more than $63°.5$. It should be emphasized that one assumption is that the compressive force is horizontal.

If the fault plane dips away from the source of active pressure—that is, the block is pushed downhill—the equations become:

$$s = \frac{b\rho g(\mu - \tan \theta)}{(1 + \mu \tan \theta)} \tag{11}$$

$$B = \frac{s(1 - \mu \tan \theta)}{\rho g(\mu - \tan \theta)} \tag{12}$$

In equation (11), if $\mu = 0.5$ and θ is greater than $26°.5$, s becomes negative. That is, under the assumed conditions, the block slides downhill under its own weight.

A series of graphs based on equations (10) and (12) are shown in Fig. 10-22. The coefficient of friction is shown on the abscissa, whereas the maximum possible size of fault blocks is shown on the ordinate. Separate curves are shown for various angles of dip. Negative dips represent faults that dip away from the active pressure.

Two facts stand out. (1) For a given dip, the maximum possible size of the fault block becomes progressively greater as the coefficient of friction becomes less. Moreover, for horizontal faults or faults dipping away from the active pressure, the maximum possible size of the fault block approaches infinity as the coefficient of friction approaches zero. (2) The steeper the dip of the fault toward the active pressure, the smaller the maximum possible size of the fault block. For example, if the coefficient of friction is 0.4, B changes from 7.8 km for horizontal faults to 2.0 km for faults dipping $40°$. (3) If the fault dips away from the active pressure, much larger blocks are possible. Blocks will move downhill under their own weight if the coefficient of friction and angle of dip are low.

We are now prepared to consider some additional factors. It has been suggested[21] that high fluid pressure in the pore spaces of the rocks may partially balance the lithostatic pressure. As indicated on page 32, λ is the ratio of pore pressure to lithostatic pressure. As λ approaches one, the effective coefficient of friction becomes zero. But it is probable that enough fractures would be present to let fluids escape and prevent the attainment of high fluid pressures.

In the preceding analysis it was assumed that the fault was already present and that coherence along the fault was lacking. But some geologists

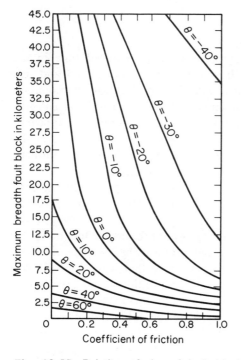

Fig. 10-22. Relation of size of fault block to coefficient of friction and dip of fault plane. Negative dips indicate fault is dipping away from source of active pressure.

believe that an extra force is necessary to break the bonds along the fault.[4] This is the "cohesive strength" or "adhesive strength." It is the value of the shearing stress crossed by the Mohr envelope (Fig. 7-20) where the normal stress is zero.

In some equations[21] the fact that the strength increases with increased thickness of the fault block has been taken into consideration.

The values shown in Fig. 10-22 would be increased if the rocks are stronger than assumed.

In summary, the maximum possible size of overthrust sheets and detachment blocks is a function of many variables including the strength of the rocks, the coefficient of friction along the fault, the angle of dip of the fault, pore pressure, and the attitude of the greatest principal compressive stress. In fact, so many variables are involved that precise mathematical analysis in any specific case is impossible. The student of overthrusts should be well aware of the problems and parameters involved. But he should realize that at best the answer will give only the order of magnitude.

Palinspastic Maps

Paleogeographic maps show the geography at some specified time in the past.[22,23] Most such maps merely show the distribution of land and sea. Others are more elaborate, showing in addition the location of mountain ranges and the thickness of the sediments that accumulated.[24] A *paleogeologic map* is a hypothetical geologic map of an area as it would have appeared in the past.[24] For example, a paleogeologic map of North America at the beginning of the Tertiary would obviously show no Tertiary or Quaternary rocks. Moreover, all rocks eroded away since the beginning of the Tertiary would be restored. A *paleotectonic map* is a map showing the tectonic conditions—folds, thrusts, normal faults, igneous rocks, etc.—at some specified time in the past.[25]

However, many such maps make no allowance for any reduction in the width of the fold belt that may have occurred, primarily by folding and thrusting. On most paleogeographic maps a geosyncline that was originally 200 miles wide but has been reduced to 150 miles by folding and thrusting is shown as 150 miles wide. But it would be more correct to "unfold" and "unthrust" the rocks before preparing the paleogeographic and similar maps. *Palinspastic maps*[26] place rocks in their presumed position before folding and thrusting. "Palinspastic" can be used as an adjective with paleogeographic, paleogeologic, and paleotectonic. Commendable as this ideal may be, it is often difficult to put into practice because of lack of adequate information on the amount of crustal shortening. But such an analysis has been made of the structure shown in Fig. 10-8. This belt, now 32 miles wide, was 70 miles wide before thrusting.[27]

References

[1] Cooper, Byron N., 1944, Geology and mineral resources of the Burkes Garden Quadrangle, Virginia, *Virginia Geol. Surv., Bull.* 60, 299 pp.

[2] Craig, Gordon Y., ed., 1965, *The geology of Scotland*, 556 pp. Hampden, Conn.: Archon Books.

[3] Peach, B. N., Horne, J., Gunn, W., Clough, C. T., Hinxman, L. W., and Teall, J. J. H., 1907, *The geological structure of the North-West Highlands of Scotland*, Mem. Geol. Surv. Great Britain, pp. 463–594, Glasgow.

[4] Hsü, K. Jinghwa, 1969, Role of cohesive strength in the mechanics of overthrust faulting and of landsliding, *Geol. Soc. Amer. Bull.* 80: 927–52.

[5] Rubey, William W., 1958, *Geology of the Bedford Quadrangle, Wyoming*, Geol. Quad. Maps of the United States, No. 109, U.S. Geol. Surv.

[6] Poulter, Glenn J., 1965, *Geology of the Georgetown Thrust area southwest of Philipsburg, Montana*, Montana Bureau of Mines and Geology, Geol. Invest., Map. No. 1.

[7] Pierce, William G., 1957, Heart Mountain and South Fork detachment thrusts of Wyoming, *Bull. Amer. Assoc. Petrol. Geol.* 41: 591–626.

[8] Pierce, William G., 1966, *Geologic map of the Cody Quadrangle, Park County, Wyoming*, Geol. Quad. Maps of the United States, Map GQ-542.

[9] Pierce, William G., 1960, The "break-away point" of the Heart Mountain detachment fault in northwestern Wyoming, U.S. Geol. Surv., Prof. Paper 400-B, pp. B 236–37.

[10] Longwell, Chester R., 1951, Megabreccia developed downslope from large faults, *Amer. J. Sci.* 249: 343–55.

[11] Noble, L. F., 1941, Structural features of the Virgin Spring area, Death Valley, California, *Geol. Soc. Amer. Bull.* 52: 941–1000.

[12] Kupfer, Donald H., 1960, Thrust faulting and Chaos structure, Silurian Hills, San Bernardino County, California, *Geol. Soc. Amer. Bull.* 71: 181–214.

[13] Greenly, E., 1919, *The geology of Anglesey*, Mem. Geol. Surv. Great Britain.

[14] Bailey, E. B., and McCallien, W. J., 1953, Serpentine lavas, the Ankara mélange, and the Anatolian thrust, *Trans. Roy. Soc. Edinburgh*, Vol. 62, Part 2 (No. 11), 403–42.

[15] Hsü, K. Jinghwa, 1968, Principles of mélanges and their bearing on the Franciscan-Knoxville Paradox, *Geol. Soc. Amer. Bull.* 79: 1063–74.

[16] Hamilton, Warren, 1969, Mesozoic California and the underflow of Pacific mantle, *Geol. Soc. Amer. Bull.* 80: 2409–30.

[17] Horne, Gregory S., 1969, Early Ordovician chaotic deposits in the central volcanic belt of northeastern Newfoundland, *Geol. Soc. Amer. Bull.* 80: 2451–64.

[18] Page, Ben M., 1963, Gravity tectonics near Passo Della Cisa, northern Appenines, Italy, *Geol. Soc. Amer. Bull.* 74: 655–72.

[19] Hafner, W., 1951, Stress distribution and faulting, *Geol. Soc. Amer. Bull.* 62: 373–98.

[20] McKinstry, H. E., 1953, Shears of the second order, *Amer. J. Sci.* 251: 401–14.

[21] Rubey, W. W. and Hubbert, M. K., 1959, Role of fluid pressure in mechanics of overthrust faulting, *Geol. Soc. Amer. Bull.* 70: 167–206.

[22] Schuchert, Charles, 1955, *Atlas of paleogeographic maps of North America*, 177 pp., New York: John Wiley & Sons.

[23] Sloss, L. L., Dapples, E. C., Krumbein, W. C., 1960, *Lithofacies maps, an atlas of the United States and southern Canada*, 108 pp., New York: John Wiley & Sons.

[24] Eardley, Armand J., 1962, *Structural geology of North America*, 2d ed. 624 pp., New York: Harper and Row.

[25] McKee, E. D., Oriel, S. S., et al., 1967, *Paleotectonic investigations of the Permian System in the United States*, U. S. Geol. Surv., Prof. Paper 515, 271 pp.; also Misc. Geol. Invest. Map I-450, 1967.

[26] Kay, Marshall, 1945, Paleogeographic and palinspastic maps, *Bull. Amer. Assoc. Petrol. Geol.* 29: 426–50.

[27] Trumpy, Rudolf, 1969, Die Helvetischen Decken der Ostschweiz, *Ecolgae Geol. Helv.* 62: 105–42.

11

NORMAL FAULTS

Introduction

As indicated on page 196, the term *normal fault* is commonly used for those faults along which the hanging wall has gone down relative to the footwall (Plates 18 and 21). The term *gravity fault* was preferred by the committee[1] that prepared a classification of faults, but this usage has not generally been followed. They preferred using the term "normal fault" for what on page 195 has been called "apparent normal fault." Moreover, as indicated in Chap. 10, low-angle faults along which there has been considerable horizontal displacement are a special group called "detachment faults."

There are many possibilities concerning the actual movement as measured from some datum, such as sea level or the center of the earth. The footwall may remain stationary and the hanging wall go down; or the hanging wall may remain stationary and the footwall go up; or both blocks may move down, but the hanging wall more than the footwall; or both blocks

244

may move up, but the footwall more than the hanging wall. In the present state of our knowledge it is impossible in most cases to determine the absolute movement, and hence an elaborate terminology is unnecessary and undesirable. In some instances, where there is evidence that the footwall has moved up along a high-angle normal fault or that the hanging wall has moved up along a reverse fault, the term *upthrust* may be used.

Size, Attitude, and Pattern

Normal faults may range from microscopic size to those that are tens of miles long and have a net slip that is measured in thousands of feet. The dip of normal faults may range from almost horizontal to vertical, but dips greater than 45° are more common. But a word of caution is necessary here. With reasonably good exposures it is possible to locate faults accurately and to determine the strike. But the dip may be much more difficult to ascertain. Of course, in mines, tunnels, highway cuts, and engineering excavations the faults can be observed; but these are usually relatively small faults. Data from bore holes may give valuable information, but even here a subjective element may be introduced; it may be difficult to determine whether or not a fault is present, and correlation of the same fault between bore holes may be very subjective. The large faults that are shown on inch to a mile and similar geologic maps are generally exposed in few places. It is also dangerous to rely on published cross sections, because the dip of the faults may be shown in accordance with some theory.

The system of faults that bounds the Triassic rocks extending from New York to Virginia on the northwest is 340 miles long. Normal faults on the Colorado Plateau are 100 to 300 miles long; in most instances, such long faults are not single fractures throughout their length but are fault zones. The *en échelon* normal faults bounding the Wasatch Range in Utah extend for at least 80 miles.

The minimum net slip along a normal fault bounding the Triassic rocks of Connecticut on the east side is 13,000 feet and perhaps more. The probable net slip along the normal fault on the west side of the Wasatch Range of Utah is 18,000 feet.

Tilted Fault Blocks

Normal faulting is accompanied in many instances by tilting of the blocks on one or both sides of the fault resulting in *tilted fault blocks*. Examples are shown in Fig. 11-1. In many cases erosion has destroyed all topographic expression of the fault blocks (Figs. 11-1A and B). In other cases the topography expresses to a greater or less degree the fault-block structure (Figs.

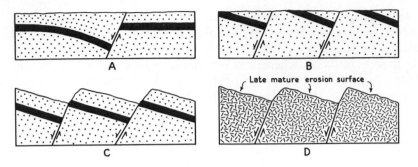

Fig. 11-1. Tilted fault blocks. Solid black, shale; dots, sandstone; diversely oriented dashes, granite. (A) and (B) Tilted fault blocks that are not expressed topographically. (C) Topographically expressed tilted fault blocks. (D) Topographically expressed tilted fault blocks developed on granite; a late mature erosion surface has been broken and tilted by the faulting.

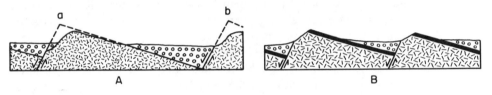

Fig. 11-2. Tilted fault blocks. Diversely oriented dashes are granite; circles are alluvium, solid black is lava. Alluvium accumulates in valleys from erosion of higher parts of fault blocks. (A) At locality *a* movement along fault ceased, and alluvium overlaps onto eroded granite. At *b* faulting is still active. (B) A broken and tilted lava bed that serves as a key bed.

11-1C and D). Instead of a single fault, the block may be bounded by several faults showing *step faulting* in which the downthrown side is systematically on the same side of several parallel faults.

Erosion rapidly attacks the uplifted area, and the debris is deposited in the adjacent valley (Fig. 11-2A). If no more faulting takes place, the valley is progressively filled up as the mountain front is eroded back (locality *a*, Fig. 11-2A). The fault is not exposed at the surface. But if, as at locality *b*, Fig. 11-2A, faulting continues during erosion, the valley fill abuts against the fault and the fault is exposed at the surface as the contact between the alluvium and bedrock of the mountain. Of course, faulting and sedimentation may interplay in a very complex way. Streams may remove all the debris and even the downthrown block may be subject to erosion.

The mountains that rise above the valleys are one variety of *fault-block mountains*. Some fault-block mountains are horsts (p. 249).

Tilted fault blocks are most readily recognized by the tilting of one or

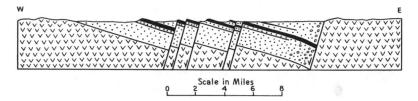

Fig. 11-3. Tilted fault block of the Connecticut Valley, Connecticut. Checks are pre-Triassic crystalline rocks; dots are arkose, sandstone, and shale; circles are conglomerate; solid black represents volcanic rocks.

more datum planes, such as the bedding (Figs. 11-1A, B, and C). In some cases a late-mature or old-age erosion surface may be broken and tilted (Fig. 11-1D). Such erosion surfaces are particularly useful in those regions that are underlain by granitic rocks, or in those areas where the strata were highly deformed prior to the normal faulting. In some mountains a thin lava flow may serve as the key bed that shows rupture and tilting (Fig. 11-2B).

The Triassic rocks of eastern North America display tilted fault blocks that have little topographic expression. In Connecticut (Fig. 11-3), the easterly-dipping Triassic rocks rest unconformably on the older crystalline rocks to the west. On the east a large, westerly-dipping normal fault, with a net slip of at least 13,000 feet, separates the Triassic rocks from the crystalline rocks to the east. The Triassic strata dip toward the east at an angle of 20° except near the fault, where the easterly dip reaches a maximum of 40°. Coarse conglomerates are well developed in the Triassic rocks directly west of the fault but pinch out westward. This indicates that an active fault scarp was present during sedimentation. A number of northeasterly-trending normal faults lie within the Triassic rocks.

The Huachuca Mountains of southeast Arizona[2] are an excellent example of a complex fault-block mountain (Figs. 11-4 and 11-5). The rocks in the mountains consist of Precambrian granite, Paleozoic and Mesozoic sedimentary and volcanic rocks, and a Jurassic quartz monzonite. The strata are highly folded and cut by thrust faults as the result of a late Mesozoic or early Tertiary period of deformation. During Late Tertiary uplift a large normal fault developed on the east side of the range and the mountains were uplifted relative to the valley. The magnitude of the uplift is indicated by the fact that Permian limestones are in contact with Precambrian granite along the Nicksville fault; the minimum stratigraphic throw is 5000 feet. The fact that the fault is no longer expressed by a scarp (Fig. 11-5) and the fact that the mountain front is deeply eroded show that considerable time has elapsed since the last significant faulting. Moreover, the Quaternary gravels extend across the fault without displacement.

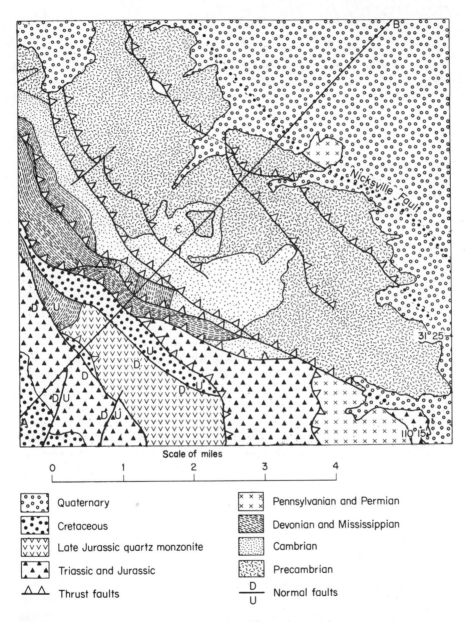

Scale of miles

0   1   2   3   4

Quaternary

Cretaceous

Late Jurassic quartz monzonite

Triassic and Jurassic

Thrust faults

Pennsylvanian and Permian

Devonian and Mississippian

Cambrian

Precambrian

Normal faults

Fig. 11-4. Geologic map of part of Huachuca Mountains, southeastern Arizona. A fault-block mountain with complex internal structure. For cross section along line *AB* see Fig. 11-5. (After Hayes and Raup.[2])

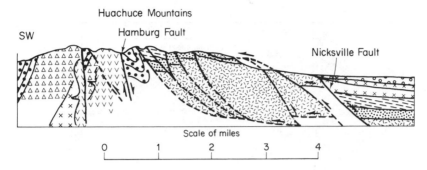

Fig. 11-5. Cross section of Huachuca Mountains, southeastern Arizona. (After Hayes and Raup.[2])

Horsts and Graben

Many fault blocks are bounded on both sides by normal faults along which the net slip is more or less equal, and, consequently, there is little or no tilting. A *horst* is a block, generally long compared to its width, that has been raised relative to the blocks on either side. A *graben* is a block, generally long compared to its width, that has been lowered relative to the blocks on either side. The faults bounding horsts and graben are usually steep normal faults. Some horsts and graben are expressed topographically, others are not.

The Black Hills in central Arizona are a horst,[3] with normal faults on the east and west (Figs. 11-7 and 11-8). One thousand feet of flat-lying Paleozoic rocks overlie the Precambrian rocks with a pronounced angular unconformity. Flat-lying Late Tertiary rocks, 500 feet thick, overlie these Paleozoic rocks disconformably. The horst is bounded on the east by the Verde fault zone. The master fault here is the Verde fault, which dips about 55°ENE; the vertical slip and stratigraphic throw of the contact of the Paleozoic and

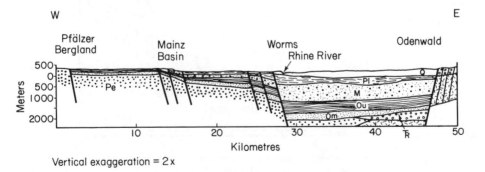

Vertical exaggeration = 2x

Fig. 11-6. Rhine graben. Black with white dashes is Pre-Permian basement; *Pe*, Permian; *Tr*, Triassic; *Om*, Middle Oligocene, *Ou*, Upper Oligocene; *M*, Miocene; *Pl*, Pliocene; *Q*, Quaternary.

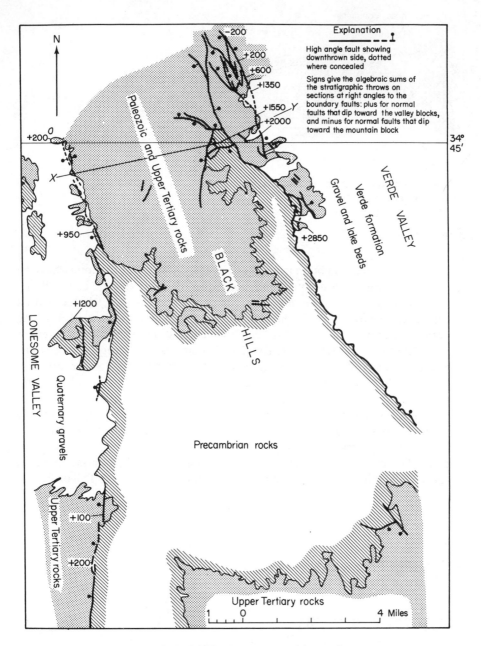

Fig. 11-7. Geologic map of Black Hills, Jerome area, Arizona. A horst. For cross section along line *XY* see Fig. 11-7. (After C. A. Anderson and S. C. Creasey.[3])

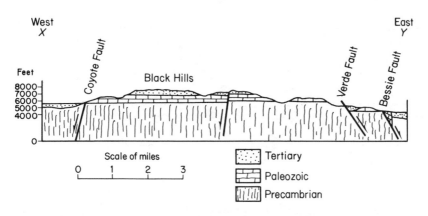

Fig. 11-8. Cross section of Black Hills, Jerome area, Arizona. (After C. A. Anderson and S. C. Creasey.[3])

the Tertiary is about 1500 feet. Although much of the displacement is younger than the Tertiary rocks, there was about 1000 feet of displacement during Precambrian time. The total net slip, bearing S.85°E. with a plunge of 50°E, totals 3300 feet. Another fault zone bounds the horst on the west. The Coyote Fault, labeled in Fig. 11-8, dips very steeply west; the stratigraphic throw of the contact of the Paleozoic and Tertiary rocks is about 1500 feet.

It is worthwhile emphasizing the vast length of time characterized by a lack of orogeny in this area. From the Early Cambrian to the Late Tertiary (over 500,000,000 years) there was no deformation except for vertical movements. Then in Late Tertiary and Early Quaternary time the region underwent normal faulting.

The Rhine graben in Europe[4] is 180 miles long and 20 to 25 miles wide (Fig. 11-6). The eastern border fault, where it is crossed by a tunnel, dips about 55° toward the west, occupying a crushed zone about 50 feet wide.

The rift valleys of Africa are a complex system of *en échelon* graben, horsts, and tilted fault blocks. The eastern rift zone lies east of Lake Victoria and extends for 3000 miles from Mozambique to the Red Sea. The western rift zone swings west of Lake Victoria and is 1000 miles long. The net slip on the border faults ranges from a few thousand to a maximum of 8000 feet at Lake Malawi.[5,6]

Rift systems are not only present on the continents, but are extensively developed on the ocean floors.[7]

Most cross sections of graben are based on surface data, and little information is available as to what happens at depth. Petroleum geology has made great contributions to this phase of the problem. Figure 11-9 is a section of a complex graben based on numerous drill holes.[8] The following points deserve special mention. (1) The graben is located on the crest of an anticline. (2) The graben is about 1½ miles wide. (3) The throw on none of the faults exceeds 150 feet. (4) Many of the lesser faults terminate against larger faults.

251

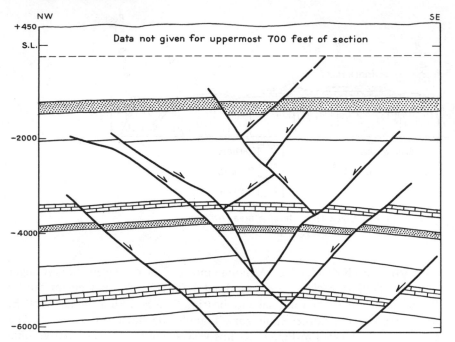

NW   SE

Data not given for uppermost 700 feet of section

Fig. 11-9. Graben on anticline. Quitman Oil Field, Texas. Horizontal and vertical scales are the same, shown in feet on left side of section. (After E. R. Scott;[8] permission of American Association of Petroleum Geologists.)

Modern Faults

Late on the night of August 17, 1959, southwestern Montana and adjacent areas were shaken by a major earthquake of magnitude 7.1. The shock was felt as far away as 350 miles from the epicenter and throughout an area of 350,000 square miles. The epicenter was near Hebgen Lake a few miles northwest of Yellowstone Park.[9]

The shoreline of the lake and the highways were deformed, and consequently it was possible to gather some very precise data on the amount of deformation (Fig. 11-10). Several faults formed during the earthquake are shown: Hebgen, Red Canyon, and Madison Range fault scarps (Plate 21). The scarps attained a maximum height of 20 feet. The isobase lines, based on deformation of the shoreline and highways, show the amount of subsidence. In the northern part of Hebgen Lake the subsidence on the southwest side of the Hebgen fault scarp was more than 19 feet. Similarly, the subsidence on the southwest side of the Red Canyon fault scarp exceeded 15 feet. In principle, the deformation is similar to that shown in Figs. 11-1C and D.

A major earthquake shook Alaska on March 27, 1964.[10] The deformation

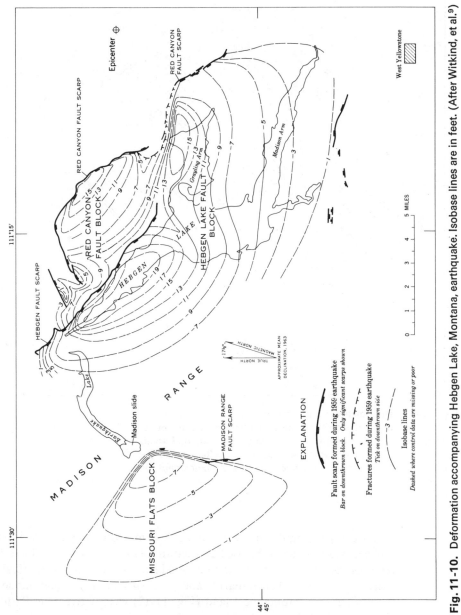

Fig. 11-10. Deformation accompanying Hebgen Lake, Montana, earthquake. Isobase lines are in feet. (After Witkind, et al.[9])

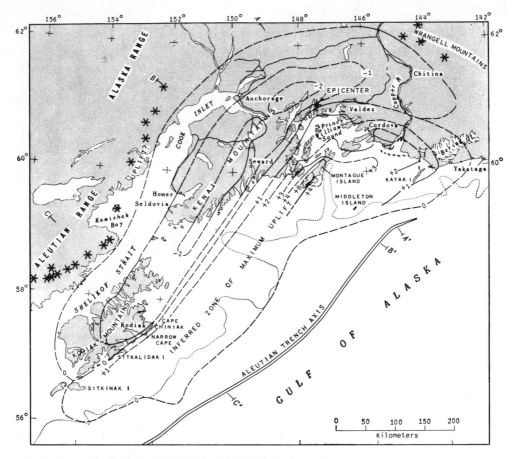

Fig. 11-11. Deformation accompanying the 1964 Alaska earthquake. Uplift or depression shown by structure contours in meters. Dotted line at −200 meters is outer edge of continental shelf. Asterisks are active or dormant volcanoes. (After Pflaker;[10] permission of American Association for the Advancement of Science.)

tion accompanying this quake could be determined with great precision inasmuch as sea level could be used as a datum plane (Fig. 11-11). The following features were displaced: (1) intertidal sessile marine organisms; (2) tidal bench marks; (3) highway bench marks; and (4) docks and other man-made structures. Faults on Montague Island (Fig. 11-12) were active. The Patton Bay Fault, which strikes northeast, could be traced for 16 kilometers, and may extend another 28 kilometers southwestward under water. The dip is vertical; the net slip is 5.2 meters. with the northwest wall uplifted. The shorter fault, the Hanning Bay Fault, also strikes northeast, dips 70 to 80 degrees north-west and the net slip, which is down dip, is 4 to 5 meters.

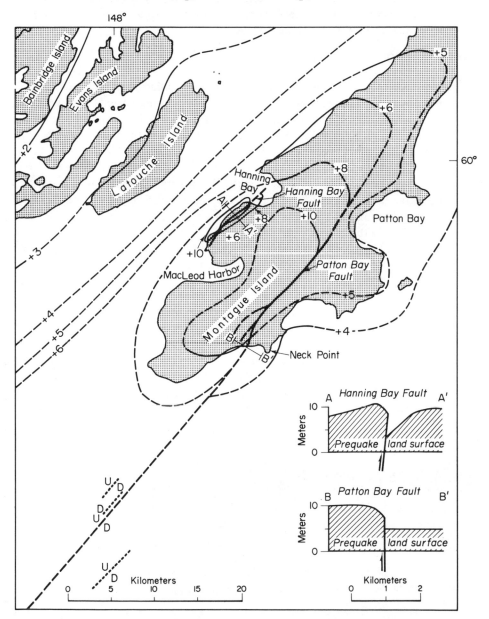

Fig. 11-12. Faulting on Montague Island during 1964 Alaska earthquake. (After Pflaker,[10] permission of American Association for the Advancement of Science.)

But a vast area in Alaska, covering 170,000 to 200,000 square kilometers, was deformed (Fig. 11-11). A central area trending northeasterly was uplifted a maximum of 8 meters, whereas a similarly trending belt to the northwest was lowered as much as 2 meters.

Renewed Faulting

Until recently it has been assumed that actual displacement along faults took place within a few seconds and was separated by long periods of quiescence, lasting for years, decades, and centuries. In general this is true, but in some instances faulting is a slow continuous process, at least in the sense that some movement occurs every year. This aspect of the problem is discussed in the next chapter.

But here we are concerned with displacements that took place along the same fault hundreds of millions of years apart. Some excellent examples have been found in the Grand Canyon area of Arizona.[11] In the central part of Fig. 11-13 the normal faults displace the Algonkian formations but not the Paleozoic. Three of these faults are clearly truncated by the unconformity at the base of the Paleozoic. The most northeasterly of these faults shows a more complex relation. The Algonkian rocks on the southwest side of the fault have been dropped down against the Archean. But the base of the Paleozoic on this same side of the fault has been uplifted. That is, the Precambrian movement was the reverse of the Paleozoic.

Figure 11-14 shows two faults on which the movement has also been different at different times. Along the fault to the southwest the Algonkian on the northeast side has gone down but the Paleozoic has gone up. The northeasterly fault shows the opposite. Hundreds of millions of years separated these two oppositely directed displacements.

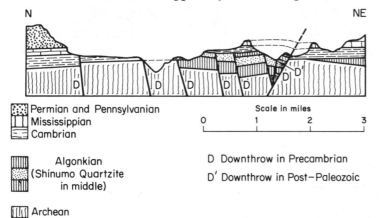

N   NE

Permian and Pennsylvanian
Mississippian
Cambrian

Scale in miles

0   1   2   3

Algonkian
(Shinumo Quartzite
    in middle)

D Downthrow in Precambrian
D′ Downthrow in Post–Paleozoic

Archean

Fig. 11-13. Faults in Grand Canyon of the Colorado, Arizona. (After John H. Maxson;[11] permission Grand Canyon Natural History Association.)

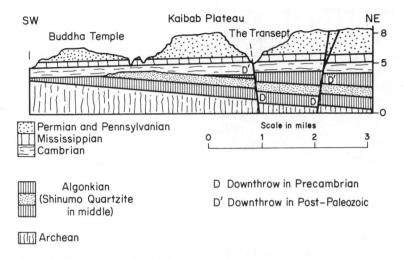

Fig. 11-14. Faults in Grand Canyon of the Colorado, Arizona. (After John H. Maxson;[11] permission Grand Canyon Natural History Association.)

Mechanics of Normal Faulting

Figure 11-15 shows the relation of normal faults to the principal stress axes. The two sets of normal faults intersect in the intermediate principal stress axis and make an angle of 30° with the greatest principal stress axis. For

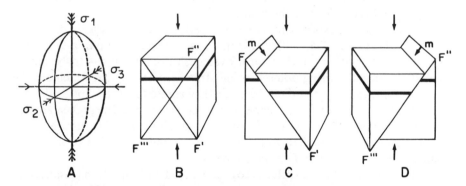

Fig. 11-15. Relation of normal faults to stress ellipsoid. (A) Stress ellipsoid: σ_1, greatest principal stress axis; σ_2, intermediate principal stress axis; σ_3, least principal stress axis. (B) Orientation of shear fractures. (C) Along FF' the right-hand block has moved relatively down. (D) Along $F''F'''$ the left-hand block has moved relatively down.

Fig. 11-16. Edge of an artificial graben. Produced by subjecting clay to tension. The master fault and many subsidiary faults dip to the right. Antithetic faults dip to the left. All the faults are of the gravity type. (After H. Cloos.[12])

normal faults dipping 60° the greatest principal stress axis is vertical, whereas the intermediate and least principal stress axes are horizontal.

In some experiments the least principal stress axis has been an active extension. A pad of clay of proper consistency is placed on two adjacent flat steel plates lying end to end.[12] When one of the plates is pulled, the clay is subjected to tension and a graben forms (Fig. 11-16). *Antithetic faults* dip toward the master fault, but the displacement on them is down dip—that is, they are normal faults. Such antithetic faults are common in a tunnel that cuts the eastern border fault of the Rhine graben.

Several sets of experiments have been performed to investigate the evolution of salt domes (Chap. 14). A plunger is slowly pushed up into sediments during sedimentation. Graben form over such domes. This is to be anticipated, because the rising dome is being stretched. The faults, which are the normal type, initially dip about 60°, but they rotate to lower dips as the uplift continues. The greatest principal stress axis is vertical. The graben are due to stretching over the top of the dome. The radiating normal faults are due to circumferential tension acting at right angles to the radii of the dome. This subject is discussed more fully in Chap. 14.

Large normal faults, such as those associated with the Rhine graben and the fault-block mountains of the Basin and Range Province of North America, present a problem. In both cases the least principal stress axis is essentially horizontal and trends east-west. But is this an active tension, due to stretching of the continent, or is it merely the least principal stress axis resulting from vertical compression? The latter is probably correct. Because of differential uplift the crust is warped. Many ranges in the Basin and Range Province may be broad uplifts without accompanying faults. In other

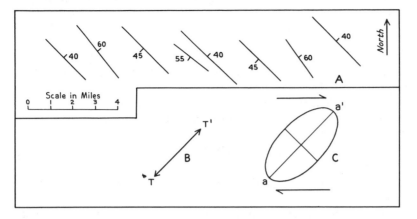

Fig. 11-17. *En échelon* gravity faults. (A) Diagrammatic map of an area of *en échelon* gravity faults. Each line represents a fault several miles long; angle of dip is shown in degrees. (B) Direction of active tension. (C) Strain ellipse, caused by a couple.

instances the fault pattern may be very complex and the faults may be of various ages. The elongation due to differential uplift is sufficient to account for the apparent stretching.

Of course, normal faults do not necessarily form with a 60° dip as shown in Fig. 11-15. There are many reasons for this. The trajectories of the principal stress axes are not necessarily vertical and horizontal. Consequently, as in the lower left-hand corner of Fig. 10-18, high-angle reverse faults and low-angle normal faults may form. Secondly, the rocks are not homogeneous and isotropic; very complex local stresses may develop. Once a fracture finally forms, the whole stress pattern may change; this aspect of the problem is more fully explored in the next chapter.

In some areas the normal faults show an *en échelon* pattern, such as that illustrated by Fig. 11-17. The individual faults strike at an angle of approximately 45° to the trend of the faulted belt as a whole. The active tension is in the direction *TT'* (Fig. 11-17), but a regional tension acting in this direction does not account for the *en échelon* pattern. The long axis of the strain ellipse is oriented northeast (*aa'* of Fig. 11-17). A couple caused by a northerly block moving toward the east relative to a southerly block would produce the type of fracture observed.

Such a belt of *en échelon* faulting in central Montana, the Lake Basin fault zone, trends east-southeast for about 56 miles.[13] More than 90 northeasterly-trending faults have been mapped, many of them over 5 miles long, and one 10 miles long. The dips of the faults range from 10 to 80°, but average about 45°; the maximum recorded stratigraphic throw is between 500 and 600 feet. Most of the faults are normal, but some are reverse, and in some there has been a strike-slip component. This belt of faulting is due to a couple, the southerly block moving east relative to the northerly block.

References

1 Reid, H. F., et al., 1913, Report of the committee on the nomenclature of faults, *Bull. Geol. Soc. Amer.* 24: 163–86.

2 Hayes, Philip T., and Raup, Robert B., 1968, *Geologic map of the Huachuca and Mustang Mountains southeastern Arizona*, U.S. Geol. Surv., Misc. Geol. Invest., Map I-509.

3 Anderson, C. A., and Creasey, S. C., 1958, *Geology and ore deposits of the Jerome area, Yavapai County, Arizona*, U.S. Geol. Surv., Prof. Paper 308, 185 pp.

4 Dorn, P., 1951, *Geologie von Mitteleuropa*, E. Schweizerbart-sche, Stuttgart, 474 pp.

5 Anonymous, 1965, *East African rift system*, Upper Mantle Committee—Unesco Seminar, Nairobi, April 1965, 115 pp., University College, Nairobi.

6 Irvine, T. N., ed., 1966, *The world rift system*, Geol. Surv. Canada, Paper 66–14, 471 pp.

7 Heezen, Bruce C., 1962, The deep-sea floor, pp. 235–88, in S. K. Runcorn, *Continental drift*, 338 pp., New York and London: Academic Press.

8 Scott, E. R., 1948, Quitman oil field, Wood County, Texas, *Structure of typical American oil fields*, 3: 419–31. Tulsa: Amer. Assoc. Petrol. Geol.

9 Witkind, Irving J., et al., 1964, *The Hebgen Lake, Montana, earthquake of August 17, 1959*, U. S. Geol. Survey, Prof. Paper 435, 242 pp.

10 Pflaker, G., 1965, Tectonic deformation associated with the 1964 Alaska earthquake, *Science*, 148: 1675–87.

11 Maxson, John H., 1961, *Geologic map of the Bright Angel Quadangle, Grand Canyon National Park, Arizona*, Grand Canyon Natural History Association.

12 Cloos, H., 1939, Hebung, Spaltung, Vulkanismus, *Geologische Rundschau*, Band 30, Zwischenheft 4A, pp. 406–527; cf. especially p. 434.

13 Hancock, E. T., 1918, Geology and oil and gas prospects of the Lake Basin Field, Montana, *U. S. Geol. Surv. Bull. 691*, pp. 101–47.

12

STRIKE-SLIP FAULTS

Introduction

Strike-slip faults are those along which the displacement is chiefly parallel to the strike of the fault. Most of them are steep and straight; crushing of the rocks in the vicinity is characteristic and the larger faults of this type are usually fault zones rather than single, clean-cut fractures.

Although numerous terms for this category of faults have been proposed, *strike-slip fault* is the most satisfactory[1]. *Wrench fault*[2] has been used by many, but this is unsatisfactory because "wrench" means a "twisting." *Transcurrent fault*,[2] a term also used by many, implies the presence of folds or other structural features to which the fault is transverse. *Flaw, lateral,* and *lateral-slip* have also all been used. *Tear-fault* has also been used for transverse strike-slip faults. The term *transform fault*, recently introduced for a special category of strike-slip fault, is discussed more fully on pages 272, 274, and 489.

To demonstrate that a fault is a strike-slip fault may often be difficult. Ideally, two or more planar structures with different attitudes should be disrupted by the fault (p. 187 and pp. 559–569) in order to calculate the net slip.

Features that have been offset by some of the strike-slip faults described in geological publications are: (1) beds or formations; (2) dikes or veins; (3) axial planes and axes of folds; (4) sedimentary basins; (5) streams; (6) ridges; (7) granitic stocks; (8) zones of metamorphism; and (9) belts of magnetic anomalies. The evidence for some supposed strike-slip faults is very weak. Sometimes a strike separation (p. 188) is erroneously assumed to demonstrate that the net-slip is strike slip, as numerous geologists have indicated.[3,4]

As indicated in Chap. 8, right-slip and left-slip may be used to describe the relative movement.

Examples of Strike-Slip Faults

SYMMETRICALLY DISPOSED
ABOUT FOLD AXES

There are some strike-slip faults that are symmetrically disposed about the fold axes. Examples of these may be cited from the Jura Mountains and from Wales.

Strike-slip faults diagonal to fold axes have been described in the Jura Mountains of northern Switzerland and adjacent parts of France.[5] Mesozoic rocks display an Appalachian type of folding. The faults (Fig. 12-1) are diagonal rather than perpendicular to the fold axes. The faults are generally vertical and some of them have coarse horizontal slickensides. Several of the faults are nearly 30 miles long. The average net slip, almost horizontal, is 0.6 mile, but along the south end of fault number 5 (Fig. 12-1) the net slip is 6 miles. Axes of folds are offset by the faults. In some instances the strata on one side of a fault are more intensely folded than those on the other side, indicating that folding continued after the fault was initiated. Most of the faults strike in a direction that is about 30° clockwise from lines drawn perpendicular to the fold axes. All these faults are left-slip. One fault and part of another strike about 30° counterclockwise from lines drawn perpendicular to the fold axes; these faults are right-slip.

Another excellent example of strike-slip faults diagonal to fold axes is found in southwestern Wales[6] (Fig. 12-2). The Paleozoic strata have been thrown into folds that strike N.80°W. Thrust faults have a similar strike and

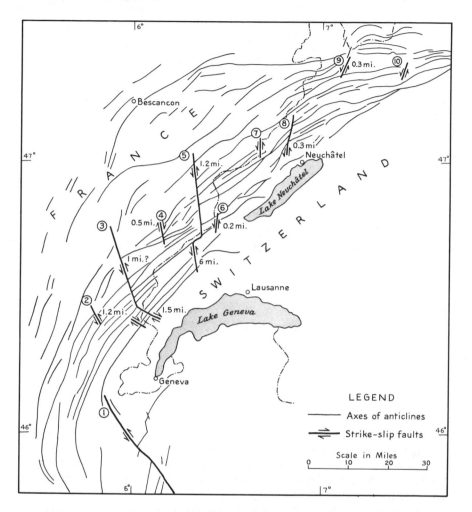

Fig. 12-1. Strike-slip faults of the Jura Mountains. Each fault is identified by a number in a circle. The relative direction of displacement is indicated by arrows and the maximum amount of the net slip, where known, is given in miles. (After A. Heim.[5])

dip south. Two sets of diagonal strike-slip faults are present, but only a few of the larger faults could be shown on the scale of Fig. 12-2. Of the 69 faults, 42 strike about N.20°W. and 27 strike about N.30°E.[2] The dips are steep. The two sets of faults thus make an angle of about 50° with each other and about 25° with lines perpendicular to the fold axes. Of the 42 faults that strike

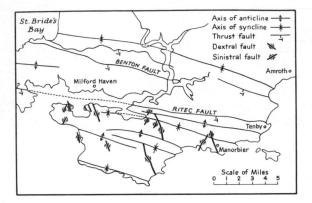

Fig. 12-2. Strike-slip faults of southwestern Wales. (Compiled from maps by E. M. Anderson,[2] and J. Pringle and T. N. George.[6])

N.20°W., 40 are right-slip; of the 27 that strike N.30°E., 22 and probably 24 are left-slip. The maximum net slip is 1200 feet.

BOUNDING A THRUST BLOCK

A block bounded by two strike-slip faults may move forward between the relatively stationary terrains on either side. In many instances the moving block is also separated by a thrust fault from the stationary block beneath it. In a sense, therefore, the moving block is bounded by one scoop-shaped fracture, the bottom of the scoop corresponding to a thrust, the sides of the scoop to the strike-slip faults.

An example is in the Piney Creek area on the northeast side of the Bighorn Mountains in Wyoming.[7] These mountains are a broad asymmetrical anticline, trending northwest, with a gentle southwest limb and a steep northeast limb. Precambrian plutonic and metamorphic rocks are exposed in the core, whereas Paleozoic and Mesozoic sedimentary rocks, about 10,000 feet thick, occupy the limbs. The dip of the sedimentary rocks on the northeast limb has a great range, but averages about 45°. In the Piney Creek area (Fig. 12-3) a block 8 miles long has been pushed forward 2.5 miles northeast of the blocks on either side of it. This block is bounded on the northeast by a thrust fault, which, although not exposed, by analogy with other faults in the area, presumably dips about 45°SW. On the northwest the block is bounded by a left-slip fault along which the net slip is 3 miles. On the southeast it is bounded by a right-slip fault along which the net slip is 2.5 miles.

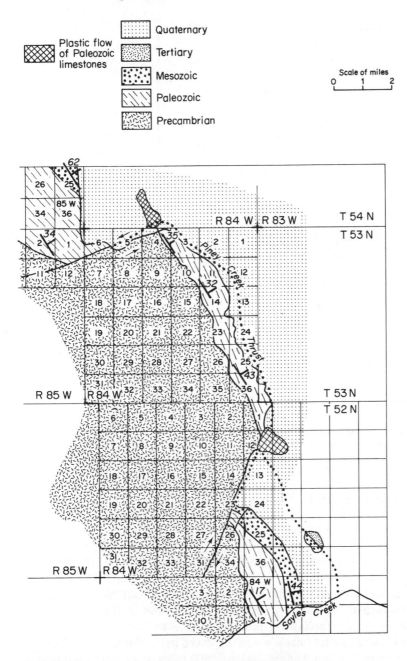

Fig. 12-3. Piney Creek thrust block, Bighorn Mountains, Wyoming. The thrust block is bounded on both sides by strike-slip faults. (After Hudson;[7] permission Geological Society of America and Robert F. Hudson.)

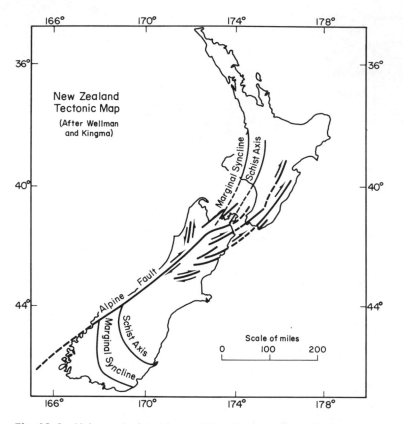

Fig. 12-4. Alpine and related faults of New Zealand. Note offset of "marginal syncline" and "schist axis."

LONGITUDINAL FAULTS

In some instances an orogenic belt has been displaced a great distance along a strike-slip fault. The northeast-trending Alpine Fault of the South Island of New Zealand[8,9] is shown in Fig. 12-4. The strike-separation of the "marginal syncline" and the "schist axis" is many hundreds of miles. The net slip is considered to be the same as this separation, that is, about 300 miles since the end of the Jurassic. Quaternary river terraces are offset by the fault.

Some strike-slip faults are essentially parallel to the folded belt in which they are located. Such faults are well displayed in California (Fig. 12-5). The largest of these, the San Andreas, extends 600 miles in a northwest-southeast direction.[10] In places the trace of the fault is marked by a trench, which has been referred to as a *rift* for many years; small ponds, called sag ponds, are one manifestation of this trench. The fault plane is steep.

Movement along this fault caused the San Francisco earthquake of April

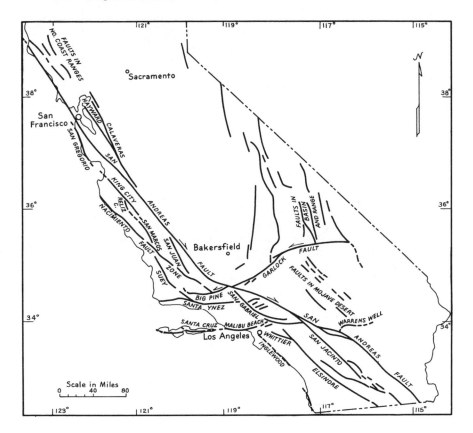

Fig. 12-5. Strike-slip and related faults of California. (After M. L. Hill and T. W. Dibblee, Jr.)

18, 1906. The block on the northeast side of the fault moved southeastward relative to the block on the southwest side. Movement occurred on 270 miles of the fault, the average net slip (essentially horizontal) was 13 feet, and the maximum net slip was 21 feet.

The total net slip along the San Andreas Fault and the length of time it has been active are matters of some conjecture. The amount of displacement in southern California since the beginning of the Miocene is indicated by the occurrence of similar rocks in the Orocopia, Solidad, and Tejon regions.[1] In this area the net slip along the San Andreas fault is 130 miles and along the San Gabriel fault (Fig. 12-5) is 30 miles. In central California the net slip during this time is 175 miles. Since the Miocene began 25 million years ago, the average rate of displacement has been 0.4 inch per year.

The Hayward fault is parallel to the San Andreas and is also a right-slip fault. But the Garlock and Big Pine faults (Fig. 12-5), which strike N.70°E, are left-slip.

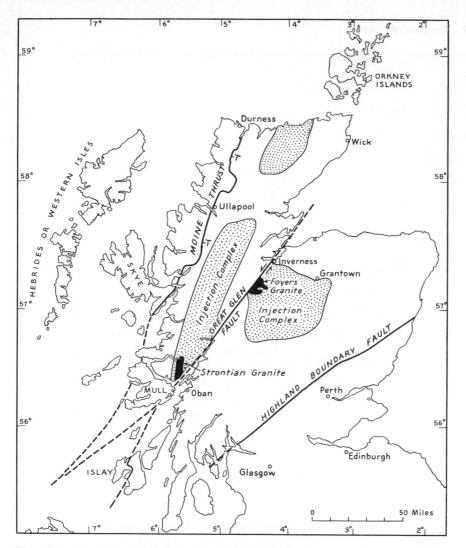

Fig. 12-6. Great Glen fault, Scotland. (After W. Q. Kennedy.)

INDEPENDENT OF FOLDING

Some strike-slip faults are younger than the folds associated with them.

There are several lines of evidence that the Great Glen fault of Scotland is a large strike-slip fault (Fig. 12-6).[11] The trace of the fault is a remarkably straight zone of crushed, sheared, and mylonitized rock, in places a mile wide. Several geological features are displaced by the fault, including: (1) a belt of highly injected schists; (2) metamorphic zones; (3) a granite stock; and (4) possibly a thrust plane. The granite stock that has been split in two by the fault has many petrological peculiarities, and the two halves undoubtedly belong to the same original body. The part northwest of the fault is known as the Strontian granite, the part southeast of the fault is known as the Foyers granite (Fig. 12-6). The block northwest of the fault has moved 65 miles southwest relative to the block southeast of the fault. There has also been some vertical movement, the southeast block having been downthrown 6000 feet. North and south of the Great Glen fault six additional faults show similar trends. Wherever data are available the block northwest of each fault has moved relatively toward the southwest from 3 to 5 miles. Although the principal movement on the Great Glen fault was in the Late Devonian or Early Carboniferous, modern earthquakes along its trace indicate it is still active.

Rate of Displacement

ELASTIC REBOUND

In the past it has generally been assumed that the movement along faults was catastrophic. This led to the *elastic rebound theory*.[12] This is illustrated by a plan in Fig. 12-7. The line aa' is a real or imaginary line prior to deformation. As a result of movement of block A relative to block B this line assumes the position of bb' (broken line). The line is stretched in the vicinity of the fault, but as yet there is no slippage on the fault. The strain eventually reaches the maximum permissible; displacement takes place along the fault, and the line snaps into positions cc' and c'' c'''. The displacement along the San Andreas fault at the time of the San Francisco earthquake of 1906 was of this type.

Initially the blocks on opposite sides of the fault are at rest. This is followed by a short interval of acceleration, subsequent deceleration, and rest. The acceleration and deceleration occupy only a fraction of a second. Some evidence is also available from seismology on the rate at which movement is propagated along the fault. Displacement must start at a point and then propagate at a speed of about 3 km/sec.

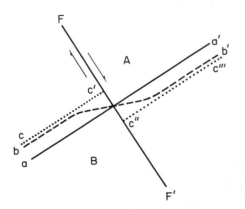

Fig. 12-7. Elastic rebound theory of faulting.

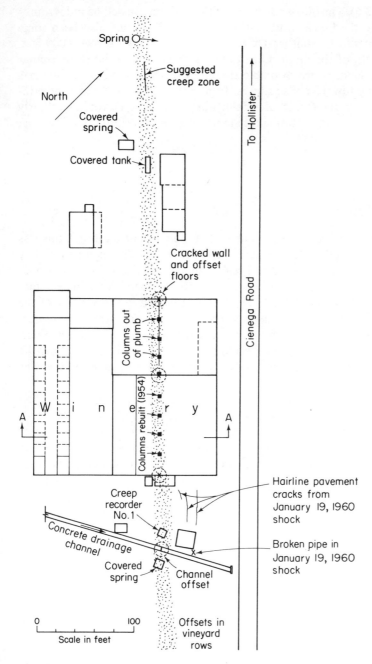

Fig. 12-8. Slow displacement along San Andreas fault, California. W. A. Taylor Winery, seven miles south of Hollister, California, showing features offset by slow movement along fault. (After Steinbrugge et al.,[11] permission Seismological Society of America.)

SLOW DISPLACEMENT

Evidence obtained in recent years shows that slow continuous movement along faults also takes place. Some of the best-studied examples are in California.

About 7 miles south of Hollister, California, a winery was built across the San Andreas fault in 1948.[13] In the ensuing 15 years many of the manmade structures were displaced in a right lateral sense as much as 6 to 12 inches. Features offset are (Fig. 12-8): (1) rows of grape vines; (2) concrete drainage ditch; (3) floor of main building; (4) walls of main building; and (5) walls and roof of spring house. Creep recorders have also been used. Between 1948 and 1960 the creep rate averaged about half an inch a year. "Ninety-two percent of the movement in a recent 371-day period accumulated in four spasms of total duration 34 days."[13]

Slow displacements along the Haywood fault (Fig. 12-5) have also been observed[14] over a length of 30 miles. Artificial structures affected include: (1) culvert under the University of California stadium at Berkeley, (2) Claremont water tunnel, (3) railroad tracks, (4) Hetch Hetchy Aqueduct, (5) chain fence, and (6) buildings. Figure 12-9 shows that a chain fence was offset

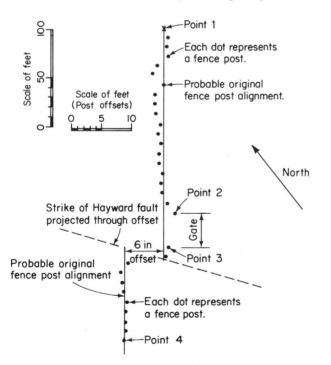

Fig. 12-9. Slow displacement along Hayward fault, California. Chain fence offset by slow movement along fault. Locality in Fremont, California. (After Cluff et al.,[14] permission Seismological Society of America.)

6 inches in 25 years. The rate of movement, determined at different places and for different intervals of time, ranges from 0.1 inch to 0.25 inch per year. An instrumental survey at the stadium lasting three months, showed an average rate of movement of 0.21 inch per year.

The slow displacement along the Buena Vista thrust fault in California may be mentioned here.[15] The fault strikes east-west, is about 10 miles long, and dips 22° to 25°N. Movement along it has been observed for several decades; oil wells as well as a specially prepared line of stakes have been displaced. The rate is 0.55 inch to 0.59 inch per year.

Fracture Zones of the Ocean Basins

Many fracture zones have been mapped in the ocean basins.[16,17] The longer zones generally trend east-west, are hundreds or thousands of miles long, and tens of miles wide; they are characterized by a topography that is rougher than that on the smoother ocean floor on either side. The shorter zones, scores or hundreds of miles long, show similar features but on a smaller scale.

Two lines of evidence indicate strike-slip movement along the fracture zones. First, magnetic anomaly belts (Chap. 22) generally trending north-south and 10 to 30 miles wide, are offset. Second, analysis of the focal mechanism of earthquakes (Chap. 23) along the shorter zones demonstrates strike-slip movement.

Transform faults[18] are one type of strike-slip fault. They are generally associated with oceanic ridges; the discontinuity of the ridge is just the opposite of the net slip along the fault. For example, in Fig. 12-10 the ridge shows a left-handed discontinuity, but the fault between *b* and *c* is right-slip. Between *a* and *b* and between *c* and *d* there is no differential movement. The evi-

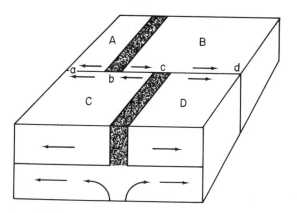

Fig. 12-10. Transform fault.

dence for the discontinuity of the ridge is topographic, whereas the evidence for the net slip is based on seismic studies of modern earthquakes.

Mechanics of Strike-Slip Faults

As shown in Fig. 10-14, strike-slip faults develop if the greatest and least principal stress axes are horizontal (Fig. 10-14C); since the intermediate principal stress axis is vertical, the intersection of the two sets of shear fractures is also vertical. As shown in Fig. 10-14D and Fig. 12-11, the fractures make angles of about 30° with the greatest principal stress axis. When displacement occurs, one of the faults is right-slip, the other is left-slip.

The principal stress axes are not necessarily vertical and horizontal. Consequently the orientation of the strike-slip faults and the displacement along them is not necessarily systematically disposed about a vertical to the surface of the earth. This is especially true of minor strike-slip faults, where heterogeneity of the rocks and preexisting fractures greatly influence the stress distribution. Also, preexisting fractures formed under different conditions may be utilized for strike-slip movements.

The orientation and displacement in the Jura Mountains and southwestern Wales conform to theory. If a series of sedimentary rocks is compressed along north-south lines, the axes of the resulting folds are oriented east-west. At times, however, the rocks may be sufficiently brittle to yield by rupture. Two sets of shear fractures may form. If the direction of easiest relief is vertical—that is, the least stress axis is vertical—two sets of shear fractures striking east-west will form; one set will dip about 30°N., the other about 30°S. If displacement takes place along these fractures the hanging wall moves over the footwall to develop thrust faults. If, however, the easiest relief is in an east-west direction—that is, the least stress axis strikes east-west and is horizontal—the two sets of shear fractures will be vertical and will strike about N.30°W. and N.30°E. Any displacements along the fractures will be strike-slip movements. The N.30°W. set will be right-slip, the N.30°E. set will be left-slip (Fig. 12-11). Not only are the strike-slip faults in the Jura and in southwestern Wales oriented in accordance with theory, but the direction of the relative movements is also correct. In the Jura, however, for some unknown reason one set of strike-slip faults is much better developed than the other.

The displacement along the strike-slip faults bounding the Piney Creek thrust blocks (Figs. 12-3) is what one would expect, but some of the faults or portions of them are perpendicular to the fold axes. These parts are presumably extension fractures.

The Alpine Fault of New Zealand is a shear fracture associated with a gigantic drag. The greatest principal stress axis was oriented east-northeast and was horizontal. The strike-slip faults of Scotland would form if the greatest principal stress axis were horizontal and oriented slightly west of south.

The great strike-slip faults of California are enigmatic, since they are essentially parallel to the fold axes. The displacements in the San Andreas and Garlock Faults suggest that the greatest principal stress axis is horizontal and oriented N.10°E.; but the interfault angle bisected by this axis is much greater than one would expect, at least 100° instead of 60°. Moreover, the trend of the folds suggests that the axis of compression was oriented northeast-southwest.

Several suggestions have been made to explain transform faults. One is given here. Convection currents rise beneath the ridges (Fig. 12-10). The rising current in the front block lies east of that in the back block. In the front block plate C moves west, plate D moves east. New magma fills the potential void as the plates move away from each other. In the back block, plate A moves west, plate B moves east. Between b and c there is right slip. But between a and b and c and d there is no differential movement. Although a compelling argument for the nature of the net slip comes from seismic studies, it is not clear why the convection currents should be discontinuous or why the faults should be at right angles to the ridge.

Shears of the Second Order

In some places a subsidiary set of strike-slip faults branches off from the main strike-slip fault at an angle of about 30° (Fig. 12-12). The displacement along the subsidiary faults is in the same sense as the master fault.[19] The acute angle between the main and subsidiary faults points in the direction in which the block containing the subsidiary fault was moving. If a slab of rock were caught between two such master faults, the secondary axis of compression could be represented by c in Fig. 12-12B. Two sets of second-order shears would theoretically develop, one set making an angle of 30° with the master fault, the other set making an angle of 90° with the master fault. But displacement along the 90° fracture would be greatly inhibited. The device of a slab between two master faults is not essential, and similar stresses and fractures would form in association with a single master fault. The compressive stress (c in Fig. 12-12) is less than the original compressive stress that produced the master fault. Hence second-order shears are not likely to be common. The suggestion that third-order shears may develop from movement on second-order shears is quite untenable.

References

[1] Crowell, John C., 1962, *Displacements along the San Andreas Fault, California*, Geol. Soc. Amer., Spec. Paper 71, 61 pp.

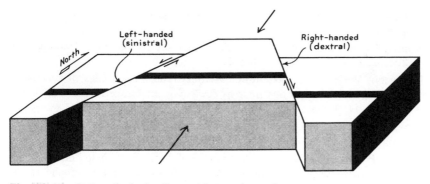

Fig. 12-11. Strike-slip faults. Formed by north-south compression with easiest relief in east-west direction. The fault striking N.30°W. shows right-separation; the fault striking N.30°E. shows left-separation.

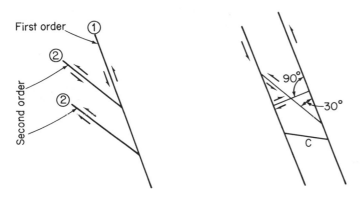

Fig. 12-12. Shears of the second order. (A) Map of the second-order shears. (B) Mechanics of second-order shears. (After McKinstry;[19] permission *American Journal of Science.*)

[2] Anderson, E. M., 1951, *The dynamics of faulting and dyke formation*, 2d ed., 206 pp., London: Oliver and Boyd.

[3] Hill, Mason L., 1959, Dual classification of faults, *Bull. Amer. Assoc. Petrol. Geol.* 43: 217–21.

[4] Kelley, Vincent C., 1960, Slips and Separations, *Geol. Soc. Amer. Bull.* 71: 1545–46.

[5] Heim, Albert, 1919, *Geologie der Schweiz*, 1: 613–26, Leipzig: C. H. Tauchnitz.

[6] Pringle, J., and George, T. H., 1948, *British regional geology, South Wales*, 2d ed., (British) Geol. Surv. and Museum.

[7] Hudson, Robert Frank, 1969, Structural geology of the Piney Creek thrust area, Bighorn Mountains, Wyoming, *Geol. Soc. Amer. Bull.* 80: 283–96.

[8] Fleming, Charles Alexander, 1970, The Mesozoic of New Zealand: chapters in

the history of the circum-Pacific mobile belt, *Quart. Jour. Geol. Soc. of London* (London) 125: 126–70.

9 Kupfer, Donald H., 1964, Width of the Alpine fault zone, New Zealand, *New Zealand J. Geol. Geophys.* 7: 685–701.

10 Dickinson, William R., and Grantz, Arthur, 1968, *Proceedings of conference on geologic problems of San Andreas Fault system*, Stanford Univ. Pub. Geol. Sci., vol. 11, 374 pp.

11 Craig, Gordon Y., 1965, *The geology of Scotland*, 556 pp, Hamden, Connecticut: Archon Press.

12 Reid, H. F., 1910, The California earthquake of April 18, 1906, in *The mechanics of the earthquake*, vol. 2, Report of the California State Earthquake Investigation Commission.

13 Steinbrugge, Karl V., et al., 1960, Creep on the San Andreas Fault, *Bull. Seismol. Soc. Amer.* 50: 389–415.

14 Cluff, Lloyd S., et al., 1966, Slippage in the Hayward Fault—six related papers, *Bull. Seismol. Soc. Amer.* 56: 257–324.

15 Wilt, James W., 1958, Measured movement along the surface trace of an active thrust fault in the Buena Vista Hills, Kern County, California, *Bull. Seismol. Soc. Amer.* 48: 169–76.

16 Menard, H. W., 1964, *Marine geology of the Pacific*, 271 pp, New York: McGraw-Hill Book Co., Inc.

17 Nafe, J. E., and Drake, C. L., Floor of the North Atlantic—summary of geophysical data. In Marshal Kay, ed., 1969, Amer. Assoc. Petrol. Geol., Mem. 12, pp. 59–87.

18 Wilson, J. T., 1965, A new class of faults and their bearing on continental drift, *Nature* 207: 343–47.

19 McKinstry, H. E., 1953, Shears of the second order, *Amer. J. Sci.* 251: 401–14.

13

DATING
OF STRUCTURAL
EVENTS

Introduction

For many decades geological events were dated primarily by their relations to the associated strata. Folds were obviously younger than the folded strata but older than flat-lying sedimentary rocks that rested unconformably on the truncated edges of the folded rocks. In the Paleozoic and younger rocks, because of their fossil content, it was possible to interpret these events within a general time scale that was only relative. But in recent decades, especially in the last two decades, radiogenic dating has become a powerful tool in unraveling tectonic history. Moreover, in recent decades a great deal of attention has been given to the interrelationship between sedimentation and tectonic activity.

Paleontology

Structural events have always been dated relative to the associated rocks. Paleozoic and younger sedimentary and volcanic rocks can in turn be correlated all over the world by means of fossils. Hence paleontology has been the fundamental basis of dating structural events and integrating them into a worldwide time table. Of course, this method gives relative ages only. Various methods were used to convert this type of timetable into one expressed in years, but it is only since the introduction of radiogenic dating that such efforts have been successful. But the importance of paleontological methods in dating events cannot be overemphasized.

Unconformities

INTRODUCTION

The structural geologist is concerned with unconformities for several reasons. Unconformities are structural features, although their origin involves erosional and depositional as well as tectonic processes. Moreover, unconformities may be confused with some kinds of faults. Most important of all, however, is the use of unconformities in dating orogenic and epeirogenic movements. Unconformities are also important to students of stratigraphy, sedimentation, and historical geology. Valuable deposits of petroleum and minerals are associated with unconformities.

An *unconformity* is a surface of erosion or nondeposition—usually the former—that separates younger strata from older rocks (Plates 24 and 25). The development of an unconformity involves several stages. The first stage is the formation of the older rock. Most commonly this is followed by uplift and subaerial erosion. Finally, the younger strata are deposited.

Rocks of various origins may participate in unconformities; sedimentary rocks, volcanic rocks, plutonic rocks, or metamorphic rocks may be involved.

In Fig. 13-1 the unconformities are labeled *ab*. In some instances, as in Fig. 13-1A, the rocks both above and below the unconformity are sedimentary. After the lower limestone was deposited, the region was uplifted and eroded; then the upper sandstones and shales were deposited. In Fig. 13-1B, the rocks beneath the unconformity are limestone, those above the unconformity are volcanic. After the deposition of the limestone, there was uplift, erosion, and, finally eruption of the volcanic rocks. In Fig. 13-1C, after the eruption of the lower volcanic rocks there was erosion—with or without a preceding uplift—and then eruption of the upper volcanic rocks. Figures 13-1D and 13-1E involve plutonic rocks. The plutonic rocks were intruded and then eroded, with or without a preceding uplift. Upon the erosion surface,

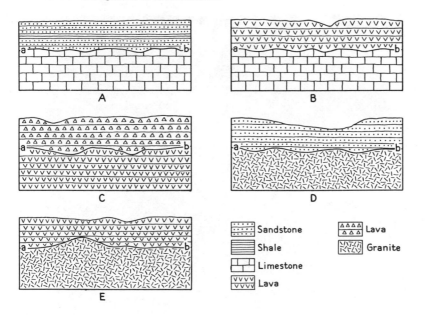

Fig. 13-1. Rocks participating in unconformities. Unconformities are labeled *ab*. (A) Sedimentary rocks both above and below the unconformity. (B) Volcanic rocks above, sedimentary rocks below. (C) Volcanic rocks above and below. (D) Sedimentary rocks above, plutonic rocks below. (E) Volcanic rocks above, plutonic rocks below.

younger sediments were deposited (Fig. 13-1D) or volcanics were erupted (Fig. 13-1E).

The relief on unconformities differs greatly. In some localities the older rocks were reduced to an extensive peneplain. In other localities, only a mature stage in the erosion cycle was reached before the younger rocks began to accumulate. The relief on the unconformity may amount to hundreds or even thousands of feet (Plate 24). The Cambrian sedimentary rocks of the Grand Canyon were deposited on a surface with a known maximum relief of 800 feet.[1] The Late Precambrian of North-West Highlands of Scotland were deposited over hills at least 2000 feet high.[2,3] The Carboniferous sedimentary rocks around Boston, Massachusetts, rest upon an unconformity with a relief of at least 2100 feet.[4]

KINDS OF UNCONFORMITIES

There are various kinds of unconformities, the distinction depending upon the rocks involved and the tectonic history that is implied. The most important varieties are: angular unconformity, disconformity, local unconformity, and nonconformity.

Plate 24 *Nonconformity.* Early Tertiary basalts (dark) and interbedded sedimentary rocks (light) resting unconformably on older gneiss. The relief on this unconformity is as much as 1000 feet. Gassefjord, East Greenland. Photo: Lauge Koch Expedition.

Plate 25 *Angular unconformity.* Yakataga District, Alaska. Photo: D. J. Miller, U. S. Geological Survey.

As is illustrated by Fig. 13-2 and Plate 25, the rocks on opposite sides of an *angular unconformity* are not parallel. Figure 13-2A is a cross section, such as an exposure in a cliff; Fig. 13-2B is a map of a different region. The first event recorded in Fig. 13-2A is deposition of sandstone and shale. These rocks were then deformed to assume dips of 70°, either by folding or tilting of fault blocks. The ensuing erosion, probably accomplished by streams, but possibly marine, reduced the region to the surface *ab.* Eventually, erosion ceased and the younger conglomerate, sandstone, and shale were deposited. Although the rocks both above and below the unconformity represented in Fig. 13-2A are sedimentary, either one or both may be volcanic.

The precision with which the period of deformation can be dated depends upon the age of the rocks on either side of the unconformity. If the rocks beneath *ab* are Upper Permian and the rocks above *ab* are Lower Triassic, the

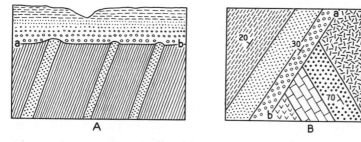

Fig. 13-2. Angular unconformity. (A) Cross section. (B) Map, not the same region as that shown in (A).

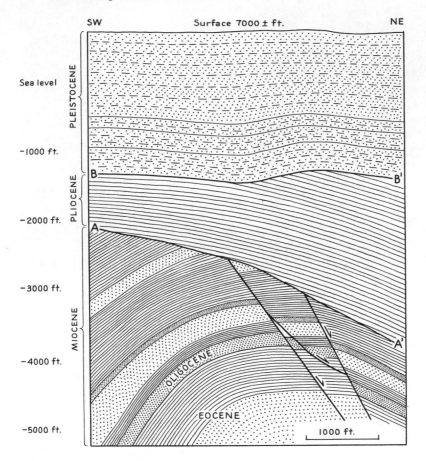

Fig. 13-3. Unconformities based on drill records. Cymric oil field, California. (After J. H. McMasters;[5] permission of The American Association of Petroleum Geologists.)

deformation was Late Permian or Early Triassic. If, however, the rocks beneath the unconformity are Upper Silurian and those above are Lower Jurassic, the deformation could have occurred at any time between Late Silurian and Early Jurassic.

An excellent example of two angular unconformities based on drill records is shown in Fig. 13-3, a cross section of the Cymric oil field, California.[5] Unconformity AA′ is between the Miocene and Pliocene. The Miocene was folded and faulted prior to the erosion that produced unconformity *AA′*. Then the Pliocene was deposited. Unconformity *BB′* is between the

Pliocene and Pleistocene. The Pliocene was gently folded prior to the erosion that produced unconformity *BB'*. The Pleistocene was then deposited and gently folded.

In a *disconformity*, the formations on opposite sides of the unconformity are parallel. A disconformity covers a large area and represents a considerable interval of time; the meaning of this somewhat vague statement will be more apparent after local unconformities have been discussed. Figures 13-1A, B, and C represent disconformities, and the history implied has been given above (page 278).

A *local unconformity* (Fig. 4-19) is similar to a disconformity, but, as the name implies, it is distinctly local in extent; the time involved is short. In the deposition of continental sediments, such as gravels, sands, and clays, the streams may wander back and forth across the basin of deposition. At time of flood these streams may scour out channels scores of feet wide and many feet deep. As the flood subsides, or some days or even years later, the channel may be filled up again.

In a sense, a local unconformity is a disconformity of small extent representing a short interval of time. Under certain conditions it may be difficult to decide which term is the more appropriate.

Although the term *nonconformity* is used in various ways in the geological literature, it may be utilized most satisfactorily for unconformities in which the older rock is of plutonic origin (Plate 24; Figs. 13-1D and E).

RECOGNITION OF UNCONFORMITIES

Unconformities may be recognized in various ways, of which *observation in a single outcrop* is the most satisfactory. The outcrop may be small and only a few feet across; it may be an artificial opening, such as a quarry; or it may be the wall of a canyon, such as the Grand Canyon of the Colorado River (Figs. 11-13 and 11-14).

If the unconformity is an angular one, the lack of parallelism of the beds on opposite sides of the contact will be readily apparent (Plate 25). This may be observed in a vertical section, such as a cliff (Fig. 13-2A), or on the surface of the outcrop (Figs. 13-2B). The lowest beds above the unconformity may consist of conglomerate with pebbles derived from the underlying formations. If the conglomerate is thin, it may be concentrated in small depressions eroded out of soft beds in the strata beneath the unconformity. But basal conglomerates are not necessarily present along angular unconformities. Faults and dikes may be truncated at the contact.

Under favorable conditions, disconformities may be readily recognized in outcrops, road cuts, and quarries. If there is a sharp contrast in color between the rocks above and below the disconformity, if the disconformity is somewhat wavy, and especially if there is a thin conglomerate just above the disconformity, the nature of the contact is apparent. Regional relations must be considered (p. 285) in order to distinguish between a disconformity and a local unconformity. But disconformities may be difficult to recognize, and in many cases paleontological evidence indicates considerable gaps in the geological record without any accompanying physical evidence.

Nonconformities must be distinguished from intrusive igneous contacts. The rocks above a nonconformity may contain fragments of the older igneous rock, either as readily recognized pebbles and boulders or as small fragments recognized only under the microscope. Some nonconformities are characterized by an arkose many feet thick, so that the plutonic rock seems to grade into the overlying strata. Along an intrusive contact, of course, dikes might be expected to penetrate the adjacent rocks; in some cases the intrusive is chilled against the older strata (see page 332).

A surface of erosion may be covered by a thick residual soil that grades into the underlying bed rock. Younger sediments deposited above this erosion surface may incorporate some of the residual soil, and a sharp contact may be lacking. Such a contact is called a *blended unconformity*.

Many unconformities are not exposed in an outcrop. This may be due

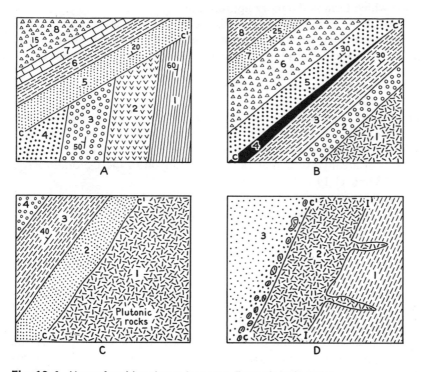

Fig. 13-4. Unconformities shown by maps. Formations in order of decreasing age: *1, 2, 3, 4, 5, 6, 7, 8. CC'* is an unconformity. (A) Angular unconformity. (B) Disconformity. (C) Nonconformity. (D) Nonconformity (*CC'*) and intrusive contact (*II'*).

to poor exposures, igneous intrusions, or faulting. In such cases other methods must be employed to detect the unconformity.

Unconformities are often demonstrated by *areal mapping*. A map showing an angular unconformity is illustrated by Fig. 13-4A. A group of older rocks, *1, 2, 3,* and *4,* strikes into the base of a group of younger rocks, *5, 6, 7,* and *8.* Formation *5* is in contact with all the older formations along the line *CC'*, and the relations can be interpreted to mean an angular unconformity. But a fault between *5* and the older formations is equally plausible. The presence of pebbles of *1, 2, 3,* and *4* in formation *5* would indicate an unconformity, but even in this case the mapped contact *CC'* could be a fault. In the last analysis, therefore, it is necessary to see the contact of formation *5* with the older rocks. The areal mapping is suggestive, but not conclusive.

Areal mapping may bring out a disconformity in a similar way (Fig. 13-4B). Although, in any one exposure, the strata on opposite sides of a contact (*CC'*) may appear to be parallel to each other, the mapping may show that the younger beds truncate the older. But just as in the case described above, the truncation may be due to faulting; consequently, a solution is obtained only if the contact is visible.

A nonconformity may be suggested by areal studies. If the sedimentary rocks, such as formation *2* of Fig. 13-4C, contain pebbles of the plutonic rock, a nonconformity must exist. But the contact *CC'*, exposed at the surface, could be a fault, and the nonconformity itself is not necessarily exposed. If dikes of the igneous rock do not cut the sediments, or if inclusions of the sedimentary rocks are not found in the plutonic rock, a nonconformity rather than an intrusive contact is implied, but not proved. As in the case of the angular unconformity and disconformity, for a satisfactory solution of the problem it is highly desirable to see the contact.

Three angular unconformities are shown in Fig. 13-5.[6] The oldest unconformity is at the base of the Fort Union formation (*Tfu*), which clearly truncates the older rocks; the Fort Union formation rests on the Meeteetse formation (*Kme*) in the northern part of the area, and on the Lance formation (*Kl*) in the southern part. A second unconformity is at the base of the Wasatch formation (*Tw*); the faults that cut the Mesaverde (*Km*), Meeteetse (*Kme*), and Fort Union (*Tfu*) formations in the northern part of the map are truncated by the base of the Wasatch formation (*Tw*). The third and youngest unconformity is at the base of the Quaternary gravels (*Q*), which are in contact with the Fort Union and Wasatch formations.

A *sharp contrast in the degree of induration* indicates an unconformity. If unconsolidated sands and clays are associated with well-cemented sandstones and compact shales, it may be presumed that the unconsolidated material is unconformable on the consolidated rocks. Some caution must be exercised, however, because an unconsolidated rock may be become indurated locally. Conversely, consolidated rocks may locally weather to loose sands and clays.

If rocks with a *distinct difference in the grade of regional metamorphism* are found in the same region, it is probable that the less metamorphosed rocks were deposited unconformably upon the more metamorphosed rocks.

The grade of metamorphism[7] is commonly defined by the nature of the minerals resulting from recrystallization; a complete discussion of this subject is beyond the scope of this book. Under conditions of regional metamorphism, a shale becomes a slate or phyllite, with such minerals as sericite and chlorite. Biotite, garnet, staurolite, and sillimanite appear successively as zones of greater metamorphic intensity are approached. Most of these minerals persist into the more intense zones of metamorphism; for example, a sillimanite schist commonly contains biotite and garnet. Staurolite is likely to disappear with the appearance of sillimanite. Thus if a sufficiently large area is studied, a single formation that was originally shale may be represented by the following different *metamorphic facies:* slate, biotite phyllite, biotite-garnet phyllite, biotite-garnet-staurolite schist and biotite-garnet-sillimanite schist. Normally a geological map will show that the change from one facies to another is gradual.

If slate and sillimanite schist are found in adjacent outcrops, an unconformity probably exists, and the more metamorphosed rock is the older. This criterion, however, must be used with some caution. A large fault, in particular, may bring together different metamorphic facies of the same formation.

Significant differences in the intensity of folding are suggestive. If some formations in an area are highly folded, whereas others are gently inclined or horizontal, the less deformed rocks are probably unconformable above the more deformed rocks. Due regard, however, must be given to the relative competency of the formations involved. Whereas thick, massive sandstone may be thrown into a few broad, open folds, thin-bedded shales deformed at the same time may be crumpled into many small folds. Moreover, even the same formation may be much more folded in some tectonic zones than in others. It is apparent that variations in the degree of folding are not a very reliable criterion of an unconformity.

The *relation to plutonic rocks* may be important. Two formations, such as *1* and *3* in Fig. 13-4D, may be separated from each other by granite (*2*) and may nowhere come into contact with each other. Formation *1* is intruded by the granite, but *3* is resting on the granite unconformably, and it contains pebbles of the granite. It is apparent, therefore, that *3* is above an unconformity and that *1* is beneath the unconformity.

The *paleontology* may indicate an unconformity. If a rock with Upper Triassic fossils is directly overlain by rocks with Lower Cretaceous fossils, even though the strata may appear to be conformable, a break representing all of Jurassic time is clearly indicated. In some instances the fossil record shows that a great hiatus is present in the midst of what is apparently a single homogeneous formation.

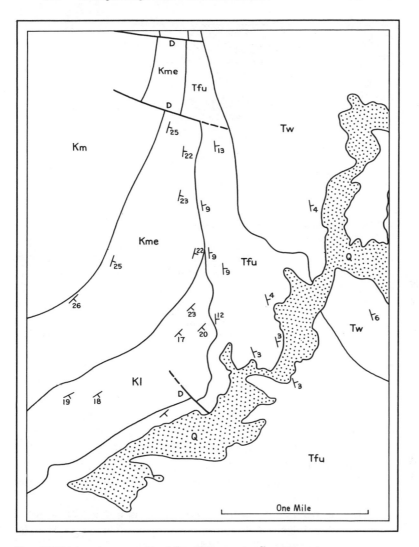

Fig. 13-5. Angular unconformities of Meeteetse Area, Wyoming. *Km*, Mesaverde Formation; *Kme*, Meeteetse Formation; *Kl*, Lance Formation; *Tfu*, Fort Union Formation; *Tw*, Wasatch Formation; *Q*, Quaternary gravels. *D*, Downthrown side of normal faults. (After D. F. Hewett.[6])

Distinguishing Faults from Unconformities

Reference has been made in several places to the danger of confusing faults with unconformities. Dip, diagonal, and transverse faults offer no difficul-

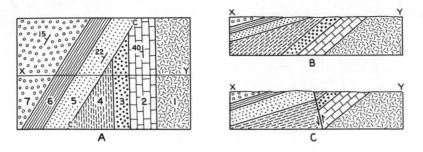

Fig. 13-6. Faults and unconformities. Formations in order of decreasing age: *1, 2, 3, 4, 5, 6, 7.* (A) Map: contact *CC'* may be either an unconformity or a fault. (B) Cross section along line *XY*; *CC'* of Fig. 13-6A interpreted as an unconformity. (C) Cross section along line *XY*; *CC'* of Fig. 13-6A interpreted as a fault.

ties. But if the bedding on one or both sides of the contact has the same strike as the contact, as in Fig. 13-6A, either an unconformity or a fault may exist. Formation *5* may lie unconformably above formations *1, 2, 3,* and *4,* or the contact may be a fault, with either group of formations the older. Even if formations *1, 2, 3,* and *4* are known to be older, the contact can be either an unconformity or a fault. In Fig. 13-6B, a cross section along *XY* of Fig. 13-6A, the contact is interpreted as an unconformity, but in Fig. 13-6C it is interpreted as a fault.

Several methods of attack may prove fruitful if the contact cannot be observed. The contact is a fault if formations *5, 6,* and *7* are older than *1, 2, 3,* snd *4* (Fig. 13-6C). Moreover, in a region of sufficient relief, it may be possible to ascertain the attitude of the contact *CC'* from its relation to topography. Under the simplest conditions, the dip of the unconformity would be essentially parallel to the dip of the beds in formation *5.* The greater the divergence of the dip of the contact from the dip of formation *5,* the greater the probability that the contact is a fault.

The presence of pebbles of formations *1, 2, 3,* and *4* in formation *5* would indicate an unconformity between the two formations, but even under such circumstances the contact *CC'* could be a fault.

If the younger beds strike or dip into the contact, a fault is indicated. This point is illustrated by Fig. 13-7A, where younger formations *5* and *6* strike into an older formation *3.*

In the final analysis, every effort should be made in the field to observe the actual contact. If it is an unconformity, small ridges of the older rock may project into the younger rocks, and a conglomerate or sandstone, with fragments of the older rock, may lie above the contact. Slickensides, gouge, and breccia would be absent from an unconformity, but would likely be present along a fault. Some faults, however, are sharp, knifelike contacts devoid of such features. Additional complexity is introduced by the fact that faults may follow unconformities, particularly angular unconformities.

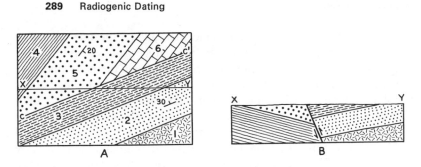

Fig. 13-7. Contact that is very probably a fault. Formations in order of decreasing age: *1, 2, 3, 4, 5, 6.* (A) Geological map. (B) Cross section along line *XY* of Fig. 13-7A.

Radiogenic Dating

The principles of radiogenic dating have been known for sixty years and some ages were determined as long ago as that. But only during the last twenty-five years have the methods been employed extensively. The lead, potassium-argon, rubidium-strontium, and fission-tracking methods are especially useful in structural problems. But for a detailed discussion of these methods the reader may be referred to a number of publications.[8,9,10,11,12]

The structural geologist should be well aware of the limitations of determining ages by radiogenic methods. The methods of chemical analysis are now relatively precise and in general the error should not be more than a few percent. The major problems are related to the geological history of the rocks and minerals. In some instances some of the daughter product, such as Ar^{40}, may have been incorporated in the rock at the time it formed; this would give too great an age. Conversely, a gaseous daughter product, such as Ar^{40}, may not accumulate until the rock is sufficiently cool, or may leak out during some subsequent later deformation. Moving solutions may remove or add some element, such as lead or uranium.

There are several ways of checking the reliability of radiogenic ages. Obviously they should be consistent with geological evidence. A dike should be younger than the rocks it cuts—or, at least, should not be older. A lava flow above an unconformity should be younger than a lava flow below the same unconformity. The radiogenic age should be consistent with the geologic time scale—a granite one billion years old should not cut a lava flow 300 $\times 10^6$ years old. Consistent ages obtained by several different methods for the same rock or mineral are obviously very reliable. Similarly, consistent ages obtained by the same method for several different minerals in the same rock are reliable.

Radiogenic ages may be used in several ways. (1) Volcanic rocks give the age of the sedimentary rocks with which they are interbedded. In some instances the age of the sedimentary rocks themselves may be determined. (2)

Plutonic rocks can be dated by radiogenic methods. A granite intruding Lower Devonian sediments but overlain by no younger rocks could range in age from Lower Devonian to the present. But radiogenic methods may give a precise age. (3) The age of metamorphism may be shown by the age of the metamorphic minerals. (4) In some instances the radiogenic ages may indicate a complex history. The zircons in a granite mass may give an age of one billion years (the time of consolidation); the biotite might give an age of 350×10^6 years (the time of regional metamorphism), and and the muscovite along shear zones might give an age of 250×10^6 years (the time of faulting).

Orogeny and Tectonism

"Orogeny" literally means "mountain genesis". More than one hundred years ago it was erroneously assumed that mountains such as the Appalachians were the direct result of folding and thrusting. It was further assumed that this meant crustal shortening due to compression. "Epeirogeny" literally means "mainland genesis" and referred to vertical movements of large areas. Thus angular unconformities are used to date orogenic events, whereas disconformities are used to date epeirogenic events. "Tectonism" refers to both orogeny and epeirogeny.

Tectonism and Sedimentation

Much information about *tectonism*—that is, deformation—may be obtained from sedimentary rocks. This evidence may bear on the movements in the sedimentary basin or in the area from which the sediments were derived.

The thickness of the sedimentary rocks indicates the amount and rate of subsidence of the basement on which they were deposited. For example, if 10,000 feet of sedimentary rocks, all deposited near sea level, are laid down on a granitic basement in 25,000,000 years, the basement has subsided, on the average, one foot every 2500 years, or at an average rate of 0.0004 foot per year. The fact that these are averages should be emphasized.

Sedimentary rocks may also give a great deal of information about the source area from which the sediments are derived. Figure 13-8 shows the distribution and thickness of Upper Cretaceous sedimentary rocks in the western United Stat s and the area from which they were derived.[13] The thickness ranges from zero on the east to nearly 20,000 feet in Wyoming. A sheet of rock averaging 5 miles in thickness was eroded from the source area. Since the Late Cretaceous lasted 35,000,000 years, the average rate of uplift and erosion was 0.7 foot per thousand years or 0.0007 foot per year.

Another example (Fig. 13-9) comes from the west flanks of the Wasatch Plateau in Utah.[14] Facies changes in the Cretaceous rocks are very striking;

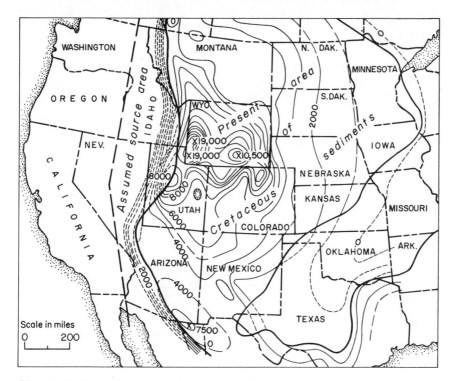

Fig. 13-8. Source of Upper Cretaceous sediments in Rocky Mountain geosyncline. Isopachs show thickness of Upper Cretaceous. Source area is further west. (Permission Geological Society of London and James Gilluly.)

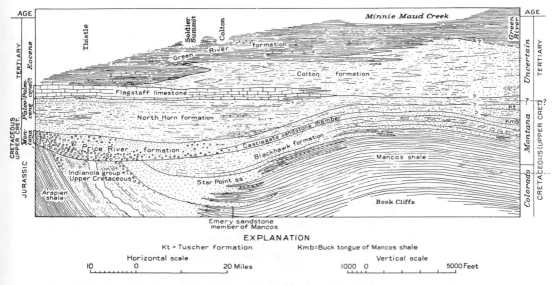

Fig. 13-9. Facies changes in Cretaceous and Tertiary rocks of Wasatch Plateau, Utah. Source area was to the west. (After Spieker.[14])

conglomerates interfinger eastward with sandstones, and still further east the sandstones interfinger with shales. The source area was clearly to the west. Moreover, conglomerates occur at three principal levels: early Indianola time, late Indianola time, and Price River time. This in turn implies that unusually rapid uplift took place in the source area at least three times.

Conglomerates do not necessarily indicate uplift. Tillites are the product of glaciation. Extensive volcanism may produce highlands, from which large quantities of sediments may be derived. Finally, deposits of limestones, dolomites, and evaporites (salt, gypsum, anhydrite) may derive their elements from far distant sources.

References

[1] Sharp, R. P., 1940, Ep-Archean and Ep-Algonkian erosion surfaces, Grand Canyon, Arizona, *Geol. Soc. Amer. Bull.* 51: 1235–70.

[2] Peach, B. N., Horne, J., Clough, C. T., Hinxman, L. W., and Teall, J. J. H., 1907, *The geological structure of the North-West Highlands of Scotland*, Mem. Geol. Surv. Great Britain, Glasgow, p. 4.

[3] Craig, Gordon Y., ed., 1965, *The geology of Scotland*, 556 pp, Hamden, Connecticut: Archon Press.

[4] Billings, M. P., Loomis, F. B., Jr., and Stewart, G. W., 1939, Carboniferous topography in the vicinity of Boston, Massachusetts, *Bull. Geol. Soc. Amer.* 50: 1867–84.

[5] McMasters, J. H., 1948, Cymric Oil Field, Kern County, California, *Structure of typical American oil fields*, 3: 38–57. Tulsa: Amer. Assoc. Petrol. Geol.

[6] Hewett, D. F., 1926, *Geology and oil and coal resources of the Oregon, Meeteetse, and Grass Creek Basin Quadrangles, Wyoming*, U. S. Geol. Surv. Prof. Paper 145, Fig. 5.

[7] Turner, Francis J., and Verhoogen, John, 1960, *Igneous and metamorphic petrology*, 694 pp. New York: McGraw-Hill Book Co., Inc.

[8] Faul, Henry, 1966, *Ages of rocks, planets, and stars*, 109 pp. New York: McGraw-Hill Book Co., Inc.

[9] Hamilton, E. I., 1965, *Applied geochronology*, 267 pp. London and New York: Academic Press, Inc.

[10] Hurley, Patrick M., 1959, *How old is the earth?*, 160 pp, Garden City: Anchor Books Doubleday & Co.

[11] Hamilton, E. I., and Farquhar, R. M., 1968, *Radiogenic dating for geologists*, 532 pp. London, New York: Wiley Interscience.

[12] Harland, W. B., Smith, A. Gilbert, and Wilcock, B., eds., 1964, The Phanerozoic time-scale, *Quart. J. Geol. Soc. London*, vol. 120S, 458 pp.

[13] Gilluly, James, 1963, The tectonic evolution of the western United States, *Quart. J. Geol. Soc. London*, 119: 133–74.

[14] Spieker, Edmund M., 1946, *Late Mesozoic and Early Cenozoic history of central Utah*, U.S. Geol. Surv., Prof. Paper 205-D, pp. 117–161.

14

DIAPIRS
AND RELATED
STRUCTURAL FEATURES

Introduction

The words "diapir," "diapirism," and "diapiric" are derived from a Greek word meaning "to pierce."[1] The words were first used to describe anticlinal folds in the Carpathian Mountains with salt cores that pierced the overlying strata.[2] The concept was originally confined to the injection of sedimentary material, but was gradually expanded to include all types of piercement, including magmatic injection. But this expansion of the term destroys its usefulness. Consequently, the term will here refer to the injection of any solid rocks, whether sedimentary, igneous, or metamorphic. In general the body cuts across the adjacent rocks, although locally it may be concordant. The rocks most commonly involved are evaporites (rock salt, gypsum, anhydrite), shale, and serpentine.

Injected rocks may range in physical properties from solid rock to liquids. For convenience in discussion we may establish the following five

293

categories: (1) solid rock, which may have a small percentage of pore space and pore liquid; (2) solid rock that is thoroughly broken up and fractured, with some liquid in the fractures; (3) solid rock that has become mobile due to partial melting, a feature that generally occurs only at considerable depth in metamorphic terranes; (4) a loose aggregate of particles, buoyed up by gases or liquids that could be derived from either magmatic or sedimentary sources; and (5) liquid.

The first two categories are diapiric, the last two are not. A strong argument could be made for classifying the third type as diapirism, but it will be considered separately rather than in this chapter.

Another point must be emphasized. The plastic or viscous flow of solid rock is not the same as diapirism. For example, in Fig. 14-1, much flowage of salt takes place before actual piercement. Thus a salt dome is not necessarily a diapir. However, the two terms are often loosely used interchangeably.

Evaporite Diapirs

INTRODUCTION

Most evaporite diapirs are composed chiefly of halite (rock salt), much less commonly of anhydrite or gypsum. They occur either as the cores of domical structural features or as the cores of anticlines. They are unusually well known because of their great economic importance, the chief product being petroleum, but also including sulfur and salt. Very little information comes from surface exposures; it is obtained chiefly from drilling. The internal structure, however, may be studied in the great underground chambers of salt mines. Geophysical methods, notably gravitational, seismic, and magnetic, contribute significant data.

The salt diapirs in the Gulf Coast of the United States are exceptionally well known; they are commonly referred to as salt domes. Over 300 such domes are known on the mainland, and an equal number are probably present on the continental shelf extending from Louisiana to Yucatán. In Northwestern Germany (Figs. 14-1 and 14-2) there are more than 200 salt diapirs.[3] About 150 are known in Iran, more than 100 in Gabon, and several hundred in the Emba and Dnieper-Donets areas of the Soviet Union.[4] Evaporite diapirs have also been found in the Arctic Islands of Canada, New Brunswick, Utah, Columbia, Peru, France, Spain, Rumania, Yemen, Pakistan, Siberia, and Australia.

SHAPE

A distinction must be made, of course, between the piercing core and the sediments dragged up with it. The cores are either circular in plan or elongate

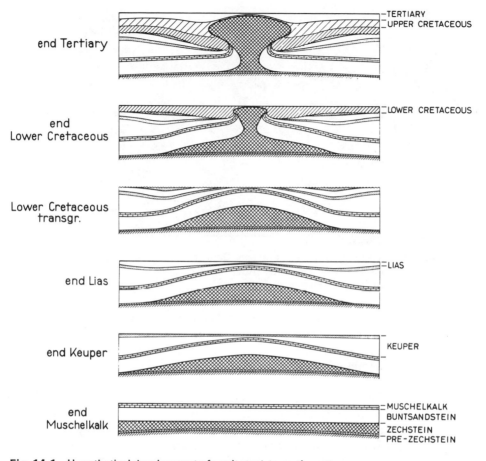

Fig. 14-1. Hypothetical development of a salt stock in northwest Germany. (After Sannemann;[3] permission American Association of Petroleum Geologists.)

(Fig. 14-2). The former have long been known as domes, and, more specifically as salt domes, because the core is largely halite.

Some salt domes in the Gulf Coast are expressed topographically, but many are not. The hills, which rise from a few feet to 40 feet above the surrounding lowlands—in exceptional cases as much as 80 feet—cover an area of more than a mile in diameter. Lakes occur in depressions above some of the domes.

The cores of most Gulf Coast salt domes are essentially circular in plan, and characteristically range in diameter from one-half to two miles, but some are as much as five miles in diameter. In many of these salt domes the walls of the core dip steeply outward; the top may be flat or domical. Some are

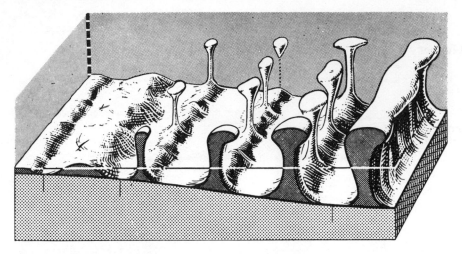

Fig. 14-2. Types of salt structures, Zechstein Basin, Germany. Surrounding country rock has been removed. (After Murray;[1] permission American Association of Petroleum Geologists.)

symmetrical, the walls dipping at the same angle on all sides; others are asymmetrical; in some the wall on one or more sides may dip inward. In some the core *overhangs* or mushrooms on several sides. Many salt domes extend to depths of several miles, as demonstrated by drilling or geophysical data.

The shapes of the German salt diapirs are illustrated in Fig. 14-2. One of these is shown as spindle-shaped, pinching out at depth. There is no evidence that any of the Gulf Coast domes do this, but some may.

In many parts of the world the salt stocks are elongate or even wall-like masses (right side of Fig. 14-2). Elongate masses are characteristic of the Paradox Basin in Utah, Coahuila in Mexico, Germany, and elsewhere.

COMPOSITION

In the Gulf Coast the core consists principally of rock salt (halite) with several percent of anhydrite. Argillaceous and potash-rich beds are interbedded with the salt in the German stocks.

A cap rock (Fig. 14-3) is generally present in the Gulf Coast domes. Absent from many domes, it may reach a maximum thickness of 1000 feet. The cap rock characteristically consists of limestone, gypsum, and anhydrite, the limestone on top and the anhydrite on the bottom. Commercial deposits of sulfur occur on some of the domes.

INTERNAL STRUCTURE

Eight of the more than 300 salt domes in the United States have mines in which the internal structure can be studied.[5,6,7] The beds, a fraction of an

NW   SE

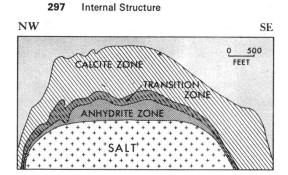

Fig. 14-3. Diagrammatic cross section of cap rock, Jefferson Island dome, Iberia and Vermilion Parishes, Louisiana. (After Murray;[1] permission American Association of Petroleum Geologists.)

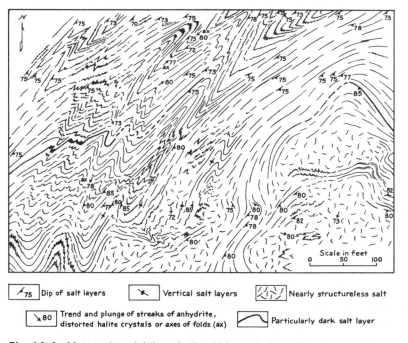

Fig. 14-4. Map to show folding of salt within a salt dome. Grand Saline salt dome, Texas. (After R. Balk,[5] *Bulletin American Association Petroleum Geologists.*)

inch to several feet thick, are displayed in different shades of gray and white; the gray beds have a greater content of anhydrite. The folding is very complex (Fig. 14-4). The bedding is vertical and is well displayed on the vertical walls of the mine workings. But the ceilings show isoclinal, attenuated, refolded and faulted folds that plunge vertically. The folds are flowage folds and resemble those produced by drawing a handkerchief through a small ring. Detailed studies suggest that the salt moved upward as lobes and spines by a series of differential movements.

SHALE SHEATH

In some domes a shale or clay sheath partially encloses the salt (Fig. 14-5). Differing greatly in size, they consist of finely divided argillaceous material that was dragged up by the rising salt. Thick breccias locally surround the salt cores in some of the Rumanian salt diapirs.

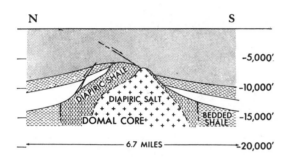

Fig. 14-5. Diapiric shale associated with a salt dome. North-south cross section of Valentine dome, Louisiana. (After Murray;[1] permission American Association of Petroleum Geologists.)

STRUCTURE OF SURROUNDING SEDIMENTARY ROCKS

The sedimentary rocks surrounding the core are uplifted into a dome or anticline. In some domes the bedding in the overlying sedimentary rocks appears to be parallel to the contact with the salt; these have been called nonpiercement domes. This feature, however, is probably due to the fact that the salt was exposed to erosion prior to the deposition of the sediments and such domes probably have cross-cutting contacts at depth. The sedimentary rocks on the top of the dome are commonly broken by normal faults. They may be radial (Fig. 14-6A) or may belong to a more or less parallel system in which one or more graben are conspicuous (Fig. 14-6B). Similar faults have been produced in experimental studies. The sedimentary rocks flanking the core dip outward at various angles, and in many instances are broken by faults. Drill records show that the core pierces the sedimentary rocks.

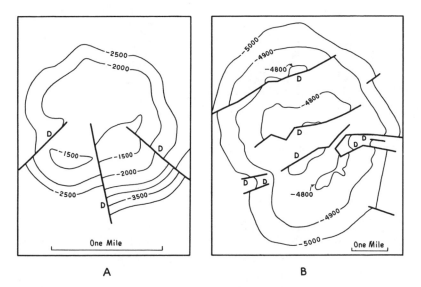

Fig. 14-6. Faulting on salt domes. Heavy black lines are faults; *D* is the downthrown side. (A) Clay Creek salt dome, Texas; structure contours on top of cap rock; contour interval 500 feet. (After W. B. Ferguson and J. W. Minton.) (B) Conroe oil field, Texas; structure contours on top of main Conroe sand; contour interval 100 feet. (After F. W. Michaux, Jr., and E. O. Buck. Data from *Bulletin American Association Petroleum Geologists*.)

ROCK GLACIERS

In some of the Rumanian diapirs the rock salt is exposed at the surface of the earth, and in some of the Iranian domes the salt has literally flowed onto the surface to form spectacular "glaciers" composed of rock salt.[4]

STRUCTURAL EVOLUTION

A vast amount of precise information has accumulated on the structural evolution of salt domes, both in America and abroad. Some, and perhaps many, of the American salt domes have been rising throughout Tertiary time. The evidence is primarily stratigraphic. An angular unconformity, such as that illustrated in Fig. 14-7A, shows that considerable uplift occurred after the deposition of formation *a*, but before the deposition of formation *b*. The salt rose up through formation *a*, truncating the bedding and doming up the sediments. Erosion followed, removing many of the younger beds in formation *a*. Formation *b* was subsequently deposited, and this was followed by renewed upward movement that slightly domed formation *b*.

In other instances a formation may become thinner over the top of the

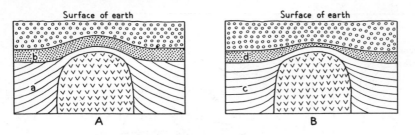

Fig. 14-7. Criteria for dating movements in salt domes. Checked area is the core of rock salt and cap rock. Rest are sedimentary rocks. (A) Unconformity between formations *a* and *b*. (B) Formation *d* thins over core of rock salt.

dome. This indicates, as illustrated in Fig. 14-7B, that the dome was actively rising throughout the deposition of formation *d*, but unconformities within formation *d* may be difficult to detect. It must be remembered that all the data are obtained from drill holes.

Topographically expressed salt domes have probably been active in relatively recent times. Moreover, if Pleistocene or Recent gravels on the dome are uplifted relative to their position in the surrounding region, it is obvious that the salt has been active during the Quaternary.

ORIGIN

Salt diapirs result from the intrusion of solid halite into the surrounding sediments. The salt is derived from some underlying source bed, usually thousands of feet thick (Fig. 14-1). In the Gulf Coast the source is very probably the Louann Salt, of Jurassic, Triassic, or Permian age, and as much as 5000 feet thick. In Germany the source is the Permian Zechstein, which is as much as 3000 feet thick.

The motivating force in the Gulf Coast results from the difference in density between the salt and the overlying sediments. Rock salt has a relatively uniform density regardless of depth. The density is about 2.2 g/cm³, but varies depending upon the amount of anhydrite and temperature. Between the surface and a depth of 2000 feet, the average density of the sediments is 1.9 to 2.2 g/cm³, but below a depth of 2000 feet the density of the average sediment increases progressively to a value of 2.46 at a depth of 20,000 feet. Thus, below a depth of 2000 feet an unstable gravitational situation exists and the salt tends to move upward in the same way that a lighter fluid rises through heavier overlying fluid (Fig. 14-8).[8] If a small anticlinal flexure exists on top of the original salt bed, upward movement starts here, and salt is drained away from the surrounding region. Eventually, the salt bed in the adjacent area may become so thin and constricted that further addition of salt is impossible.

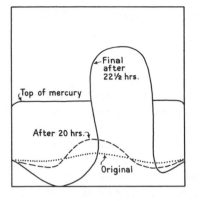

Fig. 14-8. Fluid mechanics of salt domes. At the start a layer of paraffin lies beneath the dotted line, and the mercury lies above. (After L. L. Nettleton.[8] Data from *Bulletin American Association Petroleum Geologists.*)

The elastic limit (yield stress)[9] of rock salt at 300°C and a confining pressure of 2000 bars is 100 kg/cm². But even at temperatures as low as 35° it shows creep phenomena when the confining pressure is as low as 25 kg/cm². Thus salt may flow at relatively low differential stress. Many estimates have been made of the equivalent viscosity of rock salt, some based on laboratory experiments, some based on the rate of convergence of openings in mines. The best estimates at room temperature range from 3.5×10^{16} to 4×10^{18} poises. Increasing temperature lowers the viscosity. The plastic deformation within the individual halite crystals takes place by gliding and dislocations along cube and dodecahedral planes (page 424).

The evolution of salt domes has been analyzed by computerized mathematical models.[10] Equations are prepared to calculate the manner in which the salt dome grows. Many parameters are involved, such as the density of the salt and the surrounding sediments, buoyancy in pounds per square inch (difference in density of salt and sediment at different depths), temperature, ultimate strength, time, and many others. Various values for these parameters may then be entered in the equations. Obviously, such calculations would be a very time consuming process, in fact prohibitive, if done by hand. But the computer handles a large number of substitutions in a short time. The results can be printed out on paper, as in Fig. 14-9, or they can be viewed in a closed circuit television. Figure 14-9B is a model of a salt dome that rose through shales and sandstones. Several features may be noted: (1) The vertical exaggeration is nearly six times. (2) The dome is cut off from the main part of the salt bed. (3) Overhang began when the salt dome penetrated the sandstone. In Fig. 14-9A the top of the salt has moved up in a series of spines. The average growth rate of the top of a typical dome is somewhat less than one foot per thousand years; a dome would take 10,000,000 to 20,000,000 years to develop if the rate were constant.

ECONOMIC RESOURCES

A detailed discussion of the resources of salt domes is not appropriate in this book, but salt domes are of such great economic importance that a brief discussion is desirable. Petroleum is trapped in the sediments that flank the core of rock salt, and in some instances it has been found in the cap rock. Large quantities of sulfur have been obtained from the cap rock of some salt domes. This sulfur has probably been derived from the anhydrite and gypsum normally present in the cap rock, but there is no agreement concerning the details of the process of formation. The rock salt in the core has also been exploited economically. Potash salts have been extensively mined in German salt domes, where the potash salts occur in strata that were deposited during the accumulation of the sediments.

Serpentinite Diapirs

Ultramafic rocks are common in some parts of orogenic belts. Although some are unaltered dunite (composed of olivine), peridotite (composed of olivine and pyroxene), and other similar rocks, they are commonly altered to serpentinite, a rock composed of the mineral serpentine. These ultramafic rocks have been injected into the enclosing rocks.[11] But pure dunite would crystallize at about 1700°C and peridotite at a somewhat lower temperature. Nevertheless, the enclosing rocks commonly, but not always, lack contact metamorphism; this implies that the ultramafic rock was relatively cool at the time of injection. Intense fracturing, shear planes, and slickensides in the serpentinite are consistent with the conclusion that they were injected in the solid state. The low density of the serpentinite is a factor favoring solid emplacement. Unlike the Gulf Coast salt domes, squeezing by horizontal compression is the major factor involved in emplacement.

Sedimentary Vents

The products of what, for want of a better name, have been called sedimentary volcanism, are associated with some petroleum areas, notably India, Burma, Rumania, Malaysia, Trinidad, Venezuela, and the Caucasus Mountain area.[12] Cones, identical in all respects to volcanic cones except in composition, are composed of mud. Although most of the cones are only 100 feet high, a few reach a height of over 1000 feet. The motivating force behind the eruption is gas and steam, under high pressure, derived from petroleum reservoirs.

A

B

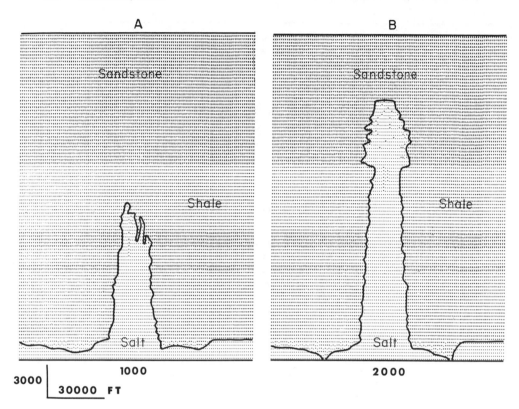

Sandstone

Shale

Salt

1000

2000

3000

30000 FT

Fig. 14-9. Salt dome simulated by a computerized mathematical model. Vertical exaggeration ×5.7. (After Howard;[10] permission University of Kansas Press.)

Vents that presumably served as feeders for sedimentary volcanoes are preserved in Miocene strata in a band 80 miles long in southeast Texas. More resistant to erosion than the surrounding sedimentary rocks, these units are circular to oval in plan, 200 to 1200 feet in diameter, and rise 10 to 125 feet above their surroundings. The filling is chiefly silicified mud-flow tuff derived from lower beds; although the mud is of igneous derivation, it is presumably much older than the emplacement of the material in the vent; moreover, the motivating force was high-pressure gas that had been associated with petroleum. Chalcedony, a form of silica, replaces much of the filling of the vent. Associated subangular blocks of quartzite, silicified tuff, sandstone, and trachyandesite range in size from an inch to a maximum of 8 feet.

Although these vents have been called diapirs, they were presumably emplaced as tiny particles surrounded by gas under high pressure.

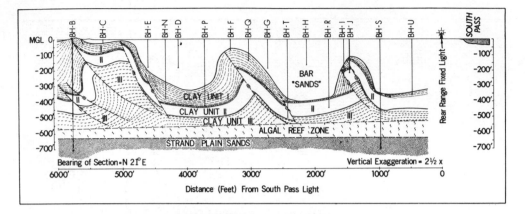

Fig. 14-10. Mud lumps in Mississippi Delta. A sand bar, extending slightly above sea level (bar "sands") made it possible to drill bore holes (labeled BH-B, etc). Note vertical exaggeration of 2½ times. Permission American Association of Petroleum Geologists. (After Murray[13])

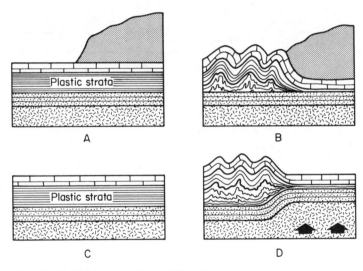

Fig. 14-11. Injection folding. (After Beloussov.)

Mud Lumps

Mud lumps are extensively developed in the Gulf of Mexico at the mouths of the Mississippi River.[13] They are, of course, ephemeral features that are rapidly destroyed by erosion. Off the mouth of South Pass there are at present about 50 islands of this type, as well as 50 submerged lumps. The islands average about 200 feet in length, but the largest is 1500 feet long. The islands rise, at the most, a few tens of feet above sea level, and wave-cut terraces are present.

The lumps are composed of clay or silt. The strata involved are Late Pleistocene and Recent, and are about 400 to 500 feet thick. A structure section is shown in Fig. 14-10, and strictly speaking, since piercement is not involved, they are not diapirs. They are the product of "injection folding" that results from deposition of an extra load of sediments (Fig. 14-11); sediments are squeezed out from beneath this extra load, but piercement is not necessarily involved. The cores of some of the mud lumps, however, have pierced the anticlines, and the clays are fractured, brecciated, and faulted.

References

[1] Braunstein, Jules, and O'Brien, Gerald D., 1968, *Diapirism and diapirs*, Amer. Assoc. Petrol. Geol., Mem. 8, 444 pp.

[2] Mrazec, M. L., 1915, *Les plis diapirs et le diapirisme en général*, Comptes rendus, Institut géologique de Roumanie, 4: 226–270.

[3] Sannemann, D., 1968, Salt-stock families in northwestern Germany, in *Diapirism and diapirs*, Amer. Assoc. Petrol. Geol., Mem. 8, 261–70.

[4] Gussow, William Carruthers, 1968, Salt diapirism: importance of temperature, and energy source of emplacement, in *Diapirism and diapirs*, Amer. Assoc. Petrol. Geol., Mem. 8, 16–52.

[5] Balk, Robert, 1949, Structure of the Grand Saline Salt Dome, Van Zandt County, Texas, *Bull. Amer. Assoc. Petrol. Geol.* 33: 1791–1829.

[6] Kupfer, Donald H., 1968, Relationship of internal to external structure of salt domes, in *Diapirism and diapirs*, Amer. Assoc. Petrol. Geol., Mem. 8, 79–89.

[7] Muehlberger, William R., and Clabaugh, Patricia S., 1968, Internal structure and petrofabrics of Gulf Coast salt domes, in *Diapirs and diapirism*, Amer. Assoc. Petrol. Geol., Mem. 8, 90–98.

[8] Nettleton, L. L., Fluid mechanics of salt domes, 1934, *Bull. Amer. Assoc. Petrol. Geol.* 18: 1175–1204; *also* Gulf Coast oil fields, 79–108, Amer. Assoc. Petrol. Geol., 1936.

[9] Odé, Helmer, 1968, Review of mechanical properties of salt relating to salt-dome genesis, in *Diapirism and diapirs*, Amer. Assoc. Petrol. Geol., Mem. 8, 53–78.

[10] Howard, James C., 1968, *Monte Carlo simulation model for piercement salt domes*, Kansas Geol. Surv. Computer Contrib. No. 22, 22–34.

[11] Oakeshott, Gordon B., 1968, Diapiric structures in Diablo Range, California, in *Diapirism and diapirs*, Amer. Assoc. Petrol. Geol., Mem. 8, 228–43.

[12] Freeman, P. S., 1968, Exposed Middle Tertiary mud diapirs and related features in south Texas, in *Diapirism and diapirs*, Amer. Assoc. Petrol. Geol., Mem. 8, 162–82.

[13] Murray, James P., et al., 1968, Mudlumps: diapiric structures in Mississippi Delta sediments, Amer. Assoc. Petrol. Geol., Mem. 8, 145–61.

15

EXTRUSIVE
IGNEOUS
ROCKS

Introduction

Igneous rocks are the product of the consolidation of magma. *Extrusive igneous rocks* form when magma pours out on the surface of the earth as lava flows or erupts to form beds of pyroclastic rocks. *Intrusive igneous rocks* form when magma consolidates beneath the surface of the earth.

This chapter is concerned with extrusive igneous rocks. In considering this subject, the exact limits of structural geology are particularly difficult to define. The petrography and chemistry of these rocks are primarily the concern of the petrologist.[1] The geomorphic forms are of interest to the physiographer.[2] All phases of the subject fall within the domain of the volcanologist.[3,4,5] The emphasis here will be on features of structural significance.

The fundamental units resulting from extrusive igneous activity are lava flows and pyroclastic beds. A large number of these units are usually associated to constitute volcanoes, which result from central eruptions, and volcanic plateaus and plains, which generally result from fissure eruptions. During the last 400 years, much more volcanic material has been erupted explosively than has poured out as lava flows.[2]

Lava Flows

Lava flows develop when magma wells out at the surface of the earth in a relatively quiet fashion, with little or no explosive activity. Lava flows are tabular igneous bodies, thin compared to their horizontal extent. The attitude corresponds in a general way to that of the surface upon which they are erupted; on flat plains, the lava flows are more or less horizontal, but, on the slopes of volcanoes, they may consolidate with a considerable inclination. Figure 15-1 shows the plan of a succession of lava flows at the top of Mauna Loa on the island of Hawaii.[6]

Individual flows differ greatly in size. Some are only a few feet thick; flows more than 300 feet thick are rare. The average thickness in the Columbia Plateaus of the northwestern United States is probably less than 50 feet; in India the average thickness of basaltic flows is less than 60 feet; in Iceland the average flow is 15 to 30 feet thick; in South Africa the flows range

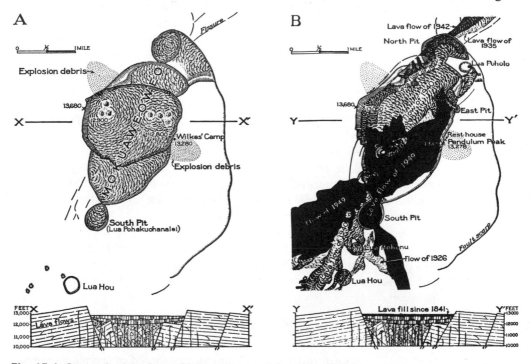

Fig. 15-1. Successive lava flows. Mokuaweoweo sink on Mauna Loa, island of Hawaii. A. In 1841. B. In 1968. (After Gordon A. Macdonald and Agatin T. Abbott, permission of University of Hawaii Press, 1970.[11])

from 1 to 150 feet in thickness; and in Hawaii the basaltic flows average 10 to 30 feet in thickness, but a maximum of 900 feet has been recorded. The area may be a few acres or many square miles. In Iceland, single flows covering over 100 square miles are known, and one flow is said to cover 400 square miles.

Tumuli (singular, *tumulus*), also called *Schollendome* and *lava blisters*, are low domical hills 20 to 60 feet long and 5 to 10 feet high. An open fracture, parallel to the long axis of the uplift, is not uncommon, and ropy masses of lava may issue from this crack. Some geologists believe that tumuli are due to the hydrostatic pressure of fluid lava beneath the crust of a gently dipping flow. Others believe that tumuli are caused by gigantic gas bubbles trapped beneath the crust of the lava.

Pressure ridges are long sharp ridges, many of which are broken by a central crack. In a flow in New Mexico, the shortest pressure ridge is said to be 130 feet long, but some pressure ridges are more than 1200 feet long; they are 10 to 25 feet high and as much as 100 feet wide.[7] Some pressure ridges are due to a compressive force imparted to the crust of a lava flow by the viscous drag of slowly moving subcrustal lava. Those cited from New Mexico are considered to be due to the collapse of the crust of a lava flow that was initially domical in cross section.

Squeeze-ups, small excrescences a few feet or tens of feet long on the surface of a flow, are due to the extrusion of viscous lava through an opening in the solidified crust. They may be bulbous or linear in form. *Driblet cones*, also called *spatter cones*, form if sufficient gas is present to cause rapid effervescence of the magma; they may be as much as 25 feet high and have steep walls. The successive layers of lava overlap one another in much the same way as does the accumulated wax on the sides of a partly burned candle. Lava flows that were very viscous at the time of eruption may possess concentric wavelike block ridges.

Lava tunnels, as the name implies, are long caverns beneath the surface of a lava flow; in exceptional cases they may be 12 miles long. They are caused by the withdrawal of magma from an otherwise solidified flow. Tunnels may be partially or completely filled with pyroclastic material or sediments that wash in through small fissures.

Depressions, circular or elliptical, are found on the surface of lava flows. Many depressions are due to the collapse of the roofs of lava tunnels. In a lava flow in New Mexico, the largest *collapse depression* is nearly a mile long and 300 feet wide, whereas the smallest is only a few feet across; one is 28 feet deep.

Pyroclastic Rocks

Pyroclastic rocks are volcanic rocks that are composed of fragments ranging in size from a fraction of an inch to many feet. Some fragments are composed entirely or partially of volcanic glass. Many of the smaller particles are glass *shards*, derived from the fragmentation of the walls of gas bubbles. Some fragments are individual crystals or parts of such crystals; many of them were crystals floating in the magma when the eruption began. Clasts of pumice are abundant in some rocks. Some fragments are composed of older volcanic rocks, either glassy or fine-grained, that were ripped from the sides of the volcanic vent. Still other fragments have been derived from older sedimentary, plutonic, and metamorphic rocks through which the magma erupted.

Indurated (cemented) pyroclastic material is classified into tuff, tuff-breccia, breccia, agglomerate, and volcanic conglomerate. *Tuff* is composed primarily of particles less than 4 millimeters in diameter; its fragmental character is usually recognized in the field and under the microscope. *Breccia*, more precisely called *volcanic breccia*, is composed mostly of angular fragments larger than 4 millimeters in diameter. If much of the matrix is less than 4 millimeters in diameter, the term *tuff-breccia* may be applied. Obviously the classification of some bodies may be arbitrary.

Much tuff and breccia is an *air-fall* deposit, that is, is composed of volcanic rock that has been blown into the air and comes to earth some distance from the vent. It may be well bedded, because of successive eruptions or because of reworking by running water. But in the last two decades it has been recognized that much tuff and breccia was erupted as ash flows.[8,9] The fragmented material was sufficiently mobile to flow like a liquid after issuing from a vent. An interstitial gas was an important factor in buoying up these fragments, many of which are glass shards. Collapsed pumice fragments may be common. These ash flows may be hundreds of feet thick or even more than 1000 feet thick; a large ash flow may contain tens or even hundreds of cubic miles of rock. Some of the sheets issued in such rapid succession that several of them collectively acted as a single cooling unit. The individually erupted sheets may show graded bedding, whereas the cooling unit controlled the thermal history. The degree of consolidation may differ greatly. The upper part of a cooling unit may be loosely consolidated. The central part may be well consolidated. The base may be welded, that is, the glass shards have welded together to produce a dense compact obsidian-like rock. The rocks resulting from ash flows have often been called *ignimbrites*.[10]

Agglomerate is composed of rounded or subangular fragments, larger than 4 millimeters in diameter, set in a finer matrix. The round or semiround shape of the fragments is not due to the action of running water; it may be original or it may result from constant attrition of the fragments. A *vent agglomerate* is the variety that is confined to a volcanic vent. *Volcanic conglomerate* is similar to breccia, but the fragments are rounded by the action of running water. Sometimes, because of torrential rains, the pyroclastic material resting on the mountain slopes becomes saturated with water and moves as a *mud flow* or *lahar*.

Fissure Eruptions

In fissure eruptions the lava is extruded through a relatively narrow crack and flows out on the surface of the earth. In Iceland some fissures are tens of miles long. The eruption of a whole succession of such flows produces a lava plateau or plain. Although pyroclastic rocks are rare, they may result from violent local explosions. Under favorable conditions, the dikes that served as feeders for the fissure eruptions are exposed by erosion.

Several of the great lava plateaus of the world have been constructed primarily by fissure eruptions. This is true of the Columbia Plateau of the northwestern United States, and of western India, South Africa, South America, and the North Atlantic volcanic field, remnants of which are found in Great Britain, Iceland, and Greenland. In each of the plateaus, which cover tens of thousands of square miles, the total thickness of the volcanics ranges from 3000 to nearly 10,000 feet. Basalts constitute 90 to 95 percent of the lavas participating in fissure eruptions.

Volcanoes

The major forms that develop at the surface of the earth as a result of central eruptions, although areally no more important than fissure eruptions, are more spectacular, partly because of their form and height, partly because of the violence that may attend their activity. These major forms may be classified as volcanoes, craters, and caldera.

Volcanoes (Plate 26) are bodies of rock built up by the eruption of magma; they rise above their surroundings. Volcanoes may be classified in several different ways, depending upon the emphasis that the investigator wishes to make. A petrographic classification is based upon the lithology. A physiographic classification is based in part upon the stage of erosion. A structural classification is based primarily upon the internal structure of the volcano and, secondarily, upon the map pattern displayed by a number of volcanoes.

On the basis of internal structure, volcanoes may be conveniently classified into pyroclastic cones, lava cones, composite cones, volcanic domes, large spines, and compound volcanoes.

The various kind of cones may be considered first. Such volcanoes are conical in shape, the apex of the cone pointing upward (Fig. 15-2). A depression, either a crater or caldera (see p. 314), is commonly present at the top of the cone, unless erosion has been so extensive as to destroy it. The slopes of the cone are concave upward, and the steepest slopes, found at the top, are controlled by the angle of repose of the material when it was erupted. Internally, the cones consist of successive layers, either lava, pyroclastic

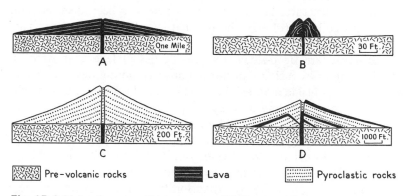

Pre-volcanic rocks   Lava   Pyroclastic rocks

Fig. 15-2. Volcanic cones. (A) Lava cone. (B) Hornito. (C) Pyroclastic cone. (D) Composite cone.

material, or both; these layers dip outward more or less parallel to the slope of the cone.

Lava cones, more commonly called *shield volcanoes*, are built chiefly of lava that was very mobile at the time of eruption; they are broad cones with low angles of slope (Fig. 15-2A). Much of the lava erupts from small vents or fissures on the side of the cone. The Hawaiian Islands are superb examples of shield volcanoes.[11] The volcanoes here rise 30,000 feet above the floor of the Pacific Ocean, attaining altitudes of 13,675 feet above the level of the sea.

Hornitos are relatively small, steep-sided lava cones constructed by the eruption of great plastic blobs of lava that were too cool to flow (Fig. 15-2B.)

Pyroclastic cones, built chiefly of pyroclastic material, may possess very steep upper slopes (Fig. 15-2C); the variety that is composed chiefly of cinders is called a *cinder cone.*

Composite cones, also called strato-volcanoes, are built of alternating layers of lava and pyroclastic material (Fig. 15-2D). Most of the lava erupts on the flanks of the cone rather than from the summit crater.

In the simplest type of volcanic cone, the pyroclastic rocks were blown out of the summit crater, from which some of the lava may issue as overflows. Some of the lava—and in many instances most of it—pours out from fractures on the sides of the mountain. *Adventive* or *parasitic* cones are subsidiary cones on the flanks of a larger volcano.

Volcanic cones differ greatly in size. Small cinder cones and hornitos may be only a few tens of feet high. On the other hand, some great giants tower five or ten thousand feet above the adjacent regions, and the base may be many miles in diameter. Mt. Rainer[12] in Washington rises to an altitude of 14,410 feet, that is, 10,000 feet above the surrounding valleys. But most active volcanoes or recently active volcanoes are not good ones in which to study the internal structure. However, where there has been deep erosion or recent caldera collapse (Pl. 27), the internal structure of the upper part of a volcano may be well exposed.[13]

Volcanic domes are steep-sided, bulbous bodies of lava that were so viscous at the time of extrusion that normal flows could not develop. Many volcanic domes form inside a crater, but this is not always true. The height of domes ranges from a few tens of feet to 2500 feet. Many domes are difficult to study because the surface is covered by a jumble of irregular blocks caused by the rupture of the original dome; this rupture is due partly to thermal contraction of the lava, partly to stresses resulting from extrusion. Moreover, the steep sides of most domes are buried under talus.

Some domes grow primarily by expansion from within. A few of these domes are characterized by concentric shells, but such a structure apparently forms most readily if a slight cover of older rocks overlies the rising magma. Most such domes are intensely fissured and brecciated. During the rise of a dome, irregular dikes that are continuously injected into the outer shell may serve as feeders to small surface flows.

A common variety of volcanic dome displays a fan arrangement of the flow layers. Such a structure may be due to the fact that the earliest lavas spread out from the vent at a low angle, forming a constricting ring that acted as a levee.[14] The later flows, because they were restricted, rose at increasingly high angles.

Guyots and *seamounts* are large conical topographic features rising above the sea floor.[15] Guyots are flat-topped seamounts. The depths of the tops of these features range from a few thousand to many thousands of feet. They are ancient volcanoes. It is generally believed that the flat tops are due to marine planation. Since this cannot extend more than a few hundred feet below sea level, guyots indicate one of the following: (1) a rise of sea level; (2) a sinking of the volcano; or (3) a sinking of a large part of the ocean floor. Fossiliferous strata resting unconformably on the surface of planation are important in dating the events. Shallow-water Cretaceous fossils above the unconformity date the volcano as pre-Cretaceous and indicate post-Cretaceous relative changes in sea level.

Compound volcanoes are those that consist of two or more of the types described above. Thus a composite volcano that has an associated volcanic dome, either in its crater or on its flanks, would be a compound volcano.

Craters, Calderas, and Related Forms

CRATERS

A *crater* (Plates 26A and B) is the normal depression at the top of a volcanic cone, and it is directly above the pipe that feeds the volcano. A crater, in its

Plate 26. *Volcano and crater.* El Cotopaxi, Ecuador. Altitude
19,498 feet. Said to be highest active volcano in world. (A)
View northeastward. El Cotopaxi in foreground, El Antisana
in right-middle distance. (B) Crater looking southeastward.
Photo: G. E. Lewis, U. S. Geological Survey.

simplest form, is a flat-bottomed or pointed, inverted cone more or less circular
in horizontal section. Immediately after an eruption, the diameter of the
bottom of the crater (probably the same as the diameter of the conduit that

fed the volcano) is seldom over 1000 feet. Subsequent landslides from the walls, however, may partially fill the bottom of the crater. The walls of a crater are composed of interbedded lava and pyroclastic rocks, but the ratio of the two depends upon the type of volcano; some crater walls are composed exclusively of lava, others are composed exclusively of pyroclastic material. An unusually strong explosion may blow away part of the crater wall, thus exposing the layers that compose the top of the cone.

Craters are primarily due to explosions at the top of the pipe feeding the volcano. Fragmental rocks are blown into the air, and the largest material, landing within some hundreds of feet, builds a circular wall. Whenever magma rises in the pipe it melts and dissolves whatever may be directly above it, thus helping to maintain the depression.

CALDERA

Large volcanic depressions may be classified as *calderas* (Plates 27 and 28). Circular or irregular in form and miles in diameter, calderas[16] are much larger than the pipe feeding the volcano. Calderas may be classified into four types: collapse caldera, resurgent caldera, explosive caldera, and erosion caldera.

Plate 27. *Caldera, Crater Lake, Oregon.* Wizard Island, in foreground, is a cinder cone. Photo: H. R. Cornwall, U. S. Geological Survey.

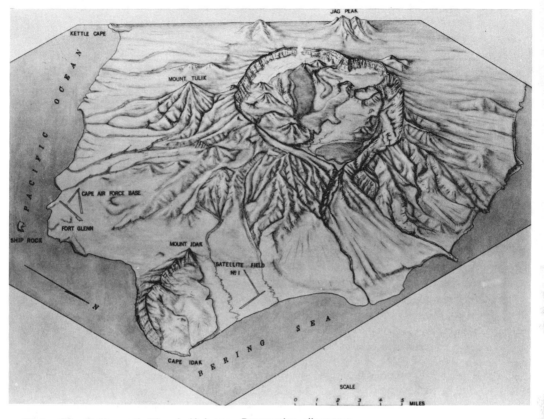

Plate 28. *Caldera of Okmok Volcano.* Perspective diagram viewed from northeast. Umnak Island, Aleutian Islands, Alaska. Credit: F. M. Byers, Jr., U. S. Geological Survey.

Collapse caldera are the result of the collapse of the superstructure of the volcano because of withdrawal of the underlying support. The rapid eruption of great volumes of ash or pumice may lower the level of the magma in the main reservoir to such an extent that a potential void is left (Fig.15-3). Collapse may be piecemeal, as shown in Fig. 15-3C, or the subsiding block may sink *en masse.*

Classic examples of caldera are Krakatau in Indonesia, which erupted violently in 1883, Crater Lake in Oregon, and Newberry Caldera, also in Oregon.[17] Of 76 known active and extinct volcanoes in the Aleutian Islands and Alaska Peninsula, at least 17 are caldera.[18] Plate 28 is a perspective block diagram of Umnak Island, one of the Aleutian chain.[19] The caldera is 6 miles in diameter. Three stages in the evolution of the volcano are recognized. (1) Eruption of basaltic flows and tuffs that built up a composite cone that rose 7000 to 10,000 feet above sea level. (2) A great erup-

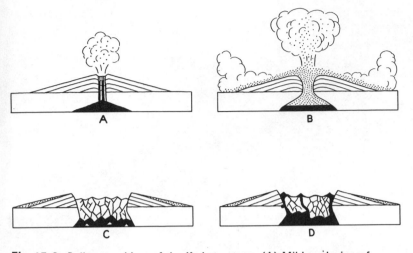

Fig. 15-3. Collapse caldera of the Krakatau type. (A) Mild explosion of pumice. (B) Violent explosions: part of pyroclastic material is thrown into air, part rushes down slopes as *nuées ardentes*. (C) Collapse of the top of the cone. (D) Renewed volcanic activity. (After H. Williams.[16])

tion, during which the caldera formed; simultaneously, ash deposits were strewn over the volcano and far beyond. (3) Cinder cones, basalt flows, and lake beds partially filled the caldera. That this is still going on is indicated by a flow as recent as 1945.

The withdrawal of support may happen in one of several ways. Large quanities of ash or pumice may be erupted from the magma reservoir directly beneath the volcano. The formation of a caldera on Mt. Katmai in Alaska during the 1912 eruption was apparently more complex.[20] Much of the ash was erupted from Novarupta, 7 miles to the west. The 7.35 cubic miles of ash erupted is equivalent to 3.4 cubic miles of solid rock. But the caldera on Mt. Katmai represents only 1.5 cubic miles of rock. Presumably collapse must also have taken place elsewhere in the region.

Resurgent calderas develop where collapse is followed by the doming of the central block. An example is the Valles Caldera[21] in New Mexico, which is 12 to 14 miles in diameter and 500 to 2000 feet deep (Fig. 15-4). In the center is a broad domical mountain, Redondo, composed of outward-dipping tuffs. Rhyolite domes occupy a circular band, 2 to 3 miles wide, 6 miles from the center of the caldera. The sequence of events is illustrated in Fig. 15-5.

Explosive calderas are due to a violent explosion that blows out a huge mass of rock. A classic example is Bandai-San in Japan, which blew up within a minute. The volcano had not erupted for more than a thousand years, but on the morning of July 15, 1888, 15 or 20 explosions occurred within little more than a minute. Although much rock was blown into the

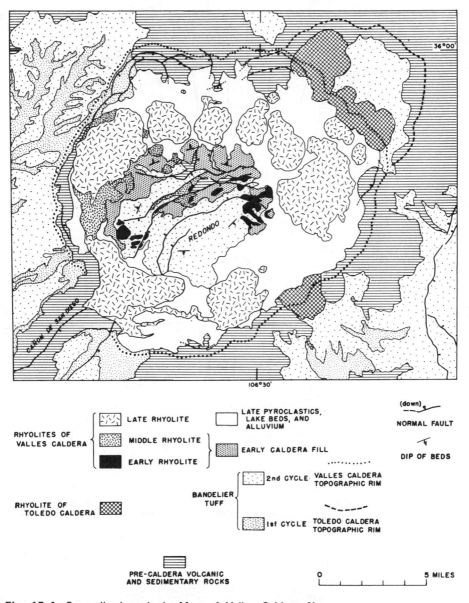

Fig. 15-4. Generalized geologic Map of Valles Caldera, New Mexico. (Memoir 116, Geological Society of America.) (Permission Geological Society of America, R. L. Smith and R. A. Bailey.[21])

air, the last explosion, which was horizontal, initiated a great avalanche containing $1\frac{1}{4}$ cubic kilometers of rock. The summit and much of the northern side of the volcano were blown away, leaving a great amphitheater $1\frac{1}{2}$ square miles in area, with walls more than 1200 feet high. No lava appeared, and

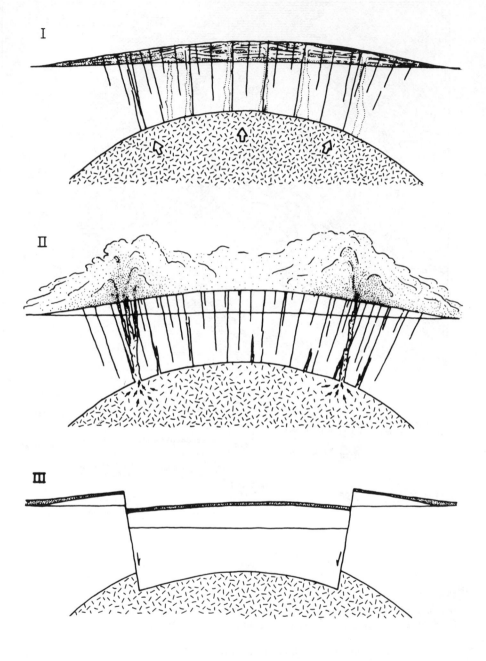

IV

V

VI

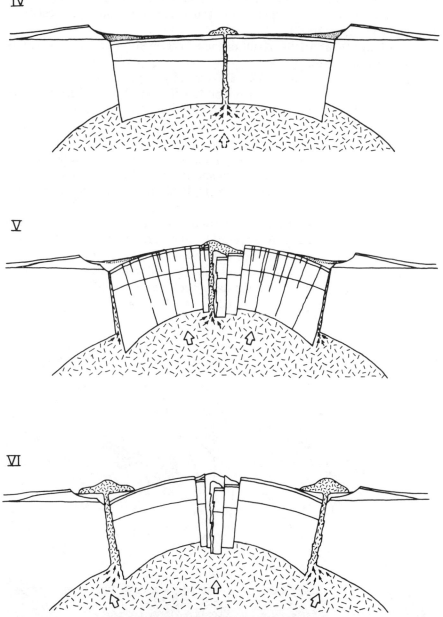

Fig. 15-5. Evolution of a resurgent caldera: Valles Caldera, New Mexico, (a) Upward pressure by magma, formation of ring fractures, and eruption of tuff, (b) Major eruption of pumices (c) Collapse, (d) Minor volcanism and sedimentation. (e) Resurgent doming. (f) Eruption of domes along ring fracture. (Permission Geological society of America, R. L. Smith and R. A. Bailey.[27])

the cataclysm was caused by a *phreatic explosion;* that is, ground water was converted to steam by volcanic heat until such a great pressure was generated that an explosion ensued.

Erosion calderas are the result of the enlargement of craters or calderas by erosional processes.

Pit craters or *volcanic sinks* are typically exposed in the Hawaiian Islands. They range in diameter from 50 feet to several miles. The walls are very steep and the depth may be hundreds of feet. In periods of quiescence the pit is floored by solid lava, but at times of igneous activity the pit may be partially or wholly filled with molten lava. At some place along a crack the magma works upward by stoping and fluxing to form a cylindrical chamber occupied by magma. When the magma subsides the roof of the reservoir collapses. The pit may be widened later by landsliding and concentric faulting.

Volcano-tectonic depressions are similar to caldera, except that they occupy much larger areas. A large graben-like area many tens of miles in length subsides as vast volumes of tuff and pumice are erupted. Lake Toba in Indonesia occupies such a depression.[22] Similar features in Australia, some of them 25 to 70 miles long, have been called cauldrons.[23] Ancient deformed volcano-tectonic basins occur in Precambrian areas.[24]

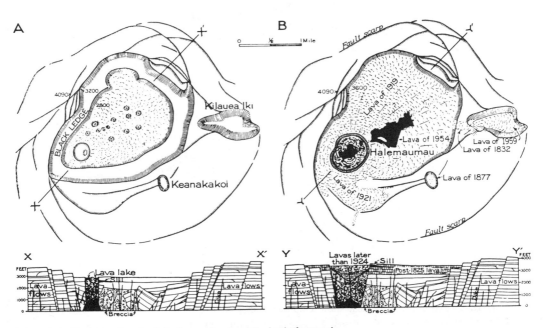

Fig. 15-6. Kilauea volcanic sink. Approximate altitudes in feet; ash beds omitted. (A) In 1825. (B) In 1960. The structure beneath the caldera is hypothetical. After Gordon A. Macdonald and Agatin T. Abbott, permission of University of Hawaii Press, 1970.[11]

Inflation and Deflation of Volcanoes

Volcanoes inflate and deflate like balloons. The magnitude of the displacement is very small and can be detected only by delicate instruments. Volcanic

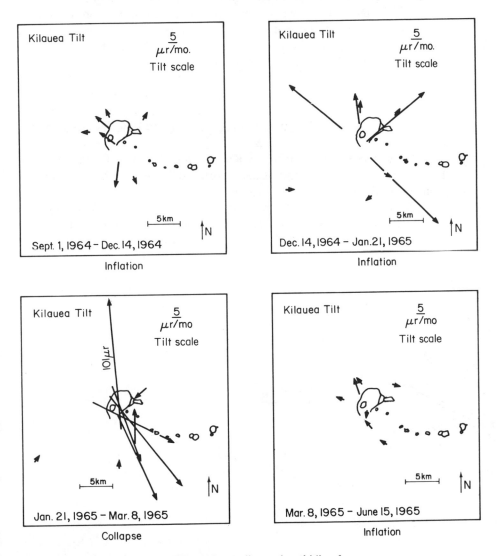

Fig. 15-7. Tilting of Kilauea. Kilauea is small area in middle of map; compare with Fig. 15-6. Smaller pit-craters are also shown to southeast. Vector shows tilting of long-base water tube. Tilt in microradians per month. (After Decker, Hill, and Wright.[25] Reprinted from *Bulletin Volcanologique*, Vol. 29, with permission from the editor.)

eruptions and small earthquakes are related to these processes. One of the most significant studies has been made at the Volcano Observatory at Kilauea on the island of Hawaii.[25,26]

Kilauea (Fig. 15-6) is a sink 3 kilometers (2½ miles) in diameter and 130 meters (420 feet) deep. Eccentrically located on the south side of the sink is Halemaumau, a lava pit, the floor of which varies considerably in altitude, sometimes being many meters below the floor of the main sink, sometimes flooding over the floor of the main sink. The lavas are all basalt.

The horizontal distance across Kilauea differs from time to time. Such measurements can be done by: (a) triangulation; (b) tellurometer, which uses a 10,000 megacycle radio-wave carrier; and (c) geodimeter, which uses a light-beam carrier. Tilting of the ground surface can be measured by precise leveling or by the use of tiltmeters; long-base liquid tiltmeters are 50 meters long.

A line 3098 meters long across Kilauea was lengthened by as much as 12 centimeters prior to eruption and shortened 29 centimeters after the beginning of an eruption on the flank of the volcano.

Figure 15-7 shows the tilting between September 1, 1964 and January 15, 1965. The scale is given in microradians per month rather than degrees. A radian is 57.3 degrees; a microradian is a millionth of a radian, that is, 0.0000573 degrees. The tail of the arrow is located at the station where the

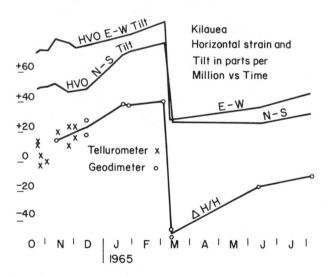

Fig. 15-8. Deformation of Kilauea from October, 1964 to August, 1965. The upper two curves give the tilt, the lower curve gives the horizontal distance. Prior to late February the volcano was inflating. In late February there was sudden collapse. (After Decker, Hill, and Wright.[25] Reprinted from *Bulletin Volcanalo-gique*, vol. 29, with permission from the editor.)

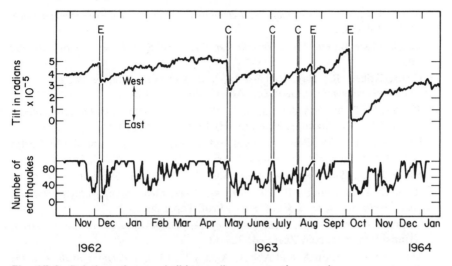

Fig. 15-9. Relation of ground tilting, collapses, eruptions, and small earthquakes at Kilauea. Upper curve shows tilting. Periods of inflation were followed by collapse or eruption. There is a tendency for daily earthquakes to become more common as the amount of tilt increases. The flat top of the curve merely indicates earthquakes in excess of 100 a day. (After U. S. Geological Survey, Moore and Koyanagi.[26])

measurement was made. An arrow pointing outward means that Kilauea was rising and inflating. From September 1 to January 21 Kilauea was rising. Then in the period from January 21 to March 8 it subsided; then it inflated again. Deflation was essentially contemporaneous with flank eruptions. Figure 15-8 shows that uplift, as shown by tilting, continued for many months. At the same time the distance between the opposite walls became greater. Then in March, 1965, there was sudden deflation. This deflation was essentially contemporaneous with a flank eruption. The eruption of 29×10^6 cubic meters of basalt is nearly identical with the 31×10^6 cubic meters of deflation. This seems fortuitous, because the correlation is usually not so good.

Figure 15-9 shows the correlation between small earthquakes, tilt, deflation, and eruption.

References

[1] Turner, Francis J., and Verhoogen, John, 1960, *Igneous and metamorphic petrology*, 2d ed., 694 pp, New York: McGraw-Hill Book Co., Inc.

[2] Cotton, C. A., 1944, *Volcanoes as landscape forms*, 416 pp., Christchurch, New Zealand: Whitcombe and Tombs, Ltd.

[3] Rittman, A., 1652, *Volcanoes and their activity*, 305 pp., New York: Interscience, John Wiley & Sons.

[4] Eckel, Edwin B., ed., 1968, *Nevada test site*, Geol. Soc. Amer., Mem. 110, 290 pp.

[5] Coats, Robert R., Hay, Richard L., and Anderson, Charles A., eds. 1968, *Studies in volcanology*, Geol. Soc. Amer., Mem. 116, 678 pp.

[6] Stearns, H. T., and Macdonald, G. A., 1946, *Geology and ground-water resources of the island of Hawaii*, Hawaii Div. Hydrogr. Bull. 9, 363 pp.

[7] Nichols, R. L., 1939, Pressure ridges and collapse depressions on the McCartys basalt flow, New Mexico, *Trans. Amer. Geophys. Union*, part 3, 432–33.

[8] Smith, R. L., 1960, Ash-flows, *Geol. Soc. Amer. Bull.* 71: 795–841.

[9] Ross, C. S., and Smith, R. L., 1961, *Ash-flows, their origin, geologic relations, and identification*, U. S. Geol. Surv., Prof. Paper 366.

[10] Marshall, Patrick, 1935, *Acid rocks of the Taupo-Rotorua volcanic district*, Trans. Roy. Soc. New Zeal., 64: 323–66.

[11] Macdonald, Gordon A. and Abbot, Agatin T., 1970, *Volcanoes in the sea, the Geology of Hawaii*, 441 pp., Honolulu: Univ. Hawaii Press.

[12] Fiske, Richard S., Hopson, Clifford A., and Waters, Aaron C., 1963, *Geology of Mount Rainer National Park, Washington*, U. S. Geol. Surv., Prof. Paper 444, 43 pp.

[13] Williams, Howell, 1942, *The geology of Crater Lake National Park, Oregon*, Carnegie Inst. Washington, Pub. 540, 162 pp.

[14] Williams, Howell, 1932, *The history and character of volcanic domes*, Univ. California, Dep. Geol. Sci. Bull. 21: 51–146.

[15] Menard, H. W., 1964, *Marine geology of the Pacific*, 271 pp, New York: McGraw-Hill Book Co. Inc.

[16] Williams, Howell, 1941, Calderas and their origin, Univ. California, Dept. Geol. Sci. Bull. 25: 239–346.

[17] Higgins, Michael W., and Waters, Aaron C., 1968, Newberry Caldera field trip, Oregon Dep. Geol., Mineral Ind., Bull. 62, pp. 59–77.

[18] Coats, Robert R., 1950, Volcanic activity in the Aleutian Arc, *U. S. Geol. Surv. Bull. 974*, pp. 35–49.

[19] Byers, F. M., Jr., et al., 1946, Volcano investigations on Umnak Island, pp. 19–53 *in* Robinson, G. D., et al., *Alaska volcano investigations*, Report No. 2, Progress of Investigations in 1946, U. S. Dept. Interior, 1947, 105 pp.

[20] Curtis, G. H., *The stratigraphy of the ejecta from the 1912 eruption of Mount Katmai and Novarupta, Alaska*, pp. 153–210 *in* Coats, Robert R., et al., 1968, Mem. 116, Geol. Soc. Amer., 678 pp.

[21] Smith, Robert L., and Bailey, Roy A., *Resurgent cauldrons*, pp. 613–62 *in* Coats, Robert R., et al., 1968, *Studies in volcanology*, Mem. 116, Geol. Soc. Amer. 678 pp.

[22] Van Bemmelen, R. W., 1939, The volcano-tectonic origin of Lake Toba, northern Sumatra, Ingenieur in Nederlandsch-Indie 6 (9), pp. 126–40.

[23] Branch, C. D., 1966, *Volcanic cauldrons, ring complexes, and associated granites of the Georgetown Inlier, Queensland, Australia*, Bur. Mineral Resources, Geol., Geophys., Bull. 76, 158 pp.

[24] Goodwin, A. M., and R., Shklanka, R., 1967, Archean volcano-tectonic depressions: form and pattern, *Can. J. Earth Sci.* 4: 777–95.

[25] Decker, R. W., Hill, D. P., and Wright, T. L., 1966, Deformation measurements on Kilauea volcano, Hawaii, *Bull. Volcanologique*, 29: 721–32.

[26] Moore, James G., and Koyanagi, Robert X., 1969, *The October, 1963, eruption of Kilauea Volcano, Hawaii*, U. S. Geol. Surv. Prof. Paper 614-C.

16

INTRUSIVE
IGNEOUS
ROCKS

Introduction

Intrusive igneous rocks form when magma consolidates beneath the surface
of the earth. Magma may be emplaced at depths ranging from near the
surface of the earth to ten or more miles.

A *pluton*[1] is a rock body composed of intrusive igneous rock. The term
may be used for bodies of all sizes, from those a few feet long to those that
are exposed over hundreds of square miles. Thus the term encompasses a
whole series of terms, such as dike, sill, laccolith, lopolith, and batholith.
But in practice, if the shape and relations of the body are readily observed
in a single outcrop, or readily inferred from exposures in several outcrops,
such terms as dike or sill are used. Thus the term pluton is commonly used
for larger bodies, especially if the shape and structural relations are not clear.

It is by no means easy to define the limits of petrology and structural
geology in this particular subject. The petrologist is primarily concerned
with the mineralogy, chemistry, and origin of the rocks in these bodies.[2]

The structural geologist is interested in their shape because they play an important role in the architecture of the crust of the earth. Moreover, he is concerned with the emplacement of these bodies and their relation to deformation. The internal structure of the rocks in plutons is significant because it offers evidence on the emplacement of the rocks and on later forces that have operated upon an already consolidated rock. Ideally the student of plutons should be both a petrologist and a structural geologist; a good field geologist and petrographer should possess a thorough knowledge of the physical chemistry of rocks. However, this book is concerned only with the structural phases of the problem.

A structural study of plutons involves a consideration of their internal structure, shape, and size, as well as the structural and chronological relations to the adjacent rocks. But a discussion of the mechanics of emplacement of larger plutons is reserved for Chap. 17.

Texture and Internal Structure

The rocks in plutons may range from fine-grained to very coarse-grained. The grain size is related to the size of the pluton, the cooling history, the volatiles present during consolidation, and the crystallizing power of the various mineral species.

Plutonic rocks are coarse-grained igneous rocks, that is, those in which the grains range in size from 5 millimeters to 3 centimeters. Many are *equigranular*, with all the grains essentially the same size. Others are *porphyritic*, with phenocrysts from 1 to 10 centimeters long. Since to many geologists the term phenocryst has a genetic implication—that is, a crystal formed during slow cooling, followed by more rapid cooling to form the groundmass—some prefer *inset* as a nongenetic descriptive term. Some insets may be the result of the rapid growth of new minerals after the rock has consolidated.

Pegmatites are very coarse rocks in which the grains exceed a diameter of 3 centimeters. The coarse grain results from the presence of volatiles during the crystallization, thus permitting large crystals to grow; consequently pegmatites may occur in small bodies.

Many plutonic rocks are *massive*, showing no preferred orientation of the constituent minerals (Fig. 16-1A). Others are characterized by *foliation* resulting from the parallel arrangement of platy or ellipsoidal mineral grains (Fig. 16-1B and 16-1D). *Primary foliation* forms during the flowage of a partially crystallized magma; it is discussed further on page 378. *Secondary foliation* forms in an already consolidated rock (Chapters 18 and 20). *Layered rocks* are those that consist of alternating sheets of different composition.[3] The layers may range from a fraction of an inch to several feet in thickness. In Fig. 16-1C the dark minerals are abundant in layers 1, 3, and 5, but are rare in layers 2 and 4. Such layering may result from lamellar flow, from settling of minerals from a crystallizing magma, or from successive injections. If the platy minerals in a layered rock are parallel to one another, the rock is foliated as well as layered (Fig. 16-1D).

Lineation is the result of the parallelism of some directional property in the rock, such as the long axes of hornblende crystals (Fig. 16-2A). Platy minerals or spherical grains may be strung out in lines to produce a lineation (Fig. 19-1C). A rock may have a lineation without foliation (Fig. 16-2A) or it may possess both foliation and lineation (Fig. 16-2C). Like foliation, lineation may be primary, secondary, or inherited. Primary lineation is discussed further on page 373.

Inclusions are fragments of older rock surrounded by igneous rock Plate 29). They may be angular, subangular, or round. A *xenolith* is an inclusion that has obviously been derived from some older formation gene-

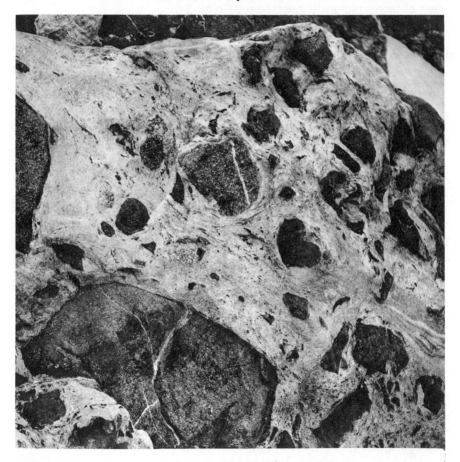

Plate 29. *Magmatic breccia.* Xenoliths of gabbro, hornblende schist, and calc-silicate hornfels in aplite-granite intrusion. Trubinasca Valley, Switzerland. Photo: J. Haller.

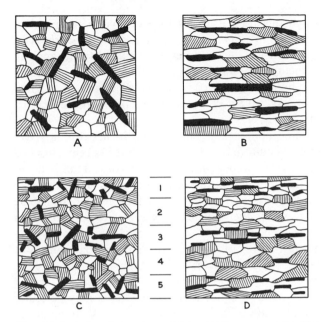

Fig. 16-1. Internal structure of intrusive igneous rocks. Natural scale. Solid black represents platy dark minerals, such as biotite; lined pattern represents feldspar; white represents quartz. (A) Massive rock. (B) Foliated rock. (C) Layered rock. (D) Layered and foliated rock. In (C) and (D), layers 1, 3, and 5 are richer in dark minerals than layers 2 and 4.

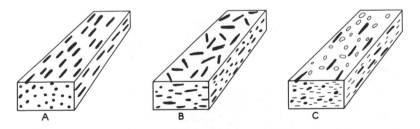

Fig. 16-2. Lineation. Solid black is a prismatic mineral, such as hornblende; dashes and circles represent a platy mineral, such as mica. (A) Lineation due to parallelism of crystals of a prismatic mineral; no platy flow structure. (B) Foliation due to orientation of crystals of a prismatic mineral parallel to the top of the block; no lineation. (C) Lineation due to parallelism of crystals of a prismatic mineral, associated with a foliation due to parallelism of a platy mineral.

tically unrelated to the igneous rock itself, as for example, a fragment of sandstone in granite. An *autolith*, sometimes called a *cognate xenolith*, is an inclusion of an older igneous rock that is genetically related to the rock in which it occurs. Thus an inclusion of diorite in granodiorite is called an autolith if it can be shown that both were derived from a common parent magma. Whether or not the two rocks have a common parentage is, however, primarily a problem of petrology.

A *segregation* is a round or irregular body, a few inches to many feet in diameter, and in some cases hundreds of feet across, that has been enriched in one or more of the minerals composing the igneous rock. Thus a hornblende granite may contain clots that have much more hornblende than the surrounding granite. These clots are segregations if they formed while the granite was consolidating, and if their formation was due to concentration of the atoms of which hornblende is composed. Petrographic methods may be necessary to distinguish a segregation from an inclusion, particularly if the latter has been modified by reaction with the magma.

Schlieren are somewhat wavy, streaky, irregular sheets, usually lacking sharp contacts with the surrounding igneous rocks. Schlieren may be either darker or lighter than the rock in which they occur. Some are disintegrating, altered inclusions, some may be segregations, and some may represent concentrations of residual fluids into layers in a rock that had otherwise crystallized.

Age Relative to the Adjacent Rocks

An intrusive igneous rock can be either older or younger than the adjacent formations. If the intrusive rock is older, the adjacent rocks must rest on it unconformably (Fig. 16-3A). The bedding in the sedimentary rocks above

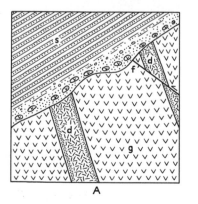
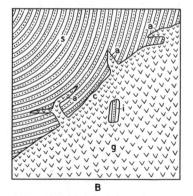

A   B

Fig. 16-3. Relative age of pluton and adjacent rocks. *s*, sandstone and shale; *g*, granite; *d*, dike of diorite; *f*, fault; *i*, inclusion; *a*, apophyses; *c*, chilled contact of granite. (A) Unconformity. (B) Intrusive contact.

the unconformity is essentially parallel to the contact. Debris from the igneous rock, either in the form of pebbles or mineral fragments, may generally be found in the overlying formations. The igneous rock is coarse up to the contact, and faults and dikes (see p. 350) in the igneous rock may be truncated at the contact.

An *intrusive contact* (Plate 30) is present wherever magma intrudes the surrounding rocks (Fig. 16-3B). Small dikes or sills (p. 335) of igneous rock may cut the adjacent formations, and inclusions of the latter may be found in the intrusive. The igneous rock may become finer-grained adjacent to the older rocks; it would give a *chilled contact*. The bedding in the adjacent formations may or may not be truncated by the contact. Conclusive data may be difficult to obtain along many contacts, but assiduous search will usually reveal critical information. Along some contacts there may be a zone of metasomatically replaced rocks, which grade in one direction into sedimentary rocks and in the other direction into typical igneous-looking rocks (page 382).

Basis of Classification of Plutons

INTRODUCTION

Plutons are classified on the basis of several factors: (1) structural relations to adjacent older rocks, (2) size, and (3) shape.

STRUCTURAL RELATIONS TO ADJACENT OLDER ROCKS

Many of the rocks into which plutons are injected are bedded or foliated. A contact is said to be *concordant* if it is parallel to the bedding or foliation (Fig. 16-4A). It is said to be *discordant* if it cuts across the bedding or foliation

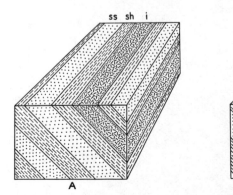
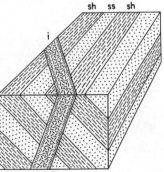

Fig. 16-4. Concordant and discordant intrusions: *ss*, sandstone, *sh*, shale, *i*, igneous rock of pluton. (A) Concordant pluton. (B) Discordant pluton.

Plate 30. *Granite intrusion.* Cliff about 3000 feet high. Devonian granite is white rock to right. Bedded rocks are quartzites and quartz schists of Lower Eleanore Group (Precambrian). Ardencaple Fjord, East Greenland. Photo: Laugh Koch Expedition.

(Fig. 16-4B). Similarly, a pluton is said to be concordant or discordant, depending on the attitude of its contacts.

In some instances a contact may be concordant in some places but discordant in others. Similarly, one contact may be entirely concordant, whereas the other may be entirely discordant. Nevertheless, a pluton may be primarily concordant or discordant, and the ensuing twofold classification of plutons is based mainly on this distinction.

If the older rock into which the pluton is intruded is massive, the terms concordant and discordant cannot be applied. But, as will become apparent below, this is not particularly important. The only instance concerns tabular bodies, which are called *dikes* if they intrude massive rocks.

SIZE AND SHAPE OF PLUTONS

The shape and size of small plutons may be observed directly. Many dikes and sills are smaller than the outcrops in which they occur. But the complete shape of the entire original body can never be determined. The part above the surface of the earth is forever lost to erosion. The part below the surface can be determined only by very expensive programs involving drilling and excavation.

If the pluton is larger than an outcrop, two vital pieces of information are necessary: (1) the ground plan and (2) the attitude of contacts.

Let us suppose that the body of igneous rock in Fig. 16-5 is too large to be observed in a single outcrop and that exposures cover only 20 percent of the area. Figure 16-5A shows that in plan the igneous body is long and narrow, trending in a northeasterly direction. The contacts are rarely exposed, but at *b* the southeast contact may be seen and dips 40°SE parallel to the bedding. Similarly, at *d* the northwesterly contact may be observed to dip 40°SE parallel to the bedding. Also at *a* and *c* the bedding dips 40°SE. Thus the pluton is a tabular concordant body dipping 40°SE, as shown in Fig. 16-5B; it is a sill.

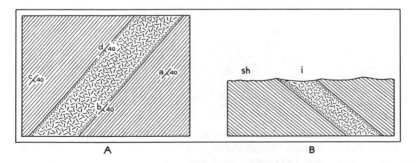

Fig. 16-5. Concordant pluton, the contacts of which are exposed: *sh*, shale; *i*, igneous rock of pluton. (A) Geological map; letters are mentioned in text. (B) Inferred cross section.

By a similar method of analysis the pluton in Fig. 16-6 is a tabular discordant body dipping 80°NW; it is a dike.

The method of analysis for larger plutons is much the same. Moreover, several indirect methods are available to ascertain the attitude of contacts that are not exposed. One of these methods is based upon topography. In Fig. 16-7A, the strata dip 40° to the southeast. The contacts of the pluton are not exposed, but the outcrops are good enough to permit location of the contact within a few feet. The contact bends downstream where it crosses a valley. If the altitudes of the southeastern contact at *c*, *d*, and *e* are known, the dip of the contact may be calculated by the three-point method (pp. 559–560). Similarly, the attitude of the northwestern contact may be determined if the altitude of points *f*, *g*, and *h* are known. If both contacts dip 40° to the southeast, the pluton is concordant.

Another method utilizes whatever primary foliation may be present in the igneous rock. In Fig. 16-8A, the bedding dips 40° to the southeast, but the actual contact of the pluton is nowhere exposed. The primary folia-

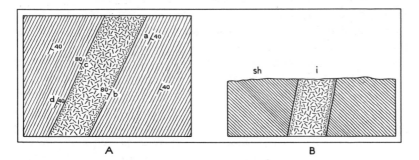

Fig. 16-6. Discordant pluton, the contacts of which are exposed: *sh*, shale; *i*, igneous rock of pluton. (A) Geological map; letters are mentioned in text. (B) Inferred cross section.

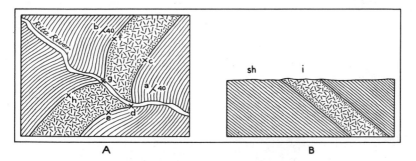

Fig. 16-7. Concordant pluton; attitude of contacts deduced from relations to topography: *sh*, shale; *i*, igneous rock of pluton. (A) Geological map; letters referred to in text. (B) Inferred cross section.

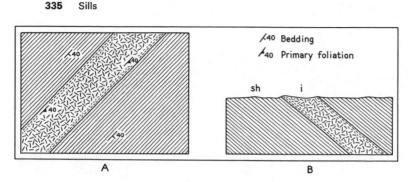

Fig. 16-8. Concordant pluton; attitude of contacts deduced from attitude of foliation : *sh*, shale ; *i*, igneous rock of pluton. (A) Geological map. (B) Inferred cross section.

tion near both contacts also dips 40° to the southeast. Observations in many parts of the world show that the primary foliation is commonly parallel to the adjacent contacts. This subject is discussed more fully on p. 373. It is inferred, therefore, that the contact in this case dips 40° to the southeast and that the pluton is concordant.

Geophysical methods may be used to determine the attitude of contacts. The principles are discussed in Chaps. 22 and 23.

Geophysical methods may also be utilized in a somewhat different way; that is, they may give a clue to the depth to which plutons extend. For example, many granites have a lower specific gravity than the rocks into which they are intruded. The magnitude of the negative gravity anomaly over the granite is an indication of the depth to which the "gravity contrast" extends. Magnetic and heat flow anomalies may indicate the depth to which the "contrast" extends. This type of evidence is discussed more fully in Chaps. 22 and 23.

Concordant Plutons

SILLS

Sills, sometimes called *sheets,* are tabular plutons that are parallel to the bedding or schistosity of the adjacent rocks (Figs. 16-5, 16-8 and Plate 31). The rock in the sill is younger than the rock on either side of it. Sills are horizontal, vertical, and inclined; the concept that a sill must be horizontal is quite erroneous. Sills are relatively thin compared to their extent parallel to the structure of the adjacent rocks.

Sills range in size from tiny sheets less than an inch thick to large bodies many hundreds of feet thick. The smaller sills can usually be traced for only a few feet or scores of feet, but the larger sills cover thousands of square miles.

A *simple sill* results from a single injection of magma (Fig. 16-9A). A

Plate 31. *Sill and discordant sheet.* The white rock is Carboniferous sandstone. A large black diabase sill is in lower part of
cliff. A large cross-cutting diabase sheet on right end of cliff
becomes concordant at top of mountain. Victoria Land, Antartica.
Photo: W. B. Hamilton, U. S. Geological Survey.

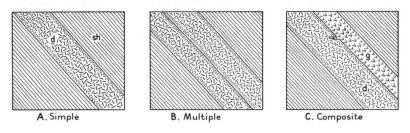

Fig. 16-9. Cross sections of sills: *sh*, shale; *d*, diorite; *g*, granite.
(A) Simple sill. (B) Multiple sill. (C) Composite sill.

multiple sill results from two or more injections of the same kind of magma
(Fig. 16-9B). Ideally, a simple sill is chilled only at the margins. A multiple sill,
although possessing chilled margins, also possesses a medial zone of finer-
grained rock, the position of which depends upon the relative thickness of
the individual sills. The greater the number of separate injections comprising
the multiple sill, the greater will be the number of chilled zones. If the injections followed one another in relatively rapid succession, the chilled zones
may be obscure. A *composite sill* results from two or more injections of

magma of differing composition. Figure 16-9C illustrates a sill formed by separate injections of diorite and granite. A small offshoot of the granite cuts the diorite, indicating that the granite is younger than the diorite.

Differentiated sills (Fig. 16-10) are of particular interest to petrographers because of their bearing on the problem of *magmatic differentiation*.[2] Such sills are generally hundreds of feet thick, and they are initiated by the injection of a horizontal sheet of magma. The magma cools slowly and, under the influence of gravity, separates into layers of different composition. As crystallization proceeds, any minerals that are heavier than the magma tend to sink, whereas crystals lighter than the magma tend to rise. Rising gases play a role, the importance of which is as yet undetermined. In the ideal case, such as is illustrated in Fig. 16-10, a differentiated sill will have at both the top and the bottom relatively thin layers, *a* and *d*, of fine-grained rock, representing the rapidly cooled original magma. Above the bottom layer are heavier rocks, *b*, which contain minerals of relatively high specific gravity. Still higher is a layer of lighter rocks, *c*, containing minerals with relatively low specific gravity. The combined chemical composition of *b* and *c* should equal the combined composition of *a* and *d*. In the simplest case, the contacts between the four layers are gradational.

Many complications may exist in detail in differentiated sills. For example, one would suppose that if any density stratification occurred within layer *b*, the layers would be progressively heavier downward. A detailed study of a similar body at Shonkin Sag, Montana (Fig. 16-11),[4] showed that the heaviest rocks were many feet above the bottom of layer *b*. Moreover, the contacts of layer *c* were relatively sharp rather than gradational; this was due to movement of the partially-liquid magma after differentiation took place.

To distinguish a differentiated sill from a composite sill may be difficult. (1) Obviously, a sill in which the lighter rocks are at the bottom cannot be due to differentiation in place under the influence of gravity. (2) Differentiation in place is implied if the various layers are gradational into one another. But this is by no means an infallible criterion because under some condi-

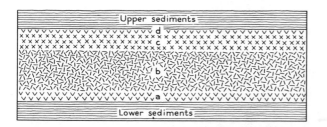

Fig. 16-10. Cross section of differentiated sill: *a* and *d*, rock with chemical composition of original magma; *b*, part of rock richer in dark minerals than *a* and *d*; *c*, part richer in light minerals than *a* and *d*.

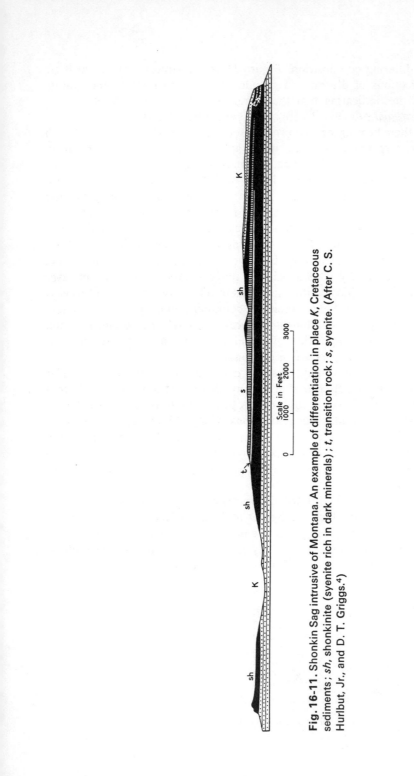

Fig. 16-11. Shonkin Sag intrusive of Montana. An example of differentiation in place *K*, Cretaceous sediments; *sh*, shonkinite (syenite rich in dark minerals); *t*, transition rock; *s*, syenite. (After C. S. Hurlbut, Jr., and D. T. Griggs.[4]

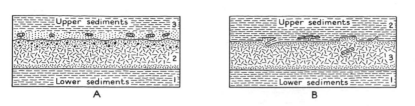

Fig. 16-12. Criteria for distinguishing sill from a flow. Diversely oriented dashes represent igneous rock; black dots are gas bubbles. (A) Lava flow. (B) Sill.

tions two intrusives of different ages show gradational rather than sharp contacts. A younger intrusive may react with an older rock to produce a transition zone that is many feet wide. (3) Although sharp contacts seem to imply successive injections, the liquid resulting from differentiation in place may move several scores of feet and produce sharp contacts.

It is quite possible, of course, for a differentiated sill to be deformed in a later orogenic episode and even to assume a vertical or overturned attitude.[5]

Lava flows as well as sills are tabular bodies of igneous rock that are parallel to the bedding of the overlying and underlying formations. It is essential therefore, to establish criteria to distingush the two. These criteria are based primarily on the fact that sills are younger than the rocks above and below, whereas lava flows, although younger than the underlying rocks, are older than those above.

The essential differences between flows and sills are illustrated by Fig. 16-12. A sill (Fig. 16-12B) is characterized by a relatively smooth, fine-grained top, in which vesicles (gas bubbles) are rare. In particular, the overlying rocks may be penetrated by apophyses of the sill, and they may occur as inclusions in the sill.

A lava flow (Fig. 16-12A), on the other hand, has a rolling, vesicular top. In particular, fragments of the lava may occur in the overlying rocks. Lava flows are likely to be cut by irregular dikes of sedimentary or pyroclastic material, representing debris washed into cracks in the lava.

LACCOLITHS

A *laccolith* is an intrusive body that has domed up the strata into which it has been inserted (Fig. 16-13). The type locality is in the Henry Mountains of Utah. Originally studied nearly 100 years ago[6], they have been restudied more recently.[7] These laccoliths, composed mostly of diorite porphyry, are 2 to 4 miles in diameter and usually a few thousand feet thick at maximum.

It is apparent that all transitions may exist between laccoliths and sills. In a typical laccolith, the diameter is only a few times greater than the thickness. In a typical sill, the diameter is many times the thickness. The body

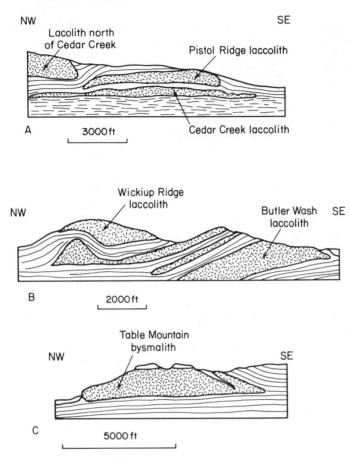

Fig. 16-13. Laccoliths in Henry Mountains, Utah. (After U. S. Geological Survey, Hunt et al.[7])

may be termed a *laccolith* if the ratio of the diameter to the thickness is less than ten; if the ratio is greater than ten, the body should be called a *sill*. This is an arbitrary figure, but in all geological classifications continous series must be artificially assigned to compartments.

Most laccoliths, especially in the western United States, are found in regions where the sedimentary rocks are essentially horizontal, except around the laccolith. Figure 16-14A is a map of an ideal laccolith. A central core of igneous rock is surrounded by outwardly dipping sedimentary rocks. The igneous rock is younger than the overlying sediments and the contacts are concordant. But in this instance the presence of a horizontal floor is based on inference.

But as shown in Fig. 16-13A the contacts are not necessarily everywhere concordant. The roof, instead of bending upward, is uplifted along faults.

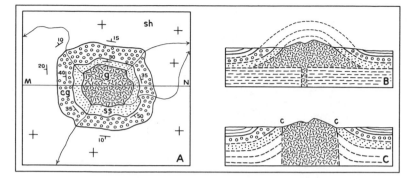

Fig. 16-14. Intrusive rock in the center of domed-up sediments:
g, granite porphyry; *ss*, sandstone; *cg*, conglomerate; *sh*, shale;
c, contact of intrusive. The granite porphyry intrudes the sandstone.
(A) Geological map. (B) Interpreted as a laccolith. (C) Interpreted as a "bottomless" stock.

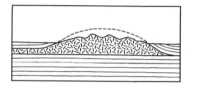

Fig. 16-15. Laccolith intruded from the side.

The fault becomes the outer contact which is thus an intrusive contact rather than a fault contact.

The nature of the conduit way through which magma enters a laccolith is problematical. In Fig. 16-14B a small, vertical pipe is shown beneath the center of the laccolith. Originally[6] it was concluded that the magma spread laterally in all directions from such a feeder. Restudy of the Henry Mountains[7] has shown that many of the laccoliths were fed from the side (Fig. 16-15). As shown by Fig. 16-16 the magma rose from depths through a centrally located cross-cutting stock. On reaching favorable stratigraphic horizons, the magma moved out laterally to form the laccoliths.

BYSMALITHS

A *bysmalith* is a body of which the roof was uplifted along a circular or arcuate fault. Such intrusions have been described from the Black Hills of South Dakota. Some of the bodies in the Henry Mountains are bysmaliths. Obviously some bodies may combine features of both laccoliths and bysmaliths, that is, the roof may be bent up in places, faulted up in others.

The Skaergaard intrusion in Greenland[3] is remarkable for the layering it displays, both on the scale of hundreds or thousands of feet and of inches

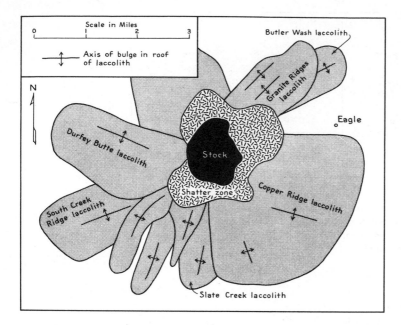

Fig. 16-16. Laccoliths fed from central stock. General plan of intrusion around Mt. Ellen stock, Henry Mountains, Utah. (After C. B. Hunt et al.[7])

(Fig. 16-17). The rocks are chiefly various kinds of gabbro, composed of plagioclase, pyroxene, and olivine. The layering, which is inferred to have been originally nearly horizontal, is due to differential settling of minerals under the influence of gravity. Because the contacts converge downward, it has been assumed that the body is shaped like an inverted cone. On the other hand, the attitude of the layering suggests that the floor is esentially flat. The intrusive was injected at the unconformity between the Precambrian gneisses and the overlying Tertiary rocks. The roof moved relatively upward and the floor moved downward along a circular fault, all trace of which has been eliminated by the intrusion.

Whether a laccolith or a sill develops when magma spreads along horizontal bedding planes depends upon several factors, of which the most important is the viscosity of the magma. Low viscosity will cause thin, widespread intrusions such as sills; high viscosity, because it prevents lateral spreading of the magma, causes laccoliths.

The roof of the laccolith is lifted by the hydrostatic pressure of the magma. The overlying sedimentary rocks, compressed by the force of the magma, are elongated parallel to the bedding. The amount of the elongation depends, of course, upon the shape of the laccolith, but in some cases it is as great as

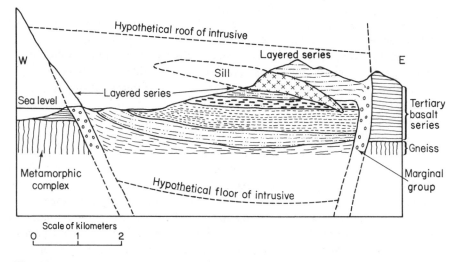

Fig. 16-17. Skaergaard Intrusive Body, Greenland. (Based on Wager and Brown[3], with permission of Oliver and Boyd, Ltd., and G. M. Brown.)

10 percent and even more. Similarly, the thickness of the overlying sedimentary rocks is reduced. Tension cracks may form because of this "stretching" of the roof, and the roofs of some experimental laccoliths are broken by radiating and concentric cracks. This is partly because the roof rocks have generally been destroyed by erosion; but even where the roof is preserved, such fractures are rare—apparently because the rocks were sufficiently plastic to yield without rupture.

LOPOLITHS

A *lopolith* has been defined as a large concordant intrusion in a structural basin (Fig. 16-18). It is essentially a sill, but differs in that it is very large and in that it is associated with a basin. In plan, lopoliths are arcuate or circular.

The type example is the Duluth lopolith[8,9] in Minnesota (Fig. 16-19). In plan this is a great arcuate body nearly 200 miles long and, at maximum, 25 miles wide. A large sheet or lens dipping southeasterly on the northwest flanks of the Lake Superior Basin, it is as much as 5 miles thick. Two major units have been distinguished in Fig. 16-19, the Duluth gabbro complex and the granophyre ("red rock"). The Duluth gabbro complex consists of multiple intrusions, which, although the average composition is gabbro, show notable differences in the percentages of the minerals. The rocks consist chiefly of olivine, plagioclase, pyroxene, and some potash feldspar. Lack of chilled contacts between the various units indicate that they were closely associated in time, that is, one unit had not cooled off before the next was emplaced. Some of the rocks are layered, due to differential crystal settling. The granophyre was originally considered a differentiate of the gabbro, but is now considered a separate intrusion. It could be a differentiate that has moved somewhat. Or the differentiation may have occurred at depth. The association of the intrusion with a basin may be due to the withdrawal of magma from an underlying reservoir.

The Bushveld igneous complex[10,11,12] is one of the most spectacular igneous bodies in the world, because of both the petrology and the mineral resources. Extending 270 miles in an east-west direction and 150 miles in a north-south direction, it underlies 24,000 square miles. It is 1950 ± 50 million years old. The complex is essentially a huge composite sheet (Fig. 16-20). The underlying rocks belong to the "older granite" and sedimentary rocks of the Transvaal system. The roof rocks consist of the upper part of the Transvaal system and the Roiberg felsite, which is not separately distingusihed from the Transvaal system in Fig. 16-20.The Karroo system is younger than the emplacement of the Bushveld complex and rests unconformably on all the other rocks.

The intrusive body is essentially a huge composite lopolith, composed in turn of two major units. An upper lopolith is composed of granite. The lower lopolith consists of a great galaxy of rocks called norite in Fig. 16-20. This unit actually consists of layered mafic and ultramafic rocks, including norite, pyroxenite, harzburgite, gabbro, anorthosite, olivine diorite, and granodiorite; there are also layers of chromitite, magnetite, and platinum. The great diversity of rocks is due in part to successive intrusions, in part to gravitational settling of minerals. Each of the two lopoliths is several miles thick.

Some geologists believe that there is evidence that the complex was emplaced from four centers of eruption and that the basal contact is discordant. Partly for these reasons it has been suggested that the Bushveld is not a lopolith. This is partly a matter of definition. The contacts of sills may be locally discordant. The number of centers of emplacement is irrelevant. The gentle inward dip of the layering, including the magnetite, chromitite, and platinum layers, invalidates the suggestion that the body really consists of a series of funnel-shaped intrusions.

At Sudbury, Ontario, a large intrusion with associated nickel ores has been interpreted by some as a lopolith, by others as a cone sheet or ring-dike.

PHACOLITHS

Phacoliths[13] are concordant intrusives confined to the crests of anticlines (Fig. 16-21B) or to the troughs of synclines. Phacoliths are not only crescentic in cross section, but also in plan because they are commonly associated

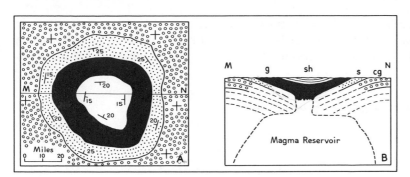

Fig. 16-18. Lopolith: *cg*, conglomerate; *s*, sandstone; *sh*, shale; *g*, gabbro of lopolith. Usual dip-strike symbols; +, flat strata. (A) Geological map. (B) Structure section.

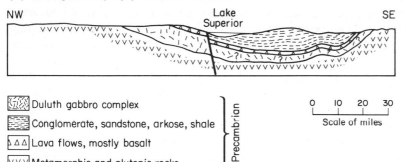

Fig. 16-19. Cross section of Duluth lopolith. (After Leith, Lund, and Leith.)

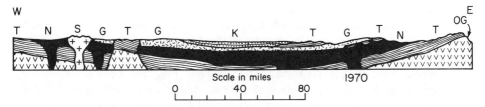

Fig. 16-20. Bushveld igneous complex of South Africa: *OG*, older granite; *T*, Transvaal system; *K*, Karroo system; *N*, norite of the lower lopolith; *G*, newer granite of the upper lopolith; *S*, younger syenites, unrelated to lopolith. (After A. L. Hall,[10] section by Willemse[11] is essentially similar.)

with plunging folds. Figure 16-21A is a geological map of a northerly-plunging anticline intruded by a phacolith, shown in solid black. The thickness of phacoliths is measured in hundreds or, at the most, several thousands of feet. In plan, phacoliths are seldom more than a few thousand feet long, as measured around the crescent. Whereas sills, laccoliths, and lopoliths characteristically force their way into place, some geologists believe that phacoliths are passively intruded, the magma filling potential cavities that form during the folding. In an anticline, for example, the upper beds may pull away from the lower beds. In seems probable, however, that the magma must be under some pressure and that it actually makes space for itself.

OTHER CONCORDANT PLUTONS

Large, more or less concordant plutons are integral parts of some orogenic belts.

In western New Hampshire a large sheet of orthogneiss, the Mt. Clough pluton, is 100 miles long and several thousand feet thick (*bg* of Fig. 16-22).[14] It is probable that this body was intruded as a horizontal sheet and subsequently folded.

Mantled gneiss domes, as commonly defined, are restricted to areas of regional metamorphism in which gneissic granite in the core of a doubly plunging anticline is overlain by a mantle of metamorphosed sedimentary and volcanic rocks with a distinctive stratigraphy. The type locality is in Finland,[15] where a paradox seemed to exist (Fig. 16-23). A domical granite gneiss is overlain by a conglomerate that contains granite pebbles; but the conglomerate is also intruded by what seemed to be the same or at least a similar granite. A sequence of sedimentary and volcanic rocks (Fig. 16-23A) was folded and intruded by granodiorite or quartz diorite (Fig. 16-23B). After erosion, a new series of sedimentary rocks was deposited (Fig. 16-23C). During a second period of deformation the granodiorite was partially transformed metasomatically into a granite, containing more potash feldspar than the original rock, and rose to form a dome. Some of the granite became sufficiently mobile to intrude the overlying sedimentary rocks.

Mantled gneiss domes are common in New England in a north-south belt 250 miles long.[16,17] The mantle of the domes in western New Hampshire consists of metamorphosed Silurian and Devonian strata, mostly quartzite, marble, lime-silicate rocks, and schists (See Fig. 16-22). Unconformably beneath these are Ordovician rocks, composed of two major units: (1) mica schist, biotite gneiss, and amphibolite; and (2) granite gneiss younger than the mica schist, biotite gneiss, and amphibolite. The foliation in the domes is related to the doming. Two factors were involved in the rise of the domes: (1) east-west horizontal compression, and (2) rise of less dense solid granite.

Irregular cylindrical bodies are not uncommon in many granitic areas.[18] Such a body is illustrated in Fig. 16-24. In plan the granodiorite of the intrusion is oval shape, except for a pronounced dimple on the northwest margin. The metamorphosed sedimentary and volcanic rocks wrap around the intrusion concordantly with the contacts, which appear to dip very steeply. The granodiorite was emplaced partly by pushing the country rocks aside and

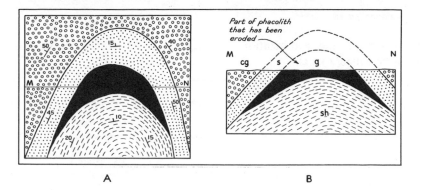

A   B

Fig. 16-21. Phacolith: *sh*, shale; *s*, sandstone; *cg*, conglomerate;
g, granite. (A) Map of granite phacolith in a northerly-plunging
anticline. (B) Cross section of the same phacolith.

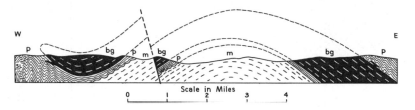

Fig. 16-22. Large concordant pluton, Mascoma quadrangle, New
Hampshire. *P*, Paleozoic schists; *m*, granitic rocks of Mascoma
group; *bg*, Bethlehem gneiss, which forms a large concordant
pluton. (After C. A. Chapman.[14])

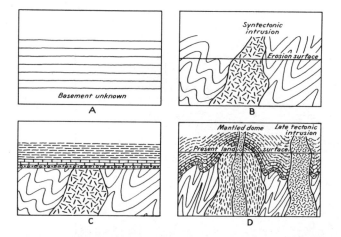

Fig. 16-23. Origin of mantled gneiss domes. (A) First sedimentation. (B) First orogeny. (C) Second sedimentation. (D) Second
orogeny. (After P. Eskola.[15])

partly by stoping and granitization (see page 381). As shown below, this
body can be called a stock.

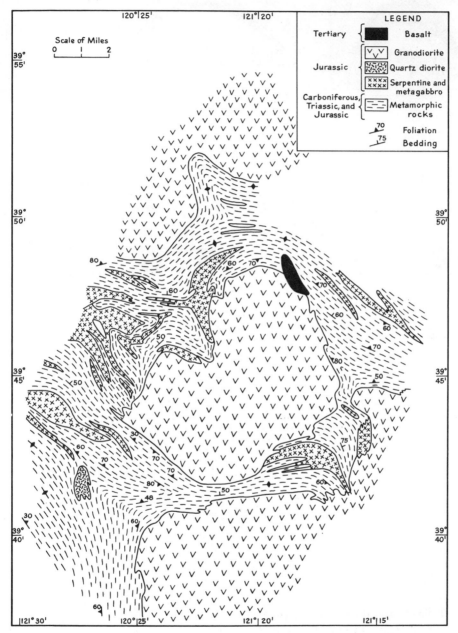

Fig. 16-24. Irregular cylindrical intrusion, composed of granodiorite. Geological map of Merrimac area, California. (After A.
Hietanen.[18])

FUNNELS

In some bodies, especially those composed of mafic and ultramafic rocks, the foliation or compositional layering may dip inward to form a funnel. The Cortlandt complex in New York[19] is such a body (Fig. 16-25). The rocks are chiefly peridotite, pyroxenite, norite, and diorite. Three funnels are present. In the eastern and northwestern funnels the foliation dips inward at 20° to 70°; the southwesterly funnel is not as well marked, but near the center the dips are 40° to 45° inward. The attitude of the foliation in the surrounding gneisses and schist conform to the foliation in the intrusive and here the contacts are believed to dip inward like the foliation.

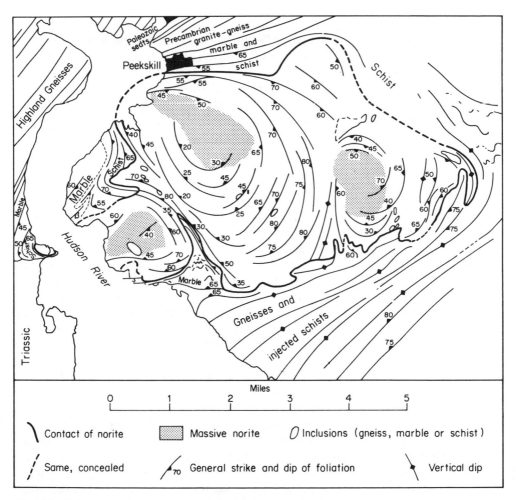

Fig. 16-25. Funnel structure shown by foliation in norite, Cortlandt complex, New York. (After Balk and Shand.[19])

Presumably each funnel occupies a small chimney at depth. The body expanded laterally as it rose, occupying a progressively bigger chamber. The plastic wall rocks were pushed aside, the foliation conforming to the walls of the intrusion. The primary foliation in the intrusive has resulted from lamellar flow.

Discordant Plutons

DIKES

Dikes (Plate 32) are tabular bodies of igneous rock that cut across the structure of the older formations (Fig. 16-4B). Tabular bodies of intrusive rock cutting massive, structureless rocks are also called dikes. Most dikes are formed by the injection of magma into a fracture; the walls may be pushed apart by the pressure exerted by the intruding magma, or the magma may quietly well up into fractures opened by tensional forces.

The terms *simple, multiple, composite,* and *differentiated* may be applied to dikes in the same sense that they are used for sills. A *simple dike* is the result of a single intrusion of magma. A *multiple dike* is the result of two or more intrusions of the same kind of magma. A *composite dike* is the product of the intrusion of two or more kinds of magma. A *differentiated dike* is one that was intruded as a homogeneous magma, but from which two or more varieties of rock have formed in situ.

Dikes may be very small, and dikes a fraction of an inch wide and a few inches long may be associated with larger igneous bodies. Most dikes are 1 to 20 feet wide, but wider and narrower dikes are not uncommon. The distance for which a dike may be followed depends in part upon the nature of the exposures. In Iceland, dikes 10 miles long are common, and some are 30 miles long; at least one is 65 miles long. In England, the Cleveland dike is 110, and perhaps 190, miles long. The Medford dike near Boston, Massachusetts, is 500 feet wide in places.

A *dike set* consists of parallel dikes; where the dikes are very abundant, the term *dike swarm* may be applied (Fig. 16-26).[20] In some areas several sets may be present, and each set may be characterized by its own peculiar petrography, indicating that the various sets are of different ages.

The fractures occupied by dikes may originate in many ways. Some are older fractures, such as joints or faults that greatly antedate the intrusion. If the hydrostatic pressure of the magma is greater than the lithostatic pressure, the walls will be pushed apart elastically or plastically. On the other hand, the older fracture may be opened up by tensional forecs, in which case

Plate 32. *Dike.* Diabase dike about two feet wide. Intruded into metamorphosed rhyolitic tuffs. Rye, New Hampshire. Photo: J. Haller.

the magma quietly wells up into the opening. In many cases the fracture is produced by the magma that fills the dike. The stresses at the edge of a wedge of magma are very great; consequently the fracture may propagate itself very rapidly. Magma in a reservoir may exert sufficiently strong pressure on its walls to form extension fractures (page 157), which are promptly occupied by the magma. Dike swarms are related to regional stresses. Anderson[29]

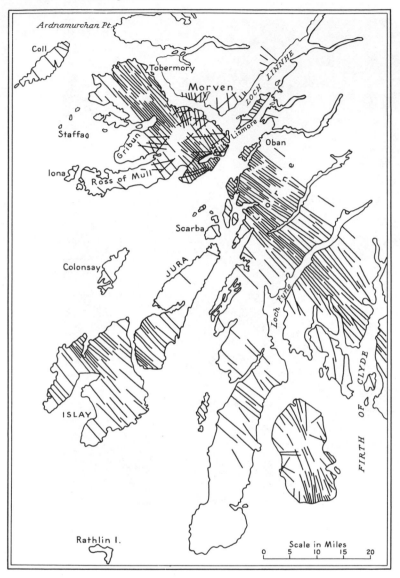

Fig. 16-26. Dike swarm. Tertiary dikes of the Southwest Highlands of Scotland. The map is diagrammatic in the sense that: (a)
each line represents 10 to 15 dikes; (b) each dike is only a few feet
wide, and not as wide as the scale implies; and (c) individual
dikes canot be traced for the long distances that the map implies.
(After J. E. Richey,[23] with permission of the Controller of Her
Britannic Majesty's Stationery Office.)

believes that such dikes occupy fractures that form perpendicular to the
least principal stress axis when the stress differences exceed a critical value.

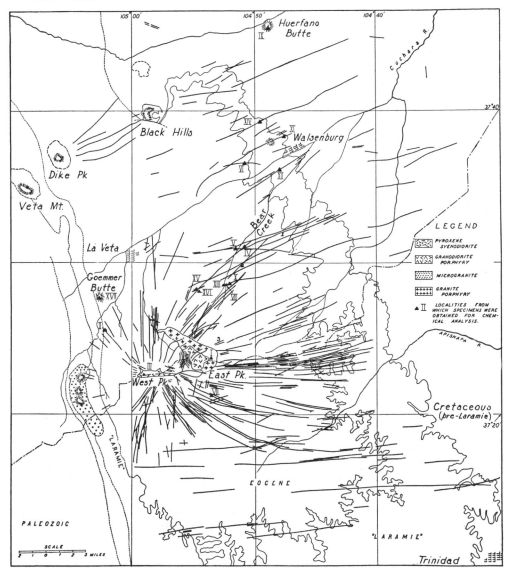

Fig. 16-27. Radiating dikes around Spanish Peaks, Colorado. (After Knopf.[21])

Radiating dikes are found around volcanic centers. The dikes around the Spanish Peaks in Colorado[21,22] are a spectacular example (Fig. 16-27). Three hundred dikes have been mapped and there may be as many as 500. Dips less than 80° are rare. Thickness ranges from 2 to 60 feet. Individual dikes may be followed for many miles and in places the dikes stand up as walls.

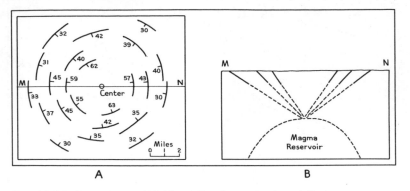

Fig. 16-28. Cone sheets. (A) Map: direction and value of dip is indicated. (B) Cross section along line *MN*.

The original fractures, now filled by the di es, are tensional in origin. The magma in the central stocks exerted an outward pressure to produce extension fractures. In addition, there was contemporaneous east-west compression, which explains the preponderance of east-west dikes.

Cone sheets are dikes that belong to a concentric, inward-dipping system (Fig. 16-28). None of the cone sheets extend all the way around the central area; in fact, a cone sheet can be followed for only a few miles at the most. The average dip is 45° but the outer cone sheets tend to dip more gently than the inner. If the Scottish cone sheets,[23] which are Tertiary in age, are projected downward, they meet at a focus approximately 3 miles beneath the present surface (Fig. 16-28B). Cone sheets, typically a few feet thick, may attain a thickness of 40 feet. In Mull, an island off the west coast of Scotland, they are concentric about two distinct centers. The cone sheets of the older center are cut by those associated with the younger center. The origin of cone sheets is discussed on page 359.

Apophyses (singular, *apophysis*) and *tongues* are dikes (in some instances rather irregular) that obviously have been derived from a nearby igneous body.

VOLCANIC VENTS

Volcanic vents, also called *necks* and *plugs*, are the roots of volcanoes that have been eroded away. They are circular, subcircular, or even irregular in plan, and are a few score feet to a mile in diameter. Some composite vents are even larger and have resulted from several successive eruptions. The contacts of volcanic vents with the surrounding country rock are typically steep; they are either vertical or inward-dipping, rarely outward-dipping.

Although it is unusually large, the vent at Cripple Creek, Colorado (Fig. 16-29) shows many of the features characteristic of vents.[24] The contacts

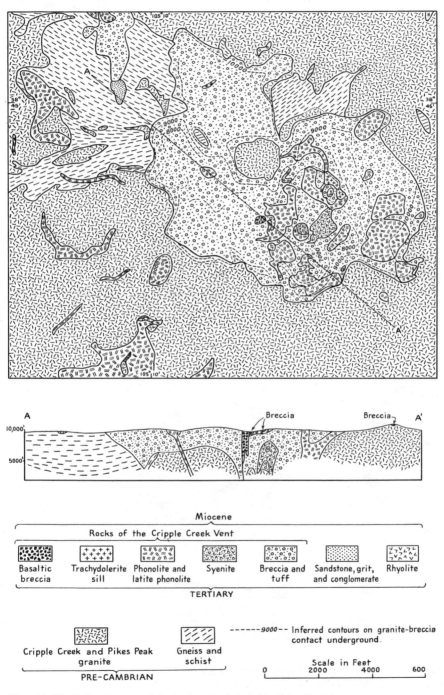

Fig. 16-29. Volcanic vent. Cripple Creek, Colorado. (After T. S. Lovering and E. N. Goddard.[24])

are discordant, and the foliation of the Precambrian schists is cut off by the material in the vent. The contacts generally slope inward. In this case the vent is composed of several separate funnels that narrow downward; this can be seen on the map by studying the structure contours that show the granite-breccia contact and by studying the structure section. Much of the rock in the vent is tuff and breccia. Lava is represented by rhyolite and phonolite. Medium-grained intrusive rocks are represented by syenite. The Precambrian rocks were blown out in series of explosions, but the voids were promptly filled by tuff and breccia. Magma pushing up through the breccia consolidated to form the rhyolite, phonolite, and syenite. In other areas some vents are filled exclusively by tuff and breccia, others are filled exclusively by lava.

RING-DIKES

Ring-dikes are oval or arcuate in plan (Fig. 16-30); the contacts are steep, either vertical or very steep. The average diameter of ring-dikes is four and a half miles, but some have a diameter of only 1000 feet, whereas others are 15 miles in diameter. The average width of ring-dikes is 1600 feet; the maximum is 14,000 feet. Down-faulted volcanics are found inside some ring-dikes; they were erupted during the same igneous cycle as the ring-dike. The classic

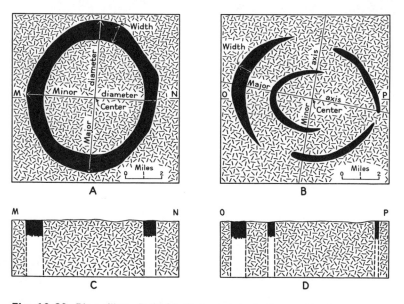

Fig. 16-30. Ring-dikes. Solid black represents igneous rock of ring-dikes. (A) Circular (complete) ring-dke. (B) Arcuate (incomplete) ring-dikes. (C) Structure section along line *MN* of diagram (A). (D) Structure section along line *OP* of diagram (B).

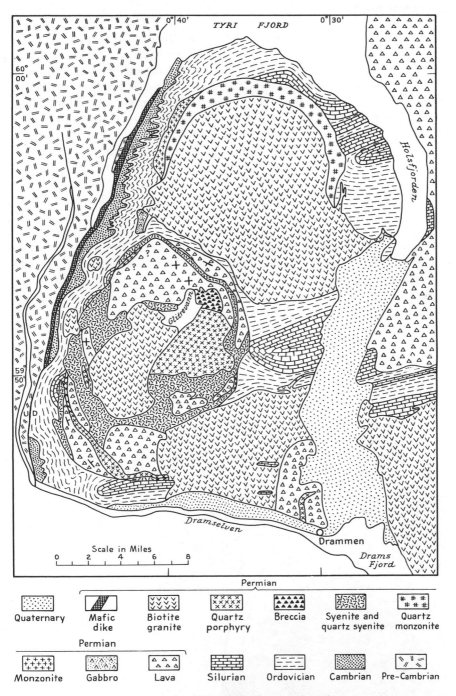

Fig. 16-31. Ring-dikes and stocks. Oslo region, Norway. Longitude is measured west from Oslo. Dip-strike symbols show attitude of volcanic rocks. *D*, downthrown side of normal fault. (Based on maps by W. C. Brögger, J. Scheltig, and C. Oftedahl.[26])

area for ring-dikes is the Tertiary Volcanic District of Scotland.[23] Some other areas are New Hampshire,[25] Norway,[26] Nigeria,[27] and Australia.[28]

Figure 16-31 is a map of a region of ring-dikes a short distance west of Oslo, Norway. The ring-dikes and stocks (see page 362) are grouped about two centers, one in the north-central part of the map, the other in the southwestern part of the map. The one in the southwestern part of the map is known as the Glitrevann cauldron (see page 360). Outside the igneous complexes are the folded Paleozoic sedimentary rocks of Cambrian, Ordovician, and Silurian age. The manner in which the folds are truncated at the igneous contact is strikingly shown, especially in the northeast corner of the map. Inside the Glitrevann cauldron are several large areas of subsided volcanics, which are of Permian age and rest unconformably on the older Paleozoic rocks. On the south and southwest sides of the cauldron a discontinuous ring-dike of quartz syenite forms the boundary between the cauldron and the older Paleozoic rocks. Also on the southwest side a second ring-dike, composed of quartz porphyry, lies chiefly within the volcanics and is nearly 10 kilometers long. In the northern part of the cauldron is another ring-dike of quartz syenite, in places forming the boundary between the volcanics and the Ordovician rocks, but in many places lying within the volcanics. This ring-dike extends almost all the way around the cauldron, although it is rather irregular in the southern part of the area. Large bodies of quartz porphyry and coarse biotite granite occupy the center of the cauldron. A body of coarse biotite granite also cuts out the southeast part of the cauldron.

The body in the north-central part of the area consists of an outer ring-dike of monzonite, an inner ring-dike of quartz monzonite, and a large central stock of biotite granite.

Anderson's theory for the formation of the fractures occupied by ring dikes and cone sheets is illustrated in Fig. 16-32.[29] Let us first assume that

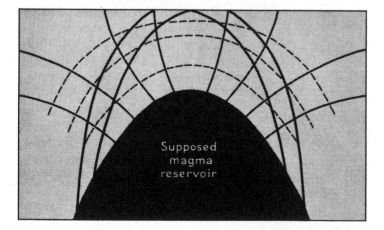

Fig. 16-32. Origin of ring-dikes. (After E. M. Anderson;[29] see text for explanation.)

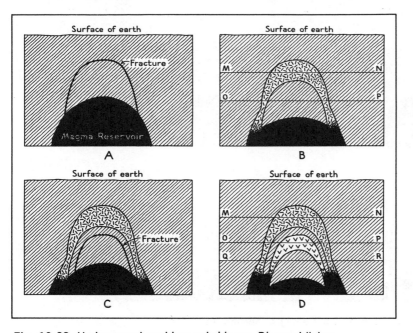

Fig. 16-33. Underground cauldron subsidence. Diagonal lining represents older country rock; diversely oriented short dashes represent one intrusion; checks represent a second intrusion. (A), (B), (C), and (D) represent successive stages of intrusion. *MN*, *OP*, and *QR* are a few of the many levels to which erosion may cut.

the hydrostatic pressure of the magma exceeds the lithostatic pressure of the surrounding rocks. The magma will be pushing on its walls. In vertical sections the trajectories of the greatest principal stress axes will radiate out from the magma reservoir; they are the lighter solid lines of Fig. 16-32. The trajectories of the least principal stress axes are shown by the broken lines. Tension fractures will form parallel to the lighter solid lines. These are the fractures that control the emplacement of cone sheets.

If the hydrostatic pressure of the magma is less than the lithostatic pressure in the surrounding rocks, the latter will push on the magma. The lighter solid lines are the trajectories of the least principal stress axes, and the broken lines are trajectories of the greatest principal stress axes. Tension cracks would form parallel to the broken lines. Two sets of shear fractures might form (page 154), making angles of 30° with these broken lines. One set of these shear fractures is shown in heavy solid lines. Anderson believes that these are the fractures that are most commonly utilized in the emplacement of ring-dikes.

When one of these fractures or sets of fractures becomes large enough, a block of country rock is isolated from the rest of the roof. It sinks because it is heavier than the magma in the reservoir (Fig. 16-33). While the central

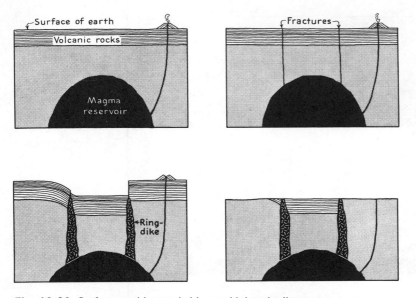

Fig. 16-34. Surface cauldron subsidence. Light stippling represents older country rock. Diversely oriented short white dashes represent intrusive rock of a ring-dike. Upper left: after eruption of volcanics. Upper right: fracture forms. Lower left: subsidence accompanied by some volcanism; drag that commonly occurs near ring-dikes has been omitted. Lower right: erosion to present topography.

block subsides, magma rises from the reservoir to fill the potential void left between the stationary walls and the sinking block. If erosion subsequently cuts deeply enough, to a level such as *OP*, a circular or oval intrusion with steeply dipping contacts is exposed. If the central block subsides several times, a number of concentric ring-dikes will form. A remnant of the older country rock left between two ring-dikes is called a screen.

In New Hampshire the contacts of many of the ring-dikes are essentially vertical.[25] It seems unlikely that they could have formed in the manner outlined above. An alternate hypothesis is that the magma (Fig. 16-34) pushes on its roof like a punch. Vertical fractures form above the reservoir. Once the overlying block is isolated by fractures it sinks into the reservoir because it is heavier than the magma. The space occupied by the ring-dike is made largely by piecemeal stoping (page 381) along the fracture.

The mechanism by which ring-dikes form is called *cauldron subsidence*. *Surface cauldron subsidence* results when the fractures extend all the way to the surface (Fig. 16-34). *Underground cauldron subsidence* results when the subsiding block does not reach the surface (Fig. 16-33).

Batholiths and Stocks

The term batholith literally means "deep rock," originally based on the assumption that coarse-grained igneous rocks, in contrast to volcanic rocks, dikes, sills, and other small bodies, congealed at relatively great depths. But for many decades controversies raged about the shape and mechanics of emplacement of batholiths. This was largely because bodies of many different shapes and origins were called batholiths, but each investigator assumed that all so-called batholiths were like the few he had studied. For this reason the term "pluton" was proposed.[1] Nevertheless, the term batho-

Plate 33. *Igneous contact.* Tertiary stock (light-colored) intrudes hornfels. Alaska. Photo: Bruce Reed.

lith is still used, generally for bodies with the following characteristics. (1) The rocks are plutonic, but most batholiths are composite, composed of several lithologic and textural varieties that were emplaced at slightly different times. (2) The term is arbitrarily restricted to bodies occupying an area of more than 40 square miles (100 square kilometers). (3) They commonly enlarge downward or at least do not get smaller (Fig. 16-35). (4) There is no visible floor; geophysical evidence indicates that many batholiths have

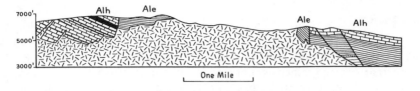

Fig. 16-35. Structure section of Marysville stock, Montana.
Diversely oriented dashes are quartz diorite. *Ale*, Empire shale;
Alh, Helena limestone. (After J. Barrell.)

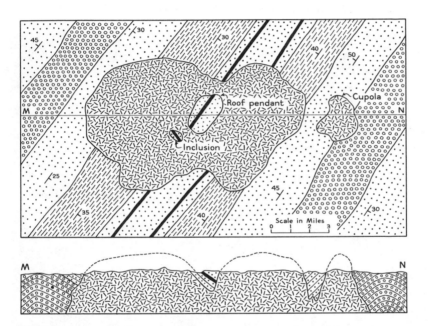

Fig. 16-36. Batholith. Circles, dots, parallel dashes, and solid
black are sedimentary rocks. Diversely oriented dashes are granite.
Map is above, structure section is below.

floors, but at depths of several miles.[30] (5) The roof may be irregular; *roof
pendants* are downward projections of the roof (Fig. 16-36). *Septa* are parts
of the roof that lie between two different intrusions forming the batholith.
The distinction between roof pendants and septa may be subjective,
depending on the geologist's concept of the history of the body. *Cupolas*
are isolated plutonic bodies that presumably connect downward with the
main batholith (Fig. 16-36). The term batholith is also often used for large,
more or less concordant bodies of plutonic rocks, in some cases with rather
diffuse contacts. *Stocks* (Plate 33) are similar to batholiths, but the surface
area is arbitrarily fixed at less than 40 square miles.

Radiogenic methods of dating have been applied very successfully in unravelling the history of batholiths. The Sierra Nevada and adjacent ranges in California and Nevada may be cited.[31] Five intrusive epochs occurred between Middle Triassic and Late Cretaceous. Each epoch lasted 10 to 15 million years, and took place at intervals of about 30 million years. These intrusive epochs were associated with regional deformation.

References

[1] Cloos, Hans, 1928, Zur Terminologie der Plutone, *Fennia*, vol. 50, no. 2, pp. 1–10.

[2] Turner, Francis J., and Verhoogen, John, 1960, *Igneous and metamorphic petrology*, 2d ed., 694 pp, New York: McGraw-Hill Book Co., Inc.

[3] Wager, L. R., and Brown, G. M., 1967, *Layered igneous rocks*, 588 pp., San Francisco: W. H. Freeman & Co.

[4] Hurlbut, C. S., Jr., and Griggs, D. T., 1939, Igneous rocks of the Highwood Mountains, Montana, pt. 1, the laccoliths, *Geol. Soc. Amer. Bull.* 50: 1043–1111.

[5] Hess, H. H., 1960, *Stillwater igneous complex, Montana*, Geol. Soc. Amer., Mem. 80, 230 pp.

[6] Gilbert, G. K., 1877, *Report on the geology of the Henry Mountains*, U. S. Geogr. and Geol. Surv., Rocky Mtn. Region, 160 pp.

[7] Hunt, Charles B., Averitt, Paul and Miller, Ralph L., 1953, *Geology and geography of the Henry Mountains region, Utah*, U. S. Geol. Surv., Prof. Paper 228, 234 pp.

[8] Grout, F. F., 1918, The lopolith, an igneous form exemplified by the Duluth Gabbro, *Amer. Sci.*, 4th series, 46: 516–22.

[9] Leith, C. K., Lund, Richard J., and Leith, Andrew, 1935, *Pre-Cambrian rocks of the Lake Superior region*, U. S. Geol. Surv., Prof. Paper 184, 34 pp.

[10] Hall, A. L., 1932, *The Bushveld igneous complex of the Central Transvaal*, Geol. Surv. South Africa, Mem. 28.

[11] Willemse, J., 1959, *The floor of the Bushveld igneous complex and its relationships, with special reference to the Eastern Transvaal*, Proc. Geol. Soc. South Africa, 62: 21–80.

[12] Willemse, J., 1964, A brief outline of the geology of the Bushveld igneous complex, in *The geology of some ore deposits of Southern Africa*, Geol. Soc. South Africa, 2: 91–128.

[13] Harker, A., 1909, *The natural history of igneous rocks*, New York: The Macmillan Company, pp. 77–78.

[14] Chapman, C. A., 1939, Geology of the Mascoma Quadrangle, New Hampshire, *Geol. Soc. Amer. Bull.* 50: 127–80.

[15] Eskola, P., 1949, The problem of mantled gneiss domes, *Quart. J. Geol. Soc. London*, 104: 461–76.

16 Thompson, J. B., Jr., et al., 1968, Nappes and gneiss domes in west-central New England, in E-an Zen et al., *Studies of Appalachian geology, northern and maritime*, New York: Interscience Publishers, pp. 203–18.

17 Naylor, Richard S., 1969, Age and origin of the Oliverian Domes, west-central New Hampshire, *Geol. Soc. Amer. Bull.* 80: 405–28.

18 Hietanen, Anna, 1951, Metamorphic and igneous rocks of the Merrimac area, Plumas National Forest, California, *Geol. Soc. Amer. Bull.* 62: 565–608.

19 Shand, S. James, 1942, Phase petrology of the Cortlandt Complex, *Geol. Soc. Amer. Bull.* 53: 409–28.

20 Richey, J. E., 1934, *Guide to the geological model of Ardnamurchan*, Geol. Surv. Scotland, Memoirs.

21 Knopf, Adolph, 1936, Igneous geology of the Spanish Peaks region, Colorado, *Geol. Soc. Amer. Bull.* 47: 1727–84.

22 Odé, Helmer, 1958, Mechanical analysis of the dike pattern of the Spanish Peaks area, Colorado, *Geol. Soc. Amer. Bull.* 68: 567–75.

23 Richey, J. E. (revised by A. G. MacGregor and F. M. Anderson), 1961, *Scotland: The Tertiary volcanic districts*, 3rd ed., Brit. Regional Geol.

24 Lovering, T. S., and Goddard, E. N., 1950, *Geology and ore deposits of the Front Range, Colorado*, U. S. Geol. Surv., Prof. Paper 223, pp. 292–98.

25 Billings, Marland P., 1956, *Geology of New Hampshire, part 2, Bedrock geology*, New Hampshire State Planning and Development Commission, 203 pp.

26 Holtedahl, Olaf, ed., 1960, *Geology of Norway*, Norges Geologiske Undersökelse, Nr. 208, 540 pp.

27 Jacobson, Reginald R. E., Macleod, William Norman, and Black, Russell, 1958, *Ring-complexes in the younger granite province of northern Nigeria*, Mem. No. 1, Geol. Soc. London, 72 pp.

28 Branch, C. D., 1966, *Volcanic cauldrons, ring complexes, and associated granites of the Georgetown Inlier, Queensland*, Australian Bureau of Mineral Resources, Geology, and Geophysics, Bull. No. 76, 158 pp.

29 Anderson, E. M., in Bailey, E. B., et al., 1924, *The Tertiary and Post-Tertiary geology of Mull, Loch Aline, and Oban*, Geol. Surv. Scotland, Memoirs.

30 Hamilton, Warren, and Myers, W. Bradley, 1967, *The nature of batholiths*, U. S. Geol. Surv., Prof. Paper 554-C, 30 pp.

31 Evernden, J. F., and Kistler, R. W., 1970, *Chronology of emplacement of Mesozoic batholithic complexes in California and western Nevada*, U. S. Geol. Surv., 47 pp.

17

EMPLACEMENT
OF LARGE PLUTONS

Introduction

The preceding chapter has been devoted largely to the geometry of plutons, that is, their shape and structural relations to the country rock. But the mechanics of emplacement of smaller plutons has been discussed. In some instances the magma fills a potential opening as fast as it develops. This is undoubtedly true for many dikes. But in some instances the magma forming a dike may force its way into the fracture, compressing the wall rocks elastically or plastically or pushing the wall rock away. According to one theory the magma forming a ring-dike fills the potential cavity left by the sinking of a central block bounded by an outward dipping fracture. But if the fracture zone is steep, the removal of fractured rock by upward-flowing magma or by downward stoping will be important. Sills, laccoliths, and lopoliths are all forcefully injected. Great energy is involved in lifting the roof. In some cases the floor may sink into an underlying reservoir, and intrusion involves only the transfer of magma from a lower level to a higher level. Volcanic vents blow out the country rock explosively.

But the mechanics of emplacement of larger plutons is difficult to determine. It is a mistake to assume that only one mechanism is involved. Much controversy has resulted from this assumption. Moreover, there are several facets to the problem: (1) time of emplacement, both relative to orogenic movements and to geologic time; (2) the depth of emplacement; (3) mechanics of emplacement; (4) stresses at time of emplacement.

Time of Emplacement

Atectonic plutons are those that are unrelated to orogenic movements. But many plutons may be genetically related to orogeny. *Pretectonic plutons* are those that are older than the folding, *syntectonic* or *synchronous plutons* are emplaced during the orogeny, and *posttectonic* or *subsequent plutons* are later than the orogeny. Obviously some very subjective elements are involved in deciding whether to distinguish an atectonic from a pretectonic pluton, or an atectonic pluton from a post-tectonic pluton. The situation is complicated, of course, if there has been more than one period of folding in the region. In such a case, an igneous body could be post-tectonic relative to the first orogeny, but pretectonic relative to the second. The ensuing discussion applies, therefore, to a specific period of folding.

Posttectonic intrusives are undeformed, the associated dikes and sills are not folded, the rocks are not granulated, and they lack a secondary foliation (p. 378). But an atectonic intrusion would be characterized by similar features. Two methods may be used to decide whether a pluton is chronologically associated with an orogeny.

One method is based on petrology. The mineralogy and chemistry of the rocks may indicate that they are comagmatic with synchronous intrusions in the same area. Thus the Concord granite of New Hampshire—a biotite-muscovite equigranular undeformed granite—is comagmatic with some of the synchronous plutons of the Middle Devonian New Hampshire plutonic series.[1] Conversely, the Early Jurassic White Mountain series, also undeformed and characterized by ring-dikes, stocks, and batholiths, is unrelated to any orogeny and is atectonic.

A second method could be based on radiogenic dating, which may show that an undeformed granite is only slightly younger than one that is synorogenic. But at present the methods are not sufficiently refined for such close dating.

Pretectonic bodies are caught in the folding. If the folding is intense, the rocks will have a cataclastic texture—that is, under the microscope the grains comprising the rock will be fractured, broken, and show strain shadows—or a granoblastic texture—that is, the grains are spherical. They

D. DE WAARD: *Tectonics of a pretectonic orthogneiss massif near St. Jean du Gard in the southeastern Cevennes, France.*

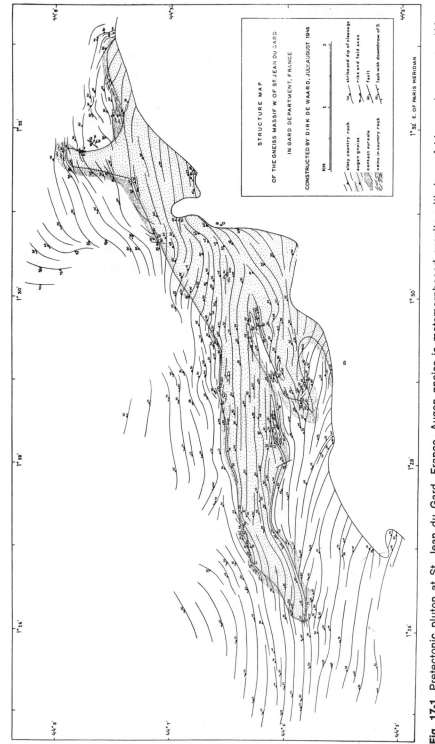

Fig. 17-1. Pretectonic pluton at St. Jean du Gard, France. Augen gneiss is metamorphosed granite with large feldspar phenocrysts. (After DeWaard.[2] Permission Kon. Ned. Akademie van Wetenschappen te Amsterdam.)

may also possess a secondary foliation. Figure 17-1 illustrates a pretectonic granite.[2] The foliation cuts indiscriminately across the granite (now gneiss), the contact metamorphic aureole surrounding it, and the country rock. Associated dikes and sills may be folded.

Some plutons are contemporaneous with sedimentation in a geosyncline.[3] They intrude some of the sedimentary rocks. But they also contribute clasts to conglomerates that are slightly younger but still part of the geosynclinal sequence.

Syntectonic plutons will show many of the features of pretectonic plutons. On the other hand contacts may cut across folds and the inclusions may contain minor folds.

So far we have been concerned with relative ages. Let us now return to the geological age, that is, the place in the geological time scale. This is deduced from the relations to the adjacent rocks—that is, whether the contact is intrusive or an unconformity (Fig. 13-4). The accuracy of the dating obviously depends on the precision with which these can be dated and the span of time involved. Of course, radiogenic dating is very important.

Depth of Emplacement

Many features of plutons, especially the larger ones, depend upon the depth of emplacement.[4] These emplacement zones have been designated the epi-, meso-, and catazones.

Plutons of the epizone are injected in the upper part of the crust, not deeper than four miles; the temperature of the country rock does not exceed 300°C. Such plutons intrude their own volcanics, that is, volcanics derived from the pluton or at least from the same reservoir—this is demonstrated by the fact that the volcanics and the rocks in the pluton are comagmatic. These bodies generally lack foliation or lineation and they may contain miarolitic cavities—that is, gas cavities, with small well-formed minerals growing in them. They are surrounded by an aureole of contact metamorphism. The Boulder batholith of Montana[5] is an example (Fig. 17-2).

Plutons of the mesozone consolidated at a depth of four to nine miles, in country rock ranging in temperature from 300°C to 500°C. Primary foliation is confined to the margins or is weak, miarolitic cavities are absent, and associated metamorphism is a combination of contact and regional. Figure 17-3 is an example of a mesozonal pluton that was forcefully injected.[6]

Catazonal plutons consolidate at depths of 7 to 12 miles, where the country rock is at temperatures ranging from 450°C to 600°C. The contacts tend to be concordant, the rocks lack miarolitic cavities, the country rock

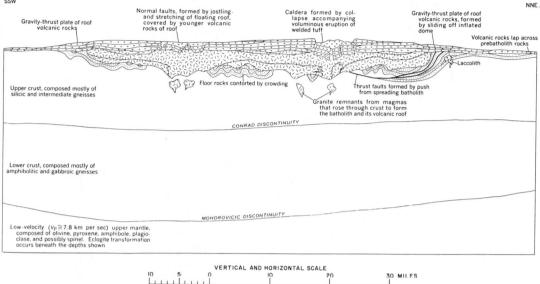

SSW   NNE.

Normal faults, formed by jostling·
and stretching of floating roof,
covered by younger volcanic
rocks of roof

Caldera formed by col-
lapse accompanying
voluminous eruption of
welded tuff

Gravity-thrust plate of roof
volcanic rocks

Gravity-thrust plate of roof
volcanic rocks, formed
by sliding off inflated
dome

Volcanic rocks lap across
prebatholith rocks

Laccolith

Upper crust, composed mostly of
silicic and intermediate gneisses

Floor rocks contorted by crowding

Thrust faults formed by push
from spreading batholith

Granite remnants from magmas
that rose through crust to form
the batholith and its volcanic roof

CONRAD DISCONTINUITY

Lower crust, composed mostly of
amphibolitic and gabbroic gneisses

MOHOROVICIC DISCONTINUITY

Low-velocity ($v_p \cong 7.8$ km per sec) upper mantle,
composed of olivine, pyroxene, amphibole, plagioclase, and possibly spinel. Eclogite transformation
occurs beneath the depths shown

VERTICAL AND HORIZONTAL SCALE

10   5·   0   10   20   30 MILES

10   5   0   10   20   30   40   50 KILOMETERS

Fig. 17-2. Epizonal batholith. Boulder batholith, Montana. Here
the batholith is interpreted to be a sheet, as much as four miles
thick, injected above Precambrian, Paleozoic, and Mesozoic
strata but beneath a cover of late Cretaceous volcanic rocks.
But geophysical evidence (p. 458) indicates that the batholith
extends to a depth of 6 to 9 miles. (After U. S. Geological Survey,
Hamilton and Myers.[5])

is regionally metamorphosed, and the plutons tend to form domes and sheets.
An example is given in Fig. 17-4.[4]

The above criteria may not always be easy to apply. An epizonal pluton
may later be buried under miles of sediments and then deformed under
catazonal conditions.

Methods of Emplacement

Basically three possible methods of emplacement must be considered: (1)
forceful injection; (2) magmatic stoping; and (3) metasomatic replacement.
The source of the magma and metasomatizing solutions will be considered
in a later section.

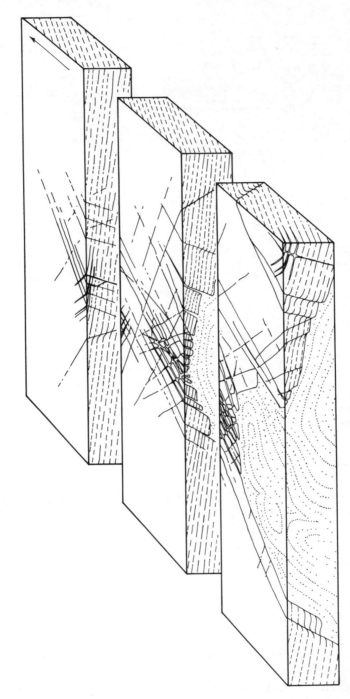

Fig. 17-3. Mesozonal pluton. Structure section of Mt. Aigoul pluton, France. Primary flow structure of the granitic pluton is shown by dotted lines. Surrounding rock is slate. After DeWaard,[6] permission Kon. Ned. Akademie van Wetenschappen te Amsterdam.

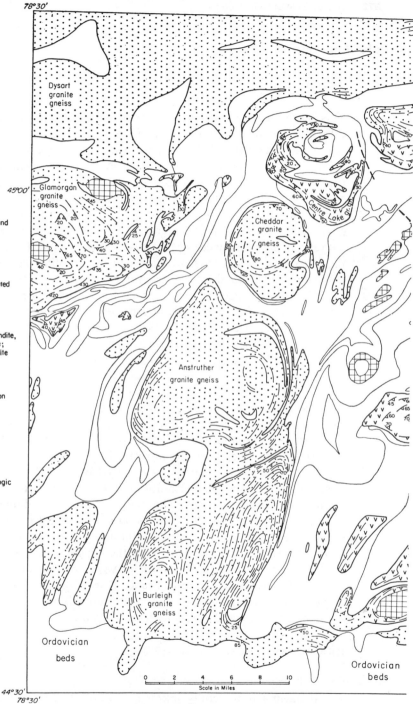

EXPLANATION

Granite, granite gneiss and
associated pegmatite

Hybrid granite gneiss,
migmatite and associated
granite pegmatite

Diorite, gabbro, hornblendite,
pyroxenite, anorthosite;
metagabbro, amphibolite

Strike and dip of foliation

Strike of foliation

Boundary lines of lithologic
units

Fault or lineament

Fig. 17-4. Catazonal pluton. Part of Haliburton-Bancroft area,
Canada. (After Buddington.[4])

371

Forceful Injection

INTRODUCTION

This interpretation assumes magma is available. The magma may be partially crystallized, and the intruding magma may be a kind of mush consisting of crystals floating in liquid. The magma pushes the older rocks aside or upward. Vertically rising magma would behave much like a rising salt dome. The rocks into which the magma rises are in many cases much more deformed and much more competent than the sedimentary rocks through which salt domes move. Hence the resulting structure would undoubtedly be more complex than around a salt dome. The movement of the magma may be entirely under the influence of gravity—that is, a magma that is lighter than the surrounding rocks tends to move upward. In other instances the magma may be pushed around by orogenic forces.

That magma may be emplaced by forceful injection is clearly demonstrated by laccoliths. Similarly, in most dikes, sills, and lopoliths the walls move apart because the hydrostatic pressure exerted by the magma exceeds the lithostatic pressure on the rocks. The manner in which the structure wraps around some stocks shows that they have been forcefully injected. Examples are given in Figs. 16-24 and 17-5.

An example of doming related to forceful injection has been described from the Black Hills of South Dakota (Fig. 17-6). The white areas are Precambrian schists, whereas the stippled areas are Paleozoic sedimentary rocks that rest with pronounced angular unconformity (see page 281) on the Precambrian.[7] The black areas are intrusive bodies of Tertiary age; these intrusive bodies make up a complex composed of a large number of separate steeply dipping dikes. The dash-dot lines are structure contours on the base of the Paleozoic formations. The highest part of the dome shown by the structure contours coincides with the area of greatest intrusion. The intrusions are the cause of the doming. Although part of the space now occupied by the intrusions was made by pushing the schists aside, much of it was made by pushing slabs of schist upward between "paired faults."

The Mt. Aigoual pluton in France (Fig. 17-3) has been cited as an example of a forcefully injected pluton. Presumably magma was inserted in a horizontal fracture and gradually lifted its roof.

GRANITE TECTONICS

Introduction

If magma is forcefully injected, a whole series of structural features may be geometrically related.[8] These features include foliation, lineation, joints, dikes, and faults. If the magma attains its destination before crystallization begins, it will be massive and will possess neither foliation nor lineation. Some of the schlieren and inclusions, however, may be oriented in the manner discussed below.

If the magma is crystallizing as it moves, and especially if the movement continues until after the rock is completely consolidated, a series of structural features develop. Even after the magma in the outer and upper parts of the intrusion has frozen, the liquid or semiliquid material below may continue to rise, subjecting the consolidated parts of the intrusion to systematic stresses. For convenience, the structures may be considered under the following headings: (1) structures of the flow stage and (2) structures of the solid stage.

Structures of the flow stage

Moving liquids are characterized by either turbulent flow or lamellar flow. In *turbulent flow*, the movement of the individual particles is irregular and unsystematic, and the particles are not confined to any one layer in the liquid. In *lamellar flow*, on the other hand, the individual particles move in parallel sheets which slide over one another like the cards in a sheared playing pack. The flow is lamellar in such a viscous substance as magma beneath the surface of the earth. The sheets tend to be parallel to nearby contacts, and at such localites the differential speed of the various layers will be much greater than in the interior of the body of the magma.

Platy flow structure, also referred to as *planar flow structure* or *planar structure*, forms during the flow stage. Platy material in a magma characterized by lamellar flow tends to become oriented with the largest face parallel to the liquid layers. Consequently, platy inclusions, such as slabs of shale, sandstone, or schist, become oriented parallel to one another. Schlieren and platy minerals, such as biotite, align themselves in the same way. Feldspar crystals, if well-formed, will tend to lie with the largest faces parallel to the layers. Prismatic minerals—those shaped like a short pencil— also become oriented so that the long axis lies in the flow layers. If inclusions, schlieren, and platy minerals are all present, they will ordinarily be parallel to one another, as shown by Fig. 17-7. *Primary foliation* (Fig. 16-1B,D) is that variety of platy flow structure that is due to the parallelism of platy minerals. Platy flow structure is thus a more inclusive term because it refers to the parallelism of slablike inclusions and schlieren as well as to the parallelism of platy minerals.

Linear flow structure, which is that variety of lineation formed in moving liquids, is also formed during the flow stage (Fig. 16-2). The long axes of prismatic minerals may be oriented parallel to one another (Fig. 16-2A). This is particularly true of a mineral such as hornblende, the longest dimension of which is ordinarily parallel to the c-crystallographic axis. Feldspar is also longer parallel to the c-crystallographic axis than it is in any other direction.

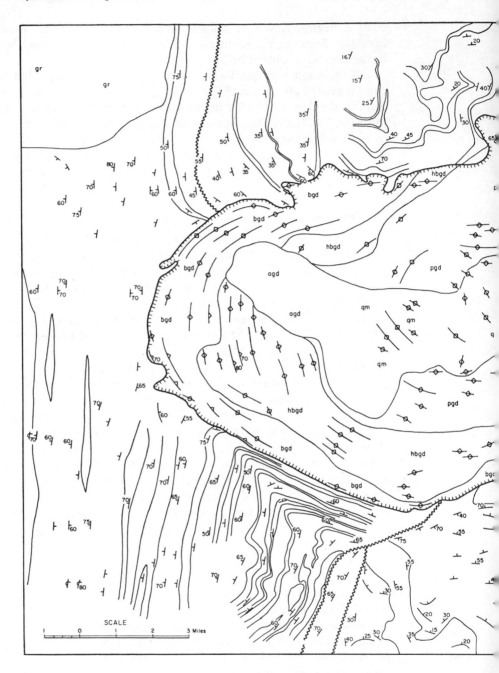

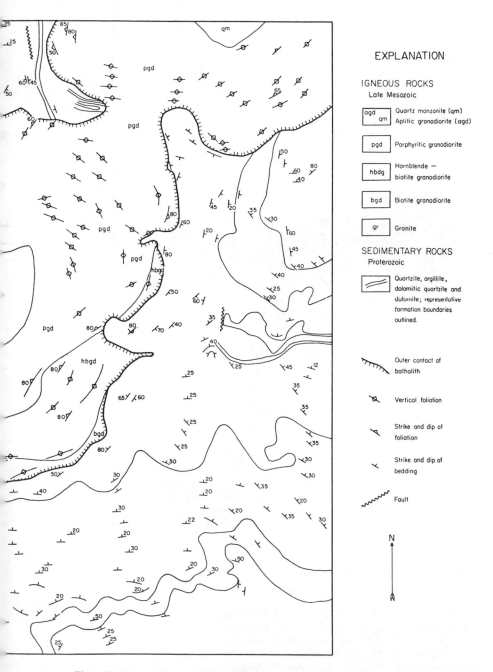

EXPLANATION

IGNEOUS ROCKS
Late Mesozoic

agd qm	Quartz monzonite (qm) Aplitic granodiorite (agd)
pgd	Porphyritic granodiorite
hbdg	Hornblende — biotite granodiorite
bgd	Biotite granodiorite
gr	Granite

SEDIMENTARY ROCKS
Proterozoic

	Quartzite, argillite, dolomitic quartzite and dolomite; representative formation boundaries outlined.

	Outer contact of batholith
	Vertical foliation
	Strike and dip of foliation
	Strike and dip of bedding
	Fault

N

Fig. 17-5. Forceful injection. Mesozonal batholith of Mesozoic age. White creek batholith, British Columbia. (After Buddington.[4])

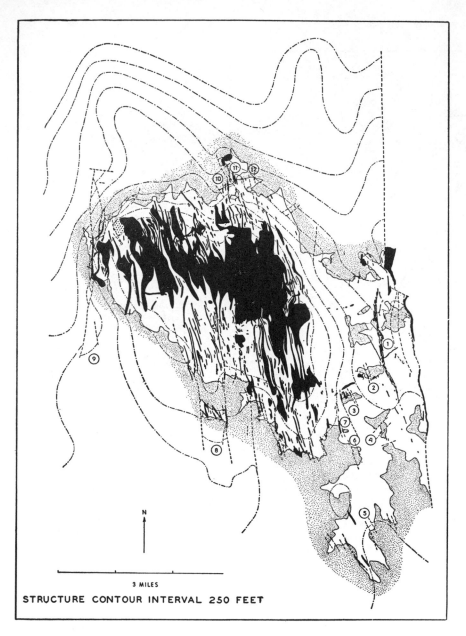

Fig. 17-6. Doming by forceful intrusion. Black Hills of South Dakota. Black represents rhyolite of Tertiary age. Stippled area is the base of the Paleozoic stata (only the base of the Paleozoic is stippled; the white area outside the stippled area is Paleozoic and Mesozoic). White inside of the stippled area represents Precambrian. Dash-dot lines are structure contours on base of Paleozoic; contour interval 250 feet; greatest uplift is in center of area. Numbers in circles refer to small horsts uplifted between paired faults (short dashes). (After J. A. Noble; permission of University of Chicago Press.)

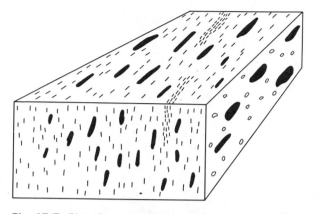

Fig. 17-7. Platy flow structure in an intrusive rock. Shown by parallel arrangement of platy inclusions (solid black), platy minerals (short dashes, except on side of block, where they appear circular), and schlieren (closely spaced dashes).

The manner of measuring the attitude of lineation and of recording it on a map is described on pp. 72, 413.

The long axes of spindle-shaped inclusions may also be parallel to one another, thus showing the lineation. In some cases, although the individual inclusions are more or less spherical, they may be strung out in lines, like the beads on a string.

The significance of the lineation depends upon the shape of the pluton. There would be no lineation if all parts of a rising body of magma were moving at the same velocity. If, however, lamellar flow were caused by friction on the walls, any elongate minerals would probably be rotated to assume an orientation parallel to the direction of flow. But if the magma were expanding upward, the elongation, and hence the lineation, would form an arch. Conversely, if magma were being forced into a progressively smaller opening, the lineation would be parallel to the direction of flow.

Structures of the solid stage

After the outer shell of the intrusion has completely consolidated, the interior may still be liquid or partially liquid. Continued movement of this interior will subject the consolidated shell to stresses that are systematically related to the structures formed in the flow stage. But the solid rock fails by rupture. In the early stages of the intrusion, the contact of the magma with the wall rock was not only a lithologic boundary, but also a dynamic boundary. Liquid magma was moving past solid rock. As the wall rock became heated and the outer shell of the intrusion consolidated, the contact became less important as a dynamic boundary. The wall rock began to participate in the movements; consequently, many of the fractures of the solid stage extend out into the country rock.

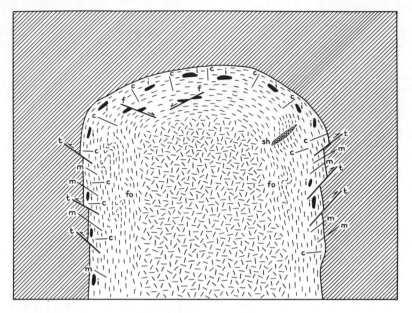

Fig. 17-8. Cross section of a hypothetical pluton. Section is parallel to strike of the linear flow structure. Short dashes are platy minerals; *i*, inclusion; *fo*, flexure; *sh*, shear filled with pegmatite or coarse granite; *c*, cross joint; *m*, marginal fissure; *t*, marginal thrust; *f*, flat-lying normal fault.

The fractures may be either joints or faults. Some of them may contain minerals similar to those in the igneous rocks. This is often taken as evidence that the fractures formed while some of the underlying magma was still liquid and consequently that the fracturing was related to the intrusion. Much later fractures, unrelated to the intrusion, would theoretically be devoid of such minerals. Some of the fractures are filled by dikes.

A detailed study of a pluton consequently involves the mapping of the joints, faults, and dikes as well as the structures of the flow stage.[8] Some of the structural features associated with a vertically intruded pluton are shown in Fig. 17-8.

Distinction between primary and secondary structures

A clear distinction between primary and secondary structures is essential for a correct interpretation of the tectonics of plutons. The distinction becomes particularly difficult if primary structures are utilized for later secondary movements.

Several criteria may be particularly useful in demonstrating that a planar structure—particularly a foliation—is primary. In regions unaffected by orogenic movements, any foliation in an igneous rock must be primary

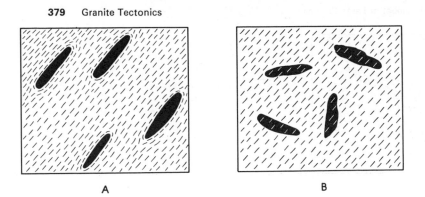

Fig. 17-9. Relation of inclusions to foliation. Solid black represents platy inclusions; black and white short dashes represent foliation due to platy minerals. (A) Platy inclusions parallel to primary foliation. (B) Platy inclusions, diversely oriented, cut by a secondary foliation.

because the forces essential for the development of secondary structures never existed. A second line of evidence is based on the attitude of inclusions. A foliation to which the inclusions are parallel (Fig. 17-9A) is probably primary. If a secondary foliation were imposed upon a rock with diversely oriented inclusions (Fig. 17-9B), the long axes of the inclusions would be unrelated to the foliation. In the example illustrated by Fig. 17-9A, the flat inclusions must have been oriented while the host rock was still molten; obviously, they would lie parallel to the foliation of the igneous rocks because the platy minerals and the slablike inclusions are oriented by the same forces.

Two exceptions to the use of these criteria may be mentioned. In one case, platy inclusions might be oriented during the flow stage, but a foliation might fail to develop in the intrusive rock. During a later orogenic movement, a secondary foliation superimposed on the igneous rock might coincide with the oriented inclusions. In this example, the orientation of the inclusions would be primary, but the foliation would be secondary. In a second case, the "stretching" during the formation of a secondary foliation could be so great that inclusions, regardless of their original orientation, might be elongated to such an extent that they become parallel to one another and to the secondary foliation.

A third line of evidence that the foliation is primary is illustrated in Fig. 17-10. The foliation is flexed, but along the axis of the flexure a dikelike body, rich in the lighter-colored minerals, grades into the main body of the rock; the material in the dike was derived from the liquid that was still uncrystallized in the rock.

In regions unaffected by orogeny, lineation in igneous rocks must be primary. The orientation of hornblende needles in a dike, sill, or laccolith injected into flat sediments is primary (Fig. 17-11).[9] In orogenic belts, on the other hand, a clear distinction between primary and secondary lineation

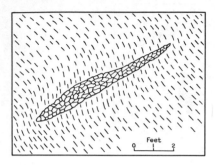

Fig. 17-10. Shear filled with pegmatite. Short dashes represent primary platy flow structure that is a primary foliation. Granular pattern is pegmatite or coarse granite.

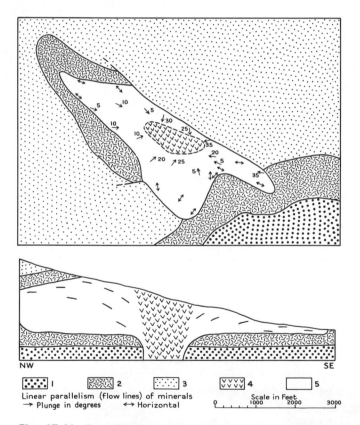

Fig. 17-11. Deer Creek laccolith. Map is above, northwest-southeast cross section is below. 1, early basic breccia; 2, early basalt flows; 3, late basic breccia; 4, decayed quartz diorite porphyry; 5, quartz diorite porphyry. (After J. T. Rouse.[9])

may be very difficult. If the lineation is obviously due to the granulation and dragging out of older minerals, it is presumably secondary. On the other hand, if the minerals in the igneous rock appear to be unstrained and ungranulated, the lineation may be primary; but the recrystallization of a highly granulated rock is a possibility that would have to be considered. No hard and fast rules can be established and each case must be considered by itself.

Magmatic Stoping

Magmatic stoping is a process whereby magma works its way up into the crust of the earth. *Piecemeal stoping*, one variety of the process, is illustrated in Fig. 17-12. The roof of the magma chamber is shattered, and blocks are surrounded by apophyses. Any block that is isolated by the magma will sink—provided, of course, that the specific gravity of the block is greater than that of the magma. The blocks may sink to great depth, where they may be reacted upon by the magma and assimilated. Thus the magma can gradually eat its way upward into the country rock. The size of the individual blocks is measured in feet, tens of feet, or even hundreds of feet. The shattering of the country rock may be due to thermal or mechanical causes. Inasmuch as the country rock is heated by the magma, it expands and cracks, especially if the heating is rapid. Mechanical shattering is probably even more important because the roof rocks will be subjected to numerous forces, particularly tensional and torsional forces.

In favor of the hypothesis of piecemeal stoping, it is possible to cite examples of the process "caught in the act." Xenoliths and autoliths attest to the actuality of stoping, and at some contacts it is possible to observe apophyses of igneous rock enveloping the country rock (Fig. 17-12, Plate 29).

On the other hand, xenoliths are surprisingly rare in many batholiths and stocks. Moreover, the specific gravity of the rocks involved must be favorable. In general, the specific gravity of magma is 7 to 10 percent less

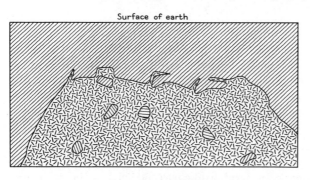

Surface of earth

Fig. 17-12. Magmatic stoping. Diagonal lines are older rocks. Diversely oriented dashes are plutonic rock.

than that of the corresponding solid rock. Thus a magma could stope its way into a rock of the same chemical composition. But the specific gravity of gabbro or diorite magma is as high as that of crystalline granite; in such a case piecemeal stoping could not be very effective. Furthermore, many sedimentary rocks have a lower specific gravity than magma. The structure at Marysville, Montana (Fig. 16-35) led to the concept of magmatic stoping for this body.

Cauldron subsidence differs from piecemeal stoping only in the size of the blocks involved. Whereas in piecemeal stoping the sinking blocks are measured in feet, the blocks involved in cauldron subsidence are measured in miles. The term *ring-fracture stoping* may be used synonymously with *cauldron subsidence*. Underground cauldron subsidence (p 360) may produce either a ring-dike or a stock, depending upon the amount the central block subsides and upon the depth of erosion. In Fig. 16-33D, for example, if erosion penetrates to the level QR, two ring-dikes will be exposed. If, however, erosion penetrates only to MN, a circular intrusive with all the characteristics of a stock will be exposed. The mechanism of cauldron subsidence explains the rarity of small inclusions in many batholiths and stocks. Moreover, this hypothesis is in accord with the smooth walls so typical of batholiths and stocks; piecemeal stoping should produce jagged, irregular walls. Specific gravities, however, would play the same role as in piecemeal stoping, and a subsiding block must be heavier than the magma in the reservoir.

The batholiths and stocks of New Hampshire[1] and Norway[10] are composite complexes made up of ring-dikes and stocks, and they have therefore attained their position in the outer shell of the earth by ring-fracture stoping. Available data indicate that such ring-dike complexes are rather exceptional in orogenic belts. Underground cauldron subsidences may have been important in other batholiths, but the individual subsidences may have been of such magnitude that only stocks rather than ring-dikes are exposed.

Metasomatic Replacement

Metasomatism is a process whereby a mineral is replaced by another of different chemical composition as a result of the introduction of material from an external source. The material may be introduced in a liquid solution or as a gas. As applied to granitic rocks, the preexisting rock is a sedimentary, volcanic, plutonic, or metamorphic rock. To convert a pure quartz sandstone into granite (in this case used in the narrow petrographic sense to mean a rock rich in potash feldspar and quartz, with lesser amounts of plagioclase feldspar and dark minerals) it would be necessary to add potash, alumina, and soda, as well as lesser amounts of lime, iron, and magnesia. If there were no change in volume, considerable silica would have to be removed.

To convert a slate into granite without any important change in volume, it would be necessary to remove iron, magnesia, alumina, and water, but add silica, soda, and perhaps potash.

Such terms as *granitization* and *migmatization* have also been used, but they have somewhat different connotations. *Granitization* has been described as the process by which a solid rock is made more like a granite in minerals, or in texture and structure, or both. *Migmatization* has been used by some in this same sense, but the term migmatite is best used as a descriptive term for a mixed rock consisting of alternating layers, each layer an inch or so thick; the darker layers are metamorphic rock, the lighter layers are granitic; no specific method of origin is implied.

The criteria for recognizing replacement granites that have not moved from the place where they originated are as follows: (1) The contact between the plutonic rock and the country rock is normally gradational. The transition zone may be only a few feet wide, but more commonly it is thousands of feet or miles wide. (2) The same structural pattern found in the country rocks continues into the plutonic rock. Thus if the country rock is characterized by gently plunging open folds, a similar structure may be shown by a weak to pronounced foliation in the plutonic rock. Even more striking are those cases where certain less readily replaceable beds can be traced for long distances into the plutonic rock. (3) Microscopic study of porphyritic rocks that have originated by replacement may show that the groundmass has the same texture as the surrounding sedimentary rocks. (4) There are also numerous petrographic and chemical criteria, but they are beyond the scope of this book.

The metasomatic replacement may take place during a period of deformation, in which case we may speak of *syntectonic granitization* (also *synkinematic* or *synorogenic granitization*). If the replacement occurs without any contemporaneous deformation, we may speak of *static granitization*.

Relative Importance of Various Mechanisms

In any specific area the field geologist should undoubtedly have an open mind concerning the mechanics of intrusion. Statistics on the relative importance of the various mechanisms are irrelevant. Moreover, in any area all three mechanisms may play a role. But he should try to evaluate the importance of the various mechanisms in the pluton he is studying.

Source of Magma

INTRODUCTION

A detailed consideration of the ultimate source of any magma that rises from below is beyond the scope of this book. Many factors other than structure must be taken into account. The petrology and chemistry of the rocks may give important clues. Trace elements will be of value. Moreover, the

Sr^{87}/Sr^{86} ratio is significant. Obviously the field geologist will get involved in this subject and in any case the problem should be clearly focused in his mind. The magma may originate in place, it may be derived from deeper parts of the crust, or it may be derived from the mantle.

ORIGIN IN PLACE

Rocks may be heated to such a high temperature that they liquify in place or at least some of the components form a partial solution. Underground atomic explosions liquify rocks, but obviously this is of no importance geologically. It has been proposed that meteor impacts may have supplied enough energy to melt rocks. Some migmatites, composed of alternating layers of mica schist and granite, each layer averaging an inch or so thick, are the product of partial melting of original siltstones. Possibly some large bodies have resulted from complete melting in place, but this might be difficult to prove on structural evidence. Chemical data might give a clue.

ORIGIN BY MELTING OF A DEEP ROOT

One idea that has been in great favor is that during orogeny the lower part of the crust is downfolded to such an extent that it liquifies. Thus older granitic rocks, sedimentary rocks, or volcanic rocks could be melted and rise to form plutons. A possible chemical test of this hypothesis is discussed below.

ORIGIN FROM THE MANTLE

It is now believed that most basaltic rock is derived from partial fusion within the mantle. That is, ultramafic rocks comprising the mantle are partially melted to give basaltic magma. Similarly, most granitic rocks might be derived from the mantle, either directly, or, more likely, by fractional crystallization (pp. 337–339) of basaltic magma that is trapped in reservoirs in the lower part of the crust. Geochemical data may help in distinguishing between granites so derived and those formed from the melting of older rocks.

Evidence indicates that the present Sr^{87}/Sr^{86} ratio in the mantle is about 0.720; this ratio has been increasing throughout geologic time because of the formation of Sr^{87} from rubidium, whereas Sr^{86} has not increased. On the other hand, a granite emplaced in the Precambrian, say 2 billion years ago, because its content of rubidium is greater than that in the mantle, would produce more Sr^{87} than the mantle. Hence the Sr^{87}/Sr^{86} ratio in this granite in the late Paleozoic would be much higher than that in the mantle. If this Precambrian granite were melted in the late Paleozoic, injected into the crust, and consolidated, it would have a higher Sr^{87}/Sr^{86} ratio than a granite of the same age derived from the mantle.

References

[1] Billings, Marland P., 1956, *Geology of New Hampshire, part II, Bedrock geology*, New Hampshire State Planning and Development Commission, 203 pp.

[2] DeWaard, D., 1950, *Tectonics of a pretectonic orthogneiss massif near St. Jean de Gard in the southeastern Cevennes, France*, Koninklijke Nederlandsche Akademie van Wetenschappen, Proceedings LIII, Nos. 4 and 5; I: 545–59; II: 662–74.

[3] Helwig, James, and Sarpi, Ernesto, 1969, Plutonic-pebble conglomerates, New World Island, Newfoundland, and history of geosynclines, *Memoir, 12*, Amer. Assoc. Petrol. Geol., pp. 443–66.

[4] Buddington, A. F., 1959, Granite emplacement with special reference to North America, *Geol. Soc. Amer. Bull.* 70: 671–747.

[5] Hamilton, Warren, and Meyers, W. Bradley, 1967, *The nature of batholiths*, U. S. Geol. Surv., Prof. Paper 554-C, 30 pp.

[6] DeWaard, D., 1949, *Tectonics of the Mt. Aigoual pluton in the southeastern Cevennes, France*, Koninklijke Nederlandsche Akademie van Wetenschappen, Proceedings LII; I: 389–402, II: 539–50.

[7] Noble, James A., 1952, Evaluation of the criteria for the forcible intrusion of magma, *J. Geol.* 60: 34–57.

[8] Balk, Robert, 1937, *Structural behavior of igneous rocks*, Geol. Soc. Amer., *Memoir 5*, 177 pp.

[9] Rouse, J. T., 1933, The structure, inclusions, and alteration of the Deer Creek Intrusive, Wyoming, *Amer. J. Sci.* 26: 139–46.

[10] Holtedahl, Olaf, ed., 1960, *Geology of Norway*, Norges Geologiske Undersökelse Nr. 208, 540 pp.

18

CLEAVAGE
AND SCHISTOSITY

Introduction

Foliation is the property of rocks whereby they break along approximately parallel surfaces.[1] In some rocks this is a primary feature, inherited from the time of their formation. Many sedimentary rocks, particularly those that are fine-grained, tend to part parallel to the stratification, and thus they possess what is often called *bedding fissility*. Bedding fissility is probably caused by platy and elongate grains more or less parallel to the stratification.[2] The primary foliation of plutonic rocks has already been discussed (pages 373–378). The present chapter is concerned exclusively with foliation that is of secondary origin, and which develops some time— often millions of years—after the original formation of the rock. Such foliation may develop in rocks of either sedimentary or igneous origin, and the product is a metamorphic rock.

The compositional layering in metamorphic rocks is also called foliation by some geologists.[3] But such usage is confusing and unnecessary. As shown in Fig. 16-1, foliation and compositional layering should be treated as separate features.

Cleavage, sometimes called *rock cleavage*, to distinguish it from *mineral cleavage*, is the property of rocks whereby they break along parallel surfaces of secondary origin (Plates 34–39). The terminology applied to the various kinds of rock cleavage has evolved over many years[4]; consequently some confusion and lack of complete uniformity is inevitable.

Much rock cleavage is inclined to the bedding, but in some instances it may be parallel to the bedding. *Schistosity* is a term applied to the variety of rock cleavage found in rocks that are sufficiently recrystallized to be called *schist* or *gneiss*. Thus the secondary foliation of a slate would be called *cleavage*, but a similar structure in a mica schist would be termed *schistosity*. Obviously, there are transitional rocks in which either term might be appropriately used.

A *schist* is a metamorphic rock that possesses schistosity, but which is not characterized by layers of differing mineral composition. A *gneiss* is a metamorphic or igneous rock characterized by alternating layers, usually a few millimeters or centimeters thick, of differing mineral composition. These bands are rich in light minerals in many cases; others are rich in dark minerals. The layers may or may not possess foliation. *Paraschists* and *paragneisses* are, respectively, schists and gneisses of sedimentary origin. *Orthoschists* and *orthogneisses* are, respectively, schists and gneisses of igneous origin. *Metasediments, metavolcanics,* and *meta-igneous* rocks are metamorphic rocks derived, respectively, from sedimentary, volcanic, and igneous rocks.

The attitude of cleavage and schistosity is measured in the same way as is the attitude of bedding. The strike is the direction of a horizontal line in the plane of cleavage; the dip, which is the angle between the cleavage and a horizontal plane, is measured at right angles to the strike.

Special symbols are employed to represent foliation on geological

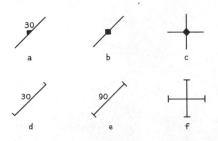

Fig. 18-1. Symbols for foliation. Upper line shows symbols for foliation in general. Lower line shows symbols for cleavage. (a) Strike and dip of foliation. (b) Strike of vertical foliation. (c) Horizontal foliation. (d) Strike and dip of cleavage. (e) Strike of vertical cleavage. (f) Horizontal cleavage.

Plate 34. *Slaty cleavage.* Slate and conglomerate of Boston Bay group, Squantum, Massachusetts. Bedding dips to left, cleavage dips to right. Photo: J. Haller.

maps. Because the symbols are not standardized, it is necessary to look at the legend accompanying the map in order to understand the meaning of the symbols. Some of the symbols used are shown in Fig. 18-1. Symbols *a, b,* and *c* may be used for foliation in general. A long line gives the strike, and a triangle, either open or solid black, shows the direction of dip; the value of the dip is given by a numeral. Symbol *a* indicates that the foliation strikes N. 45°E. and dips 30°NW. Symbol *b* indicates that the foliation strikes N. 45°E. and is vertical. Symbol *c* indicates horizontal foliation. Symbols *d, e,* and *f* may be used for those variations of foliation that are more appropriately called cleavage.

Descriptive Terminology

INTRODUCTION

Some of the terminology commonly applied to cleavage has certain genetic implications with which all geologists do not agree. Consequently, insofar as possible, the nomenclature should be descriptive rather than genetic. Nevertheless, the terms should be those established by long custom. This first section is, therefore, essentially descriptive. In a later section the genetic problems involved are discussed.

SLATY CLEAVAGE OR SCHISTOSITY

Slaty cleavage (Plates 34 and 35) and schistosity are caused by the parallel arrangement of platy minerals, such as the micas or chlorites, or by the parallel arrangement of ellipsoidal grains of such minerals as quartz and feldspar (Fig. 18-2A). Elongate minerals, such as hornblende, may impart a cleavage to the rock if the long axes lie in the same plane but are not parallel to one another. Theoretically, a rock possessing slaty cleavage can be split into an indefinite number of thin sheets parallel to the cleavage. The term *slaty cleavage* is used for less intensely metamorphosed rocks, such as slate, whereas *schistosity* is employed if the rock is recrystallized into minerals that are readily recognized by the naked eye. The term *continuous cleavage* has been suggested to embrace slaty cleavage and schistosity.[5] Cleavage and schistosity may or may not be parallel to bedding.

FRACTURE CLEAVAGE

Fracture cleavage is essentially closely spaced jointing. The minerals in the rock are not parallel to the cleavage (Fig. 18-2B). The distance between the individual planes of cleavage can be measured and is commonly a matter

Plate 35. *Cleavage.* Slaty cleavage in more argillaceous layers is nearly vertical; fracture cleavage in more arenaceous layers dips to right. Pogibshi Island, near Kodiak, Alaska. Photo: G. K. Gilbert, U. S. Geological Survey.

of millimeters or centimeters. If the distance between the fractures exceeds a few centimeters, the term "jointing" is more appropriately used. The term *spaced cleavage*[5] has been suggested to include all cleavages that are separated by a finite distance. It would include fracture cleavage, slip cleavage, and shear cleavage (Plate 36).

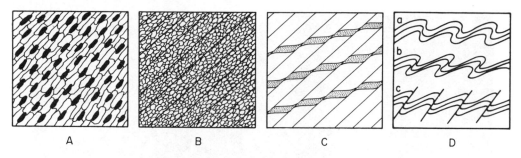

A   B   C   D

Fig. 18-2. Kinds of cleavage. (A) Slaty cleavage or schistosity. (B) Fracture cleavage. (C) Shear cleavage. (D) Slip cleavage.

SHEAR CLEAVAGE

Although this term has been used as synonymous with slip cleavage, it is more appropriately used for closely spaced fractures along which there has been some displacement (Fig. 18-2C). It is essentially a fracture cleavage along which there has been displacement.

SLIP CLEAVAGE

This feature has also been called *strain-slip cleavage* and *crenulation cleavage.* In many metamorphic terranes the schistosity may be crinkled into small folds with a wavelength of a fraction of an inch. One limb of these small folds becomes a zone of weakness (Fig. 18-2D). Eventually the mica flakes are rotated into discrete zones parallel to the axial planes of the crinkles, and displacement may take place along these zones. The rock tends to break parallel to these zones. Various terms have been used for this type of cleavage. Traditionally it has been called slip cleavage[6] or strain-slip cleavage.[7] It has been suggested that crenulation cleavage is a better term for this phenomenon.[8] At times it may be difficult to distinguish shear cleavage from slip cleavage. But slip cleavage evolves from small folds, whereas in shear cleavage small drags develop after the fracture has formed.

BEDDING CLEAVAGE

Cleavage or schistosity that is parallel to the bedding is commonly referred to as bedding cleavage or bedding schistosity. It is commonly similar to slaty cleavage in that it is caused by parallel platy minerals.

AXIAL PLANE CLEAVAGE

Cleavage or schistosity that is essentially parallel to the axial planes of the folds is called axial plane cleavage (Plates 37–39). The term is generally used in combination with one of the terms given above.

Plate 36. *Fracture cleavage.* Locally is a shear cleavage. Blue Canyon Formation, south bank of South Yuba River, $3\frac{1}{2}$ miles east of Washington. Nevada County, California. Photo: L. D. Clark, U. S. Geological Survey.

Plate 37. *Axial plane cleavage in asymmetric anticline.* Bedded tuff of the Gopher Ridge Volcanics. California. Photo: L. D. Clark, U. S. Geological Survey.

Origin

SLATY CLEAVAGE AND SCHISTOSITY

Much slaty cleavage or schistosity, but by no means all, is *flow cleavage*, a genetic term meaning that the cleavage is the result of rock flowage. The rock is shortened at right angles to the cleavage but lengthened parallel to it. The most compelling argument in favor of this interpretation is the fact that the cleavage results from the parallelism of platy and ellipsoidal grains.

Although the cleavage of this type forms perpendicular to the least strain axis—that is, at right angles to the greatest principal stress axis—it is still necessary to explain the exact mechanism whereby the mineral grains become oriented. One factor is the rotation of platy and ellipsoidal grains.

Plate 38. *Axial plane cleavage in syncline.* Washington County, Maryland, three miles west of Hancock. Photo: C. D. Walcott, U. S. Geological Survey.

Plate 39. *Axial plane cleavage in syncline.* Old quarry No. 2, Slatington, Lehigh County, Pennsylvania. Photo: E. B. Hardin, U. S. Geological Survey.

The greater the deformation, the more such grains will tend to lie perpendicular to least strain axis. A second factor is the flattening of grains. A spherical quartz grain, for example, will be flattened, partly by granulation, partly by recrystallization. Minerals with glide planes will change shape by gliding (p. 424). Thirdly, any new platy minerals will crystallize with their flat faces perpendicular to the greatest principal stress axis. The details of these processes are discussed in Chapter 20.

Rotated minerals, as described on page 432, are also indicative of considerable displacement parallel to the cleavage, but whether this means that the cleavage was initially a shear phenomenon, or is merely a flow cleavage parallel to which there was later shearing is not clear.

The condition of the rock during the formation of the cleavage has not been explicitly discussed in the preceding paragraphs. It was tacitly assumed that the rock was hard and consolidated. But it has been proposed that slaty cleavage forms while the rocks are still unconsolidated.[9,10] Under such conditions the liquid pore pressure may be very high, aiding the flowage of the clay minerals and sand grains, and facilitating the intrusion of argillaceous material into the planes of cleavage, similar to that shown in Fig. 18-3. But such a mechanism fails to explain the flattening of such minerals as quartz and feldspar in the development of slaty cleavage.

Some geologists believe that all slaty cleavage is a shear phenomenon. According to this theory the rock is cut by a vast number of shear planes along which there is slight differential movement. It is also assumed that any platy minerals will tend to be rotated so that they are parallel to the shear planes, and that any new platy minerals will form so that their flat

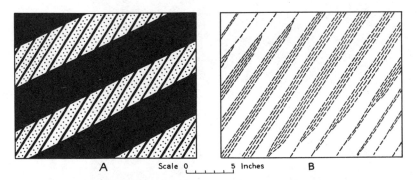

A   Scale 0 ⌊_ _ _ _ _⌋ 5 Inches   B

Fig. 18-3. Cleavage banding and segregation banding. (A) Cleavage banding. Solid black represents shale; dots represent sandstone. Bedding dips 25 degrees to left; cleavage dips 60 degrees to left. The more plastic shale has been injected along cleavage in the sandstone to produce a rhythmic alternation of shale and sandstone that simulates bedding. (B). Segregation banding. Short dashes represent bands rich in dark minerals. White areas are rich in light mineral. Bedding dips 25 degrees to the left.

faces are parallel to the shear planes. But a strong argument against this hypothesis is based on the orientation of deformed grains in schistose rocks. It will be recalled that the long and intermediate axes lie in the plane of the cleavage (Fig. 18-4A). If the slaty cleavage were a shear phenomenon, the longest axis of the deformed grains would be inclined to the cleavage. This becomes apparent from a study of Fig. 18-4. Figure 18-4B is a spherical pebble cut by cleavage planes prior to any movement. If, however, differential movement takes place along the cleavage planes, like cards slipping over one another, the sphere would be deformed into the jagged ellipsoid shown in Fig. 18-4C. But if the shear planes are very close together, the sphere would be deformed into the smooth ellipsoid represented by Fig. 18-4D. The significant point is that the long axis of the grains would be inclined to the cleavage. It is apparent that this hypothesis does not explain the common parallelism between the long and intermediate axes of deformed pebbles and the associated cleavage.

However, some schistosity is a shear phenomenon, developing parallel to one of the shear fractures of the strain ellipsoid; that is, the cleavage forms

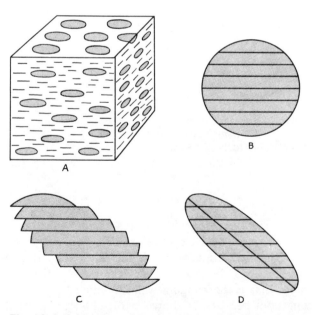

Fig. 18-4. Evidence against shear theory for origin of cleavage. (A) Characteristics of slaty cleavage; short dashes are platy minerals, stippled areas are ellipsoidal grains. (B) Circle cut by shear planes. (C) Circle sheared to jagged ellipse. (D) Same as (C), but with shear planes so close that ellipse is smooth.

at an angle of about 60° to the least principal strain axis. Because of displacement parallel to this shear cleavage some of the platy minerals may be dragged into parallelism with the cleavage. Moreover, some new platy minerals may crystallize parallel to the planes of cleavage. In this way the shear cleavage gradually becomes a schistosity. In eastern Vermont a shear cleavage becomes a schistosity in the direction of more intense deformation.[11]

FRACTURE CLEAVAGE

Fracture cleavage (Fig. 18-2B) is a shear phenomenon that obeys the laws of shear fractures; consequently, fracture cleavage is inclined to the greatest principal stress axis at an angle of about 30°. In Fig. 18-5 the fracture cleavage is parallel to the planes represented by FF' and F″F‴. Ordinarily only one set of fracture cleavage planes forms. We are thus using fracture cleavage in both a descriptive and a genetic sense.

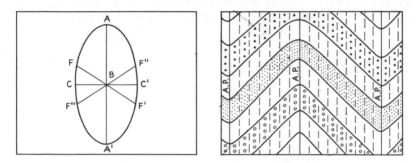

Fig. 18-5. Relation of cleavage to strain ellipsoid. Flow cleavage forms at right angles to the least strain axis (*CC'*) of the strain ellipsoid; it includes the greatest strain axis (*AA'*) and the intermediate strain axis (*B*), which is perpendicular to the plane of the paper. Fracture cleavage develops essentially parallel to the planes represented by *FF'* and *F″F‴*.

SHEAR CLEAVAGE

Shear cleavage is a fracture cleavage along which displacements have taken place. Plate 36 is an illustration of shear cleavage, which is especially well displayed on the left part of the illustration, whereas folds are well developed on the right side. The cleavage dips about 30° to the right and the displacement along it is "normal"; that is the hanging wall has moved down. At first this may seem incongruous because, as shown in Fig. 10-15, we would expect the displacement to be a "thrust"; that is, the hanging wall to move up. But in the example in Plate 36 the stress distribution was similar to that in the left side of Fig. 10-19. On any of the fractures dipping from 20 to 50° to the right the displacement is normal.

SLIP CLEAVAGE

Slip cleavage is associated with small crinkles (Fig. 18-2D). The shorter limbs of the crinkles become so stretched that they become planes of weakness. Minute thrusts may also occur along these limbs. It has already been pointed out that it is an older schistosity that is thrown into crinkles. A compressive force perpendicular to a plane normal to the schistosity may cause the crinkles. In most cases, however, a couple acting parallel to the schistosity causes asymmetrical crinkles.

BEDDING CLEAVAGE

In some metamorphic rocks the cleavage is parallel to the bedding, and hence it may be called *bedding cleavage* or *bedding schistosity*. Cleavage parallel to the bedding may be due to: (1) isoclinal folding; (2) mimetic recrystallization; (3) flow parallel to bedding; and (4) load metamorphism.

(1) The bedding on the limbs of an isoclinal fold is parallel to the axial plane. Because the slaty cleavage is parallel to the axial plane (Fig. 18-6) as shown on page 400, it will also be parallel to the bedding on the limbs. But on the nose of such folds the slaty cleavage cuts across the bedding at a considerable angle.

(2) In many localities, however, the schistosity follows the bedding and wraps around the noses of the folds. Such schistosity may be *mimetic;* during the recrystallization of the rock, the new platy minerals grew with their long dimensions parallel to the bedding fissility in the rocks. Little or no plastic deformation need accompany such recrystallization.

(3) In some localities characterized by bedding cleavage, originally spherical pebbles have been flattened so that the shortest axis is perpendicular to the bedding; conversely, the rocks must be elongated parallel to the bedding. In places, this type of deformation may be the result of stretching of the bedding on the limbs of folds. In such cases the rock is shortened perpendicular to the bedding, and flow cleavage develops parallel to the bedding.

(4) Bedding cleavage has sometimes been ascribed to *load metamorphism.* According to this hypothesis, the weight of an overlying thick column of rocks exerts vertically directed pressure on flat strata. Such a load, however, would produce a confining pressure that is essentially hydrostatic, and the proposed mechanism does not seem competent to explain bedding cleavage.

The field geologist must be exceedingly cautious in concluding that bedding and schistosity are parallel. Under some conditions of metamorphism, where shales and sandstones are interbedded, the more plastic shale may be squeezed into inclined planes of cleavage that cut the sandstones. The resulting structure is *cleavage banding* (Fig. 18-3A). The individual bands of shale are characteristically a fraction of an inch thick. In other rocks, notably those that have been thoroughly recrystallized under conditions of high metamorphic intensity, the light and dark minerals may segregate into alternate bands parallel to the schistosity (Fig. 18-3B); the individual bands are a fraction of an inch to an inch thick.[12] This may be called *segregation banding*. The original rock may have been homogeneous, but during recrystallization the various elements moving in solution tended to accumulate in different bands. Segregation banding thus differs from the plastic injection characteristic of cleavage banding.

In such cases the ordinary criteria for recognizing bedding—compositional and textural differences—cannot be applied. If, however, the layers that differ in composition are many inches or feet thick, they presumably represent bedding. In the final analysis, however, each case must be decided on its own merits.

Relation of Cleavage and Schistosity To Major Structure

INTRODUCTION

Empirical observation in the field has shown that in many localities the cleavage bears a consistent relationship to the major structure. The systematic pattern shown by folds and cleavage is of the utmost importance to the structural geologist attempting to solve complicated field problems. The constancy of this relationship is, after all, not unexpected. Inasmuch as folds and cleavage generally develop contemporaneously under the same forces, a definite correlation is to be expected. Even in those areas where the folds and cleavage are not contemporaneous, if they developed under similar forces they would be related to one another. In some regions, of course, the tectonic history was more complicated; if the various structural features under consideration developed successively under forces acting in different directions, no simple relationship would occur.

The simpler case, in which folds and cleavage developed under the same force, or under successive applications of forces acting along the same lines, will be treated first; the more complicated situation will be discussed later.

SLATY CLEAVAGE

Experience has shown that in many areas where slaty cleavage is diagonal to the bedding, it is more or less parallel to the axial planes of the folds (Plates 37-39). In some cases the slaty cleavage is confined to the incompetent beds between the competent beds, in others a regional cleavage is present throughout a thick series of argillaceous strata. Cleavage that is parallel to the axial planes of the folds is of great use in solving structural problems. Its utility is a function of the geometrical relations and is independent of any theory of origin.

The methods may first be considered from the point of view of the relations in cross sections and in natural vertical faces that are perpendicular to the strike of the bedding. From Fig. 18-6 certain generalizations are apparent. (1) If the cleavage is vertical, it follows that the axial planes of the folds are vertical, and all the beds are right-side-up (Fig. 18-6A). (2) If the cleavage dips in the same direction as the bedding but more steeply (*a, c,* and *e* of Fig. 18-6), the beds are right-side-up, and the synclinal axis is in the direction in which the beds dip. (3) If the cleavage and bedding dip in opposite directions (*b* of Fig. 18-6B), the beds are right-side-up, and the synclinal axis is in the direction in which the beds dip. (4) If the bedding is vertical, the synclinal axis is in the opposite direction of that in which the cleavage dips (*d* in Fig. 18-6C). (5) If the cleavage dips more gently than the bedding, the beds are overturned; moreover, the synclinal axis is in the opposite direction of that in which the bedding and cleavage dip (*f* in Fig. 18-6D). (6) If the cleavage is horizontal, it follows that the axial planes are horizontal; in such a case the cleavage-bedding relations cannot be used to tell which beds are right-side-up and which are overturned.

In summary, the beds are right-side-up unless the cleavage dips in the same direction as the bedding and at a gentler angle; the synclinal axis is in the opposite direction from that in which the bedding and cleavage dip. If the beds are vertical, the synclinal axis is in the opposite direction from that in which the cleavage dips.

A few examples of the use of these principles are illustrated in Fig. 18-7.

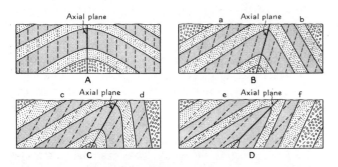

Fig. 18-6. Relation of slaty cleavage to folds in two dimensions. Cleavage represented by broken lines. Smaller letters are referred to in text. Rigorous parallelism of cleavage to axial plane is diagrammatic. (A) Symmetrical fold. (B) Asymmetrical fold. (C) Asymmetrical fold with one steep limb. (D). Overturned fold.

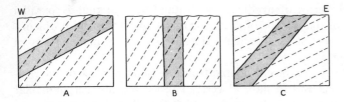

Fig. 18-7. Use of slaty cleavage to solve structure in two dimensions. Cleavage represented by broken lines. (A) Syncline is to left. (B) Syncline is to right. (C) Syncline is to right.

These are assumed to be isolated exposures of vertical faces that strike more or less at right angles to the strike of the bedding. In Fig. 18-7A the bedding and cleavage dip in the same direction, but the cleavage dips more steeply. Hence the beds are right-side-up and the syncline is to the west. In Fig. 18-7B, the bedding is vertical, but the cleavage dips to the west. Therefore, the top of the beds is to the east and the synclinal axis is toward the east. In Fig. 18-7C the cleavage and bedding both dip to the west, but the cleavage dips less steeply. It follows that the beds are overturned and the synclinal axis is toward the east.

From the illustrations given in Fig. 18-7, it is apparent that there is no evidence of the distance to the synclinal and anticlinal axes. This can be done only by a careful study of all the outcrops in the area. Suppose one outcrop shows that there is a synclinal axis to the east, but an outcrop 100 feet to the east shows that there is a synclinal axis to the west. The synclinal axis lies somewhere between the two outcrops; it can be located more precisely only if some additional information is available.

Actually, of course, the cleavage-bedding relationship should be considered in three dimensions. From such studies it is possible to determine the direction in which the folds plunge. Still assuming that the slaty cleavage is parallel to the axial planes of the folds, the following relations are geometrically inevitable. If the axes of the folds are horizontal, the strike of the cleavage and bedding are parallel (Fig. 18-8A).

If the folds plunge, the strike of the cleavage is diagonal to the strike of the bedding (Fig. 18-8). The observer faces in the direction of the younger beds. He then imagines a line drawn on the ground perpendicular to the strike of the bedding. The direction in which he would measure the *acute angle* between this perpendicular and the strike of the cleavage is the direction in which the fold plunges. Suppose this method were applied to Fig. 18-8B. On the west limb of the syncline he would face eastnortheast. The acute angle between the perpendicular to the strike of the bedding and the strike of the cleavage is to the left (north). Hence the fold plunges north. If the observer were on the east limb, he would face westnorthwest. Now the acute angle lies to the right (north). Hence the fold plunges to the right (north). A similar analysis may be applied to Fig. 18-8C and D.

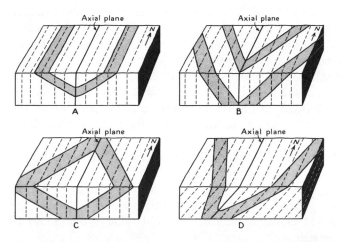

Fig. 18-8. Relation of slaty cleavage to folds in three dimensions. Cleavage represented by broken lines. Rigorous parallelism of cleavage to axial plane is diagrammatic; in many anticlines the cleavage diverges downward. (A) Symmetrical nonplunging fold. (B) Symmetrical fold plunging north. (C) Symmetrical fold plunging south. (D) Overturned fold plunging north.

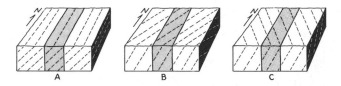

Fig. 18-9. Use of slaty cleavage to solve structure in three dimensions. Cleavage represented by broken lines. (A) Syncline to right, does not plunge. (B) Syncline to right, plunges north. (C) Syncline to right, plunges south.

A few examples of problems that might be encountered in the field are illustrated in Fig. 18-9. The normal procedure is to examine first the vertical exposure perpendicular to the strike of the bedding. This is in order to determine the direction in which the synclinal and anticlinal axes are located. Then the horizontal surfaces are studied to determine the direction in which the folds plunge.

Examination of the vertical faces in Fig. 18-9 shows that in all three cases, inasmuch as the bedding is vertical and the cleavage dips to the west, the synclinal axes are to the east. The horizontal surface in Fig. 18-9A shows that the strike of the bedding is the same as the strike of the cleavage. Therefore the plunge of the fold is horizontal. From the vertical face in Fig. 18-9B we have already determined that the younger beds lie to the east. Hence the observer, standing on the outcrop, faces east. The acute angle between a line perpendicular to the strike of the bedding and the strike of the cleavage lies to the left (north). Hence the fold plunges north. On the outcrop represented by Fig. 18-9C the observer also faces east, but the acute angle lies to the right (south). Hence the fold plunges south.

It is even possible to measure the value of the plunge. As shown in Fig. 18-10, the plunge of the folds is the same as the attitude of the trace of the bedding on the cleavage. In Fig. 18-10 the trace of the bedding on the cleavage plunges about 25° south. It follows that the fold plunges about 25° south.

The direction and value of the plunge can also be determined in another way. The attitude of the intersection of two planes, such as the bedding and the cleavage, can be determined by descriptive geometry (pages 546–550) or by the equal-area projection (pages 570–574). Hence, if we measure the attitude of bedding and cleavage, the direction and value of the plunge of the fold can be calculated.

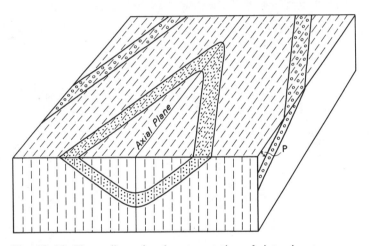

Fig. 18-10. Three-dimensional representation of slaty cleavage. Cleavage represented by broken lines. Value of plunge of fold is equal to *P,* which is measured on the cleavage; it is the angle between the trace of the bedding and a horizontal line.

FRACTURE CLEAVAGE

Inasmuch as fracture cleavage is a shear phenomenon, its relation to the deforming forces will differ from that of slaty cleavage. Fracture cleavage is characteristically developed in incompetent beds that lie between uncleaved competent beds. But in Plate 35 a slaty cleavage has developed in the incompetent beds, whereas a fracture cleavage has formed in the more competent beds, Empirical observation shows that fracture cleavage in a fold has the relations shown in Fig. 18-11A. The cleavage is inclined to the bedding

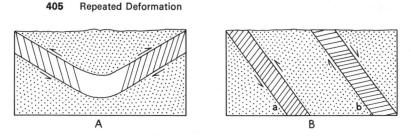

Fig. 18-11. Fracture cleavage. (A) On the limbs of a fold. (B) In isolated outcrops. Relations at *a* indicate a syncline to the right; relations at *b* indicate a syncline to the left.

and the acute angle, between the bedding and cleavage points in the direction in which the beds shear past one another. In a syncline each younger bed shears upward relative to the older bed beneath it (Fig. 6-7). In Fig. 18-11B the diagonal lining represents fracture cleavage. At locality *a* the relation of the fracture cleavage to the bedding shows that the bed to the right has sheared upward relative to the bed to the left. Hence the synclinal axis is to the right, and the beds are right-side-up. At locality *b*, however, the bed to the left has sheared upward. Therefore the synclinal axis in this case is to the left and the beds are overturned.

Inasmuch as the intersection of fracture cleavage and bedding is parallel to the axes of the folds, the trace of the bedding on the fracture cleavage may be used to determine the direction and plunge of the axes of the folds.

REPEATED DEFORMATION

In the preceding discussion it has been assumed that the cleavage was contemporaneous with the folding or, if later, that it was caused by similar forces. On the other hand, it is conceivable that the forces producing the cleavage may be later and different from those causing the folding. In Fig. 18-12, the folds formed by simple horizontal compression, H and H', are essentially symmetrical. The flow cleavage, which dips to the left, was formed by a later couple, C and C', consequently, the flow cleavage is not parallel to the axial planes of the folds. In such a case, use of the methods outlined above to determine the major structure would not necessarily give the correct solution. At *b*, one would correctly deduce that a synclinal axis lay to the right. But at *a*, one would deduce that a synclinal axis lay to the left; that is, of course, incorrect.

In Chapter 4 it has been pointed out that in some areas, especially those characterized by repeated deformation, several stages of folding may be recognized. For example, in Fig. 18-13 the recumbent anticline F_1 has been refolded by anticline F_2. Axial plane cleavage may be associated with both stages of folding. The planar surfaces are designated by the letter S. Thus the bedding is S_0. The axial plane cleavage associated with the recumbent

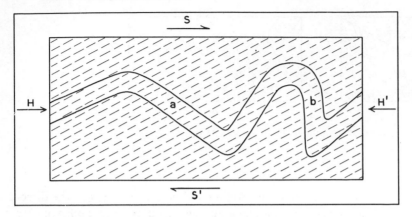

Fig. 18-12. Slaty cleavage that is not parallel to axial planes of folds. The folds resulted from simple compression, represented by H and H', and the axial planes are essentially vertical. The slaty cleavage, represented by the broken lines, was formed by a couple represented by S' and S'.

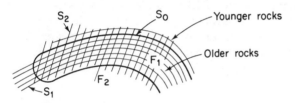

Fig. 18-13. Two stages of folding and schistosity. F_1, first stage recumbent fold; F_2, second stage open fold. S_0, bedding; S_1, schistosity of first stage of folding; S_2, slip cleavage of second stage of folding.

anticline (F_1) is designated S_1; it is parallel to the bedding on the limbs of the fold but crosses the bedding on the nose of the fold. The axial plane cleavage associated with F_2 is designated S_2; it may be a fracture cleavage, a shear cleavage, or a crenulation cleavage.

SUMMARY

Inasmuch as there are several kinds of cleavage and because successive cleavages may be superimposed on one another, it is obvious that considerable judgement is necessary in field investigations. In some areas the nature of the cleavage may be readily apparent; in others, an intensive study may be necessary. As Mead[1] has said: "No hard and fast specifications can be written for the identification of foliate structures. The problem perhaps falls more into the category of an art than a science. Experience and familiarity with a wide variety of occurrences is of value."

References

[1] Mead, W. J., 1940, Folding, rock flowage, and foliate structures, *J. Geol.* 48: 1007–21.

[2] Ingram, R. L., 1953, Fissility of mudrocks, *Geol. Soc. Amer. Bull.* 64: 869–78.

[3] Turner, Francis J., and Verhoogen, John, 1960, *Igneous and metamorphic rocks*, 2d ed., 694 pp. New York: McGraw-Hill Book Co., Inc.

[4] Dennis, John G., 1967, *International tectonic dictionary*, Amer. Assoc. Petrol. Geol. Mem. 7, 196 pp.

[5] Chidester, A. H., 1962, *Petrology and geochemistry of selected talc-bearing ultra-mafic rocks and adjacent country rock in north-central Vermont*, U. S. Geol. Surv. Prof. Paper 345, 207 pp.

[6] Pumpelly, Raphael, Wolff, J. E., and Dale, T. Nelson, 1894, *Geology of the Green Mountains in Massachusetts*, U. S. Geol. Surv. Monogr. 23, 206 pp.

[7] Wilson, Gilbert, 1961, *The tectonic significance of small scale structures, and their importance to the geologist in the field*, Annales de la Societé Géologique de Belgique, LXXXIV: 423–458.

[8] Rickard, M. J., 1961, *A note on cleavages in crenulated rocks*, Geol. Mag. XCV-III: 324–32.

[9] Maxwell, John C., 1962, Origin of slaty and fracture cleavage in the Delaware Water Gap area, New Jersey and Pennsylvania, Geol. Soc. Amer. *Petrologic Studies: A Volume in Honor of A. F. Buddington*, 281–311.

[10] Braddock, William A., 1970, Origin of slaty cleavage, *Geol. Soc. Amer. Bull.* 87: 589–600.

[11] White, Walter S., 1949, *Cleavage in east-central Vermont*, Amer. Geophys. Union Trans. 30: 587–94.

[12] Turner, F. J., 1941, The development of pseudo-stratification by metamorphic differentiation in the schists of Otago, New Zealand, *Amer. J. Sci.* 239: 1–16.

19

SECONDARY LINEATION

Introduction

Lineation, as already indicated (page 329) is expressed by the parallelism of some directional property in the rock (Fig. 16-2A). Primary lineation, which is found in both sedimentary and igneous rocks, has already been described. Some varieties found in sedimentary rocks (page 88) may be used to tell top and bottom of beds, some may be used to tell the direction of transport of sediments, and some may be used for both purposes. In igneous rocks (page 377), both intrusive and extrusive, it may be used to tell the direction in which magma was flowing or ash flows were moving.

Secondary lineation,[1,2,3] with which the present chapter is concerned, is superimposed on the rocks some time after they were originally deposited, erupted, or intruded. The time lapse may be short or long. Moreover, secondary lineation may be imposed on rocks more than once, so that several differently oriented lineations may be present in the rock. All transitions exist between rocks that possess no lineation and rocks with excellent lineation. In one rock the long axes of 90 percent of the hornblende crystals might lie within a few degrees of each other; such a striking linear parallelism would be recognized immediately. In another rock, however, the long axes of the hornblende crystals might be rather evenly oriented in all possible directions within the rock (Fig. 16-1A); in such a case no lineation would exist.

Secondary lineation may occur with or without foliation (Fig. 16-1). A rock without cleavage or schistosity may possess lineation. More commonly, however, secondary lineation is associated with foliation and lies in the plane of the foliation (Fig. 16-2C).

Kinds of Secondary Lineation

Fold axes are commonly considered to be lineation. In fact, the attitude of the fold axes is often the reference to which other lineations are compared.

Elongated or "stretched" pebbles (Plate 40) or boulders are one of the most spectacular types (Fig. 19-1A). These pebbles, which are irregular

Plate 40. *Stretched conglomerate.* Actually a garnet amphibolite derived from an agglomerate, a clastic rock of volcanic origin, Lower Paleozoic. Överuman (upper lake), Swedish Lapland. Photo: J. Haller.

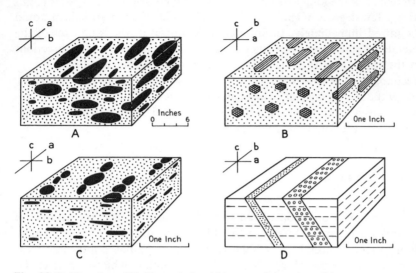

Fig. 19-1. Lineation. (A) Elongated pebbles are shown in solid black. Each pebble is an irregular ellipsoid, the longest axis of which is parallel to *a*, the shortest axis is parallel to *c*, and the intermediate axis is parallel to *b*. (B) Elongate crystals of hornblende, the long axes of which are parallel to *b* in the diagram. (C) Lineation caused by circular plates of mica, shown in solid black, strung out like beads on a string. (D) Cleavage is represented by top of block and by planes shown by broken lines. Bedding is shown by dots and open circles. Trace of bedding on cleavage gives a lineation.

ellipsoids, are associated with a foliation. The shortest axes of the pebbles are perpendicular to this foliation. The intermediate and long axes lie in the plane of foliation; the parallelism of the long axes of the pebbles produces the lineation.

Lineation is more commonly expressed by the minerals constituting the rock. As noted above, hornblende crystals, which have one long dimension, with the result that individual crystals are more or less needle-shaped, may display an excellent lineation (Fig. 19-1B). In some instances, biotite occurs as elliptical plates, the long axes of which are parallel. In still other instances, an original, more or less spherical mineral may be granulated into numerous fragments which become strung out into an ellipsoidal group.

The schistosity or cleavage may be thrown into small corrugations or crinkles, with a wave length and an amplitude that are measured in millimeters (Fig. 18-2D).

The intersection of bedding and cleavage produces a lineation, because the intersection of two planes is a line. If the rock breaks parallel to the cleavage, the trace of the bedding appears as parallel streaks on the cleavage.

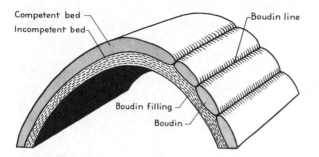

Fig. 19-2. Boudinage. In this case the boudin line is parallel to fold axis, but this is not necessarily true.

In Fig. 19-1D, the top of the block is parallel to the cleavage; the trace of the bedding on the cleavage is parallel to *b*. On the other hand, if the rock breaks parallel to the bedding, the trace of the cleavage appears as minute fractures on the bedding.

Slickensides (Fig. 9-2A) are a type of lineation. In some cases the scratches may be very obvious and clearly indicate the direction of movement. At other places a streaking caused by concentration of dark minerals along certain lines may be caused by sliding of layers past one another.

Boudinage, or "sausage structure," is illustrated in Fig. 19-2 and Plate 41. In cross section a competent bed thickens and thins in such a way as to simulate a string of sausages. Parallel to the bedding the individual units look like sausages lying side by side. The line of junction of the individual units may be called the *boudin line* and is a lineation. The boudin line may be occupied by quartz, feldspar, or some other mineral. Boudinage is clearly the result of stretching at right angles to the boudin line; it is analogous to the necking of a rod of metal under tension (Fig. 7-9). The more brittle bed is that one that shows the boudinage. Whereas the more ductile adjacent beds could yield plastically, the brittle bed first necked and then broke along a tension fracture.

Quartz *rods* are found in some areas of metamorphosed rocks.[4] In their extreme development they are long cylindrical rods of quartz. A complete study of their evolution shows that in their simplest form they are small concordant veins of quartz (Fig. 19-3A) that may have been derived from silica secreted from the country rock. However, as the individual veins become more irregular and discontinuous, they assume forms not unlike those of boudins, but far more irregular (Fig. 19-3B); many are rodlike.

Mullion structure consists of a series of parallel columns. Each column may be several inches in diameter and several feet long. Each column is composed of folded sedimentary or metamorphic rocks.

Plate 41. *Boudinage.* Light-colored dolomite interstratified with dark shaly limestone. Boudin in center is about 400 feet across. Limestone-dolomite series of Precambrian Eleanore Bay Group. Kejser Franz Joseph Fjord, East Greenland. Photo: Lauge Koch Expedition.

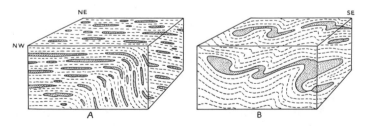

Fig. 19-3. Quartz rods. Quartz shown by dots, bedding by broken lines. (A) Quartz lenses parallel to bedding. (B) More irregular quartz lenses parallel to bedding. (Based on diagrams by G. Wilson.[7])

412

Attitude and Symbols

The attitude of secondary lineation is measured in the same way as primary lineation. The bearing is the azimuth of the horizontal projection of the lineation, whereas the plunge is the angle of inclination of the lineation.

The symbols for lineation may be shown independently of foliation or bedding, as in Fig. 19-4a, b, and c. But, as shown in Fig. 19-4g, it may be combined with a bedding symbol, or, as shown in Fig. 19-4d, e, and f, with foliation symbols. Lowercase letters may be used to indicate the kind of lineation, such as *p* for pebble, *h* for hornblende, *c* for crinkles, etc.

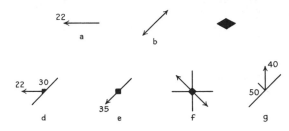

Fig. 19-4. Map symbols for lineation. (a) Lineation plunging 22°W. (b) Horizontal lineation striking NE. (c) Vertical lineation. (d) Foliation striking N. 45°E., dipping 30° NW.; lineation plunging 22°W. (e) Vertical foliation striking N.45°E., lineation plunging 35°SW. (f) Horizontal foliation, with horizontal lineation striking N.45°W. (g) Bedding striking N.45°E., dipping 50°NW.; lineation plunging 40°N.

Origin

DEFORMED PEBBLES, OÖLITES, AND MINERAL GRAINS

The orientation of the lineation relative to the major structure and the origin depend, of course, on the kind of lineation. In general, the lineation is systematically related to the major folds, but this is not necessarily the case if the lineation is the result of stresses independent of those that produced the folding.

The attitude of deformed pebbles, oölites, and grains of such minerals as quartz are generally related in some systematic way to the associated folds. But the relations may differ in different areas and in different parts of the same fold.

The attitude of deformed oölites in Maryland and Pennsylvania[5] are

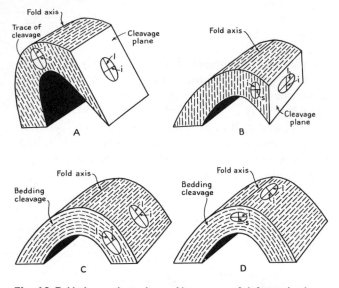

Fig. 19-5. Various orientations of long axes of deformed spheres. (A) In axial plane cleavage and perpendicular to fold axis. (B). In axial plane cleavage and parallel to fold axis. (C) In bedding plane cleavage and perpendicular to fold axis. (D) In bedding plane cleavage and parallel to fold axes.

illustrated in Fig. 19-5A. The fold axes are essentially horizontal, whereas the axial planes of the folds and the slaty cleavage dip to the southeast. The long axes (l) and intermediate axes (i) of the ellipsoidal oölites lie in the plane of the slaty cleavage. The long axis of each oölite plunges southeast essentially down the dip of the cleavage. It is apparent that in this case, because of a couple acting along northwest-southeast lines (Fig. 6-1B), the thick pile of sediments was elongating upward toward the northwest, partly by folding but especially by rock flowage.

Near Newport, Rhode Island, the Carboniferous sediments contain some very coarse-grained conglomerates. Many of the boulders are from one to three feet long. The fold axes plunge gently south. The long axes of the deformed boulders also plunge gently south (Fig. 19-5B). The intermediate axes are essentially vertical, parallel to the axial plane of the major fold; the short axes are almost horizontal, striking east-west, and are thus perpendicular to the axial plane of the major fold. Under east-west compression the sediments were thrown into folds striking north-south; the easiest relief was north-south, and thus the sediments elongated in that direction.

In much of eastern Vermont the long and intermediate axes of the pebbles on the limbs of folds lie parallel to a bedding cleavage (Fig. 19-5C). The fold axes are are essentially horizontal, and the long axes of the pebbles plunge directly down the dip. This type of deformation is the result of stretching of the limbs of the folds, accompanied, of course, by considerable thinning of the beds. Near the crests of the folds, however, the long axes of the pebbles are parallel to the fold axes (Fig. 19-5D). Here the easiest relief was parallel to the fold axes. In western New Hampshire the long axes of stretched pebbles are oriented this way.

ELONGATE MINERALS

Lineation shown by elongate minerals, such as hornblende needles or flat oval flakes of mica, likewise show various orientations. In these cases, however, the mineral was not plastically deformed into its present shape, but grew in that direction because it was the easiest direction of growth.

INTERSECTION OF BEDDING AND CLEAVAGE

A lineation that is the result of the intersection of bedding and an axial plane slaty cleavage is parallel to the fold axis. A consideration of the geometry of folds shows that this must be the case (Fig. 18-10). Similarly the intersection of bedding and of fracture cleavage that is related to the folding is parallel to the fold axes (Fig. 19-6B). The same is true of the intersection of fracture cleavage with slaty cleavage that is parallel to the axial planes of the folds.

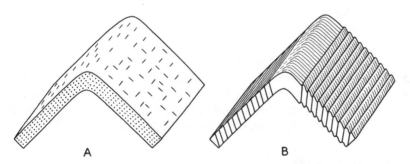

A   B

Fig. 19-6. Different kinds of lineation on folds. (A) Two lineations, one parallel to axis of fold and other at right angles. (B) Left-hand limb shows lineation due to intersection of fracture cleavage with bedding. Right-hand limb shows lineation due to intersection of shear cleavage with bedding; displacement on the cleavage causes either tiny faults or small crinkles.

CRINKLES

If a bedding cleavage is thrown into small crinkles the axes of the crinkles are likely to be parallel to the major fold axes. This is because the crinkles are essentially drag folds. After an axial plane cleavage has been produced, further compression may cause slippage along this cleavage toward the major fold axes. The axes of any crinkles will likewise be parallel to the major fold axes. In many cases, however, the rocks may be so tightly compressed that the differential movements parallel to the bedding or cleavage are unrelated to the major fold axes; consequently the crinkles may be oriented in most any direction. In some areas it is not uncommon to find several sets of differently oriented crinkles. They were produced in succession as the rocks slipped in different directions parallel to the cleavage or bedding.

SLICKENSIDES AND MINERAL STREAKS

Normally, as shown on page 90, the beds on the flank of folds slip past one another perpendicular to the fold axes. Consequently any slickensides on the bedding planes are perpendicular to the fold axes. Similarly, any mineral streaking (streaming) that is the result of differential movement (Fig. 19-1C) is likely to be perpendicular to the fold axes. Figure 19-6A shows two types of lineation. That parallel to the fold axis is a minute crinkling. That down the dip of the bedding is a mineral streaking caused by slippage of the beds past one another.

BOUDINAGE, RODDING, AND MULLION STRUCTURE

If boudinage is the result of stretching of a competent bed on the limbs of a fold, the boudin line will be parallel to the fold axis (Fig. 19-2). But if the boudinage is caused by stretching parallel to the fold axes, the boudin line will be perpendicular to the fold axis. The quartz rodding in Scotland is parallel to the fold axes. This is because some of the rods are the detached noses of folds. It is also because quartz layers are broken up into rods, the long axes of which are parallel to the fold axes. Mullion structure consists of the detached noses of folds of sedimentary or metamorphic rocks.

INCONGRUOUS ORIENTATIONS

It is obvious from the preceding discussion that even within a small body of rock some of the linear features may be parallel to the fold axes, whereas others may be perpendicular. Moreover, it is worthwhile emphasizing again that the lineation is congruous with the major structure only if it forms simultaneously with the major structure or is produced under similar conditions of stress. But if the lineation is distinctly younger or older than the folding or thrusting, it may be incongruous to the major structure. Even during a single period of deformation the rocks may be so badly squeezed that masses of rocks deform plastically quite independently of the major structure. The lineations may then be very unsystematic. Detailed study may show that an apparently erratic lineation is systematic.[6]

Successive Lineations

In earlier chapters it has been shown that more than one stage of folding may deform the strata. These successive stages of folding may be labelled F_1, F_2, F_3, etc. (page 66). These stages may affect separate parts of the area being studied, or all of them may affect the entire area, resulting in very complex geometrical patterns. Each of these stages of folding may be accompanied by a foliation, often the axial plane type, and labeled S_1, S_2, S_3, etc. (page 406). Moreover, each of these stages of folding may also be accompanied by the development of a lineation. Each of these lineations may be designated L_1, L_2, L_3, etc. If more than one type of foliation or lineation developed during any one period of folding, even more complex designations would be necessary, such as S_l', S_l'', S_l''', or L_l', L_l'', L_l'''.

A hypothetical case, based however in part on an area in Scotland, is shown in Fig. 19-7. This has resulted from three stages of folding. (a) A large isoclinal fold (F_1), the axial surface of which trends N.30°E. and dips 80°SE. The fold plunges 10°N.35°E. (b) A second stage of folding (F_2), the

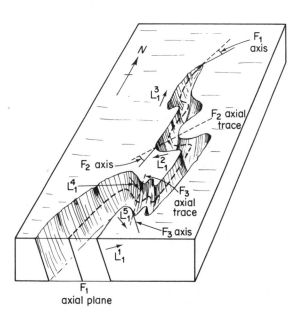

Fig. 19-7. Orientation of original lineation changed by succeeding stages of folding. Three stages of folding, F_1, F_2, F_3. In first stage of folding the axial plane of an isoclinal anticline dips steeply east and plunges 15° north. Lineation, L_1^1, forms parallel to fold axes at this time. But this lineation is reoriented to positions such as L_1^2 and L_1^3 by a second folding, and to such positions as L_1^4 and L_1^5 by a third stage of folding.

axial surfaces of which strike N.60°E. and dip 60°SE; the axis plunges 40°S.15°W. (c) A third stage of folding with the axial surfaces striking N.15°E. and dipping vertically; the plunge is also vertical. The axial plane cleavage S_1 will obviously be deformed by F_2 and F_3. Let us also assume that F_1 is accompanied by a lineation plunging 10° in a direction N.35°E. (L_1^1). But as a result of the F_2 folding, L_1^1 will be rotated to new positions, of which L_1^2 and L_1^3 are only two examples. Similarly, as a result of the F_3 folding, L_1^1 is rotated to new positions, of which L_1^4 and L_1^5 are only two examples. Methods of analyzing these problems of rotated lineations are described in Laboratory Exercises 12 to 14. If, in the example cited, new lineations were produced during foldings F_2 and F_3, it is apparent that a very complex pattern of lineation would develop.

Relation of Minor Structures to Overthrusts

A superb example of the relationship of minor structures, including lineation, to the major structure, has been described from England.[7] In the area described it has been known for some time that thrust planes dip gently northwest. Many have assumed that the relative overthrusting was from the northwest. However, an analysis of the many minor structures show that the relative overthrusting was in a direction N.20° to 30°W. Only some of the significant data are shown in Fig. 19-8. Linear features parallel to the move-

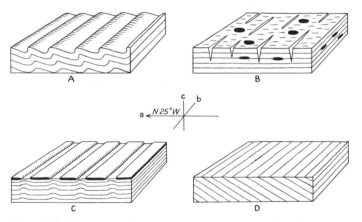

Fig. 19-8. Relation of minor structures to overthrusting in Tintagel area, North Cornwall, England. (A) Drag folds. (B) slickensides shown by short lines on top of block; deformed pillows and amygdules shown in solid black; tension cracks shown by open gashes. (C) Boudinage. (D) Fracture cleavage. (After Gilbert Wilson.[7])

ment direction, *a*, are slickensides, elongated lava pillows, and elongated amygdules (gas bubbles partially or completley filled by minerals) (Fig. 19-8B). Linear features parallel to *b* are the axes of drag folds (Fig. 19-8A), tension cracks (Fig. 19-8B), boudinage (Fig. 19-8C), and intersection of fracture cleavage with bedding (Fig. 19-8D). The dip of the axial planes of the drag folds and the dip of the fracture cleavage show that the relative overthrusting was toward the northwest.

Lineaments

A lineament is expressed on the surface of the earth as a relatively straight line and it presumably represents the trace of a fracture or fracture system on the surface. Some lineaments are readily recognized on topographic maps or aerial photographs. Others may be suggested by discontinuities of geologic structures. Still others may be suggested by the alignment of volcanoes. Most lineaments are caused by steeply dipping faults or joint systems. A long dike, eroded below the general level of the country rock, would produce a lineament. But a dike that was harder than the country rock would be readily recognized as a dike and so called. In a sense, therefore, lineament is a negative term, meaning that the exact cause is unknown. But the term may also be used even if the cause is well established. In the literature the term linear is sometimes misused for lineament.

References

[1] Turner, Francis J., and Weiss, Lionel E., 1963, *Structural analysis of metamorphic tectonites*, 545 pp., New York: McGraw-Hill Book Co. Inc.

[2] Ramsay, John G., 1967, *Folding and fracturing of rocks*, 568 pp., New York: McGraw-Hill Book Co. Inc.

[3] Cloos, Ernst, 1946, *Lineation, a critical review and annotated biliography*, Geol. Soc. Amer. Mem. 18. Supplement, Review of Literature, 1942–52, 1953.

[4] Wilson, Gilbert, 1953, *Mullion and rodding structures in the Moine Series of Scotland*, Proc. Geol. Assoc. Vol. 64, part 2, pp. 118–51.

[5] Cloos, Ernst, 1947, Oölite deformation in the South Mountain Fold, Maryland, *Geol. Soc. Amer. Bull.* 58: 843–918.

[6] Lowe, K. E., 1946, A graphic solution for certain problems of linear structure, *Amer. Mineral.* 31: 425–34.

[7] Wilson, Gilbert, 1951, *The tectonics of the Tintagel area, North Cornwall*, Quart. J. Geol. Soc. London CVI: 393–432.

20

PLASTIC DEFORMATION

Introduction

In earlier chapters some evidence has been presented to show that rocks may acquire a permanent change in shape. In Chap. 2 the experimental evidence for plastic deformation was given. Folding (Chaps. 3 to 6) is the result of plastic deformation. Certain types of cleavage result from rock flowage. But in these earlier chapters we were not especially concerned with the internal changes within the rock. How is plastic deformation possible? But before analyzing the internal movements within the rock we may discuss some methods of making quantitative estimates of the amount of strain.

Evidence of Strain

INTRODUCTION

Various types of evidence give a quantitative or semiquantitative measurement of the amount of strain that has affected deformed rocks. These include fossils, primary sedimentary structures, clasts, and oölites.

DEFORMED FOSSILS

Figure 20-1 shows fossils that have been deformed. Figure 20-1A is a brachiopod. Diagram *a* is the original shape before deformation. If there is no significant volume change and elongation is parallel to the hinge line of the brachiopod, the deformed fossil will appear as in *b*. But if the lengthening is at right angles to the hinge line, the appearance will be as in *c*.

Figure 20-1B is a deformed plant in which the elongation was as shown.

The undeformed and deformed fossils may be found in the same general area or in different parts of the same fold. The amount of deformation can be calculated, assuming no volume change. But this analysis gives only the component of the deformation within the bedding plane. The maximum elongation and shortening may not be in the bedding plane. Moreover, the fossils in many cases may be more rigid than the matrix, and hence less deformed.

DEFORMED SEDIMENTARY STRUCTURES

Deformed primary sedimentary features may be used in a similar way. In Fig. 20-2 it is assumed that an original flexure fold, with dips ranging from zero to 90°, was subjected to additional horizontal compression. But the rocks could no longer yield by flexure folding. Instead they deformed by flowage, the mass becoming shorter horizontally but elongating vertically. Figure 20A shows the effects on ripple marks. The deformation of cross-bedding is shown very diagrammatically in Fig. 20-2B. So many variables are involved in the deformation of primary sedimentary structures that great caution is necessary in utilizing them to measure the amount of strain.

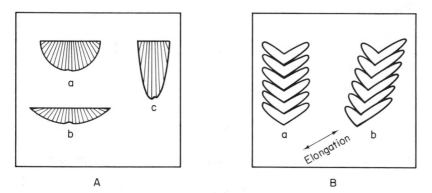

A   B

Fig. 20-1. Deformed fossils. (A) Deformed brachiopod. (a) Original shape. (b) Elongation parallel to hinge line. (c) Elongation at right angles to hinge line. (B) Deformed plant. (a) Original shape. (b) After elongation as shown.

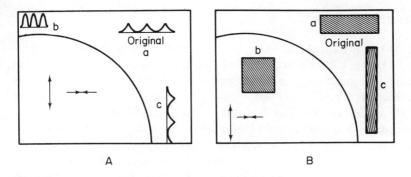

Fig. 20-2. Deformed primary sedimentary structures. Elongation is vertical, shortening horizontal. (A) Ripple marks. (B) Cross-bedding.

DEFORMED CLASTS AND OÖLITES

Bodies that were originally spherical may be deformed into ellipsoids. The fact that the bodies were originally spheres may be very clear in some cases.

In an area in Maryland and Pennsylvania[1] it is possible to trace unde-formed oölites in limestone into a region where they are deformed into ellip-soids. In many metamorphic areas "stretched conglomerates" are evidence of deformation. In some instances it is possible to trace the conglomerates into areas where the clasts are essentially spherical. But if a statistical study is made of several hundred elongate clasts, it is not necessary that they were originally spherical. It is only necessary that there was no preferred orienta-tion of the long, intermediate, and short axes of the clasts before deforma-tion.

The amount of deformation may be calculated as follows:

The volume of a sphere is

$$V_s = \tfrac{4}{3}\pi r^3 \tag{1}$$

where V_s is the volume and r is the radius of the sphere.

The volume of an ellipsoid is

$$V_e = \tfrac{4}{3}\pi \cdot a \cdot b \cdot c \tag{2}$$

where V_e is the volume of the ellipsoid, a is the half-length of the greatest axis, b is the half-length of the intermediate axis, and c is the half-length of the least axis.

If a sphere is deformed into a ellipsoid without any change in volume, then

$$V_s = V_e \tag{3}$$

or

$$\tfrac{4}{3}\pi r^3 = \tfrac{4}{3}\pi \cdot a \cdot b \cdot c \tag{4}$$

or

$$r = \sqrt[3]{a \cdot b \cdot c} \tag{5}$$

where *r* is the radius of the original sphere from which the ellipsoid was derived.

In a hypothetical example let us assume that the following facts are established in a region of deformed conglomerates: (1) the short axes of the pebbles are perpendicular to the cleavage, whereas the other two axes lie in the plane of the cleavage; (2) all the long axes of the pebbles are parallel to one another; (3) comparison with the same formation in areas where it is less deformed indicates that the pebbles were originally spherical; (4) although the pebbles are of different sizes, the ratios of the lengths of the axes to one another are similar. The axes of one representative pebble are 2, 4, and 6 centimeters. The half-axes are then 1, 2, and 3 centimeters. Using equation (5), *r* is 1.8 centimeters; the diameter was 3.6 centimeters. This means that the original sphere was shortened 1.6 centimeters perpendicular to the cleavage, but was lengthened 0.4 centimeters parallel to the intermediate axis and 2.4 centimeters parallel to the long axis. The rock mass as a whole must have deformed in a similar manner, although under some circumstances the matrix of a conglomerate may deform plastically even more than the pebbles.

Mechanics of Plastic Deformation

PROBLEM

The plastic deformation of solids[2,3] is a subject of utmost importance to structural geology. How can solid rocks change their shape without the appearance of any visible fractures? Just what happens within the rocks to permit such a change in form? The processes may be classified into intergranular movements, intragranular movements, and recrystallization.

INTERGRANULAR MOVEMENTS

Intergranular movements involve displacements between individual grains. Intrusive igneous rocks are usually composed of such minerals as quartz, feldspar, mica, and hornblende. Sandstones consist of rounded grains, usually quartz, cemented together. Limestones are composed of small interlocking crystals of calcite. If such rocks are subjected to stress, the individual crystals and grains may move independently. All the displacements, because they are between grains, may be described as intergranular. The individual grains maintain their shape and size. The deformation of such a body might be compared to the change in shape undergone by a moving mass of BB shot. Each grain can move and rotate relative to its neighbors.

In the plastic deformation of metals, such intergranular movements seem to be of subordinate importance. In rocks, particularly those of granitoid character, in which the crystals tend to interlock, more or less *granula-*

tion takes place first; that is, the larger crystals are broken into smaller spherical grains that may rotate relative to each other.

INTRAGRANULAR MOVEMENTS

Intragranular movements are very important in the plastic deformation of metals, a subject in which a large amount of experimental work has been done. Intragranular movements are also very important in the deformation of rocks. The displacements take place within the individual crystals by gliding and dislocations.[4,5,6,7]

Some minerals have no *glide planes*. In others there is one glide plane—that is, a plane parallel to which there are a vast number of additional planes along which gliding takes place. In other minerals there are several glide planes—that is, several planes parallel to each of which there are a vast number of additional planes. During deformation several glide planes may operate simultaneously.

The atomic structure controls the position and number of glide planes. Hence the glide planes are related to the atomic structure. Actually we must think in terms of the *space lattice* rather than individual atoms. The space lattice of a mineral is that assemblage of atoms that is constantly repeated to form a mineral. Gliding is of two types, translation-gliding and twin-gliding.

Translation-gliding is illustrated very diagrammatically by Fig. 20-3. Each unit of the space lattice is represented by dots, and the glide planes by heavy horizontal lines labelled g_1g_1 and g_2g_2. Diagram A shows the space lattice before gliding. Diagram B shows the arrangement of the space lattice after gliding. The shape of the figure as a whole has been changed. The space lattice shows the same pattern. The distance between glide planes depends on the substance. In gold it is 0.00045 mm, in zinc 0.00080 mm.

In *twin-gliding* the layers slide a fraction of an interatomic distance relative to the adjacent layers (Fig. 20-4). Figure 20-4A shows the space lattice before gliding, whereas Fig. 20-4B shows the pattern after gliding. In this way the lattice of the displaced part of the crystal is symmetrically

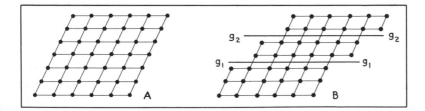

Fig. 20-3. Translation gliding. Each dot represents center of a unit of the space lattice. (A) Before gliding. (B) After gliding along planes g_1g_1 and g_2g_2.

altered with respect to the lower, undisplaced part. In the language of mineralogy, the displaced part bears a twinned relation to the undisplaced part.

The lattice cannot slip along the glide plane in all directions. There are a limited number of lines parallel to which the movement can take place, and these are known as the *glide directions*. The number and position of the glide planes and glide directions depend upon the mineral. Aluminum has four glide planes and three glide directions in each; hence there are twelve possible movements in this metal.

So far it has been assumed that entire layers of the space lattice slide past one another. But it has long been realized that far less energy is required if the movement is piecemeal. That is, small portions of the space lattice move progressively by dislocations.[7] A *dislocation* is an imperfection in the crystal structure. This imperfection can move through the crystal. In an edge dislocation the lattice slips in the direction in which the imperfection is spreading. In a screw dislocation the imperfection spreads at an angle to that in which the lattice is moving. By gliding and dislocations a grain may thus change shape (Fig. 20-5).

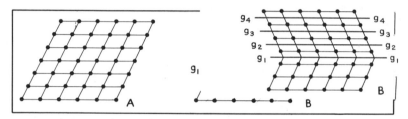

Fig. 20-4. Twin gliding. Each dot represents center of unit of the space lattice. (A) Before gliding. (B) After gliding on planes g_1g_1, g_2g_2, g_3g_3, and g_4g_4.

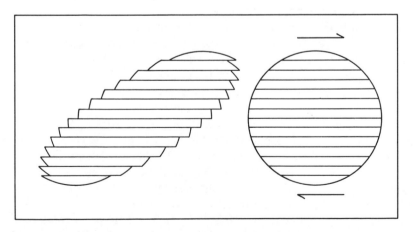

Fig. 20-5. Diagrammatic example of a circle change into an ellipse by movement along glide planes.

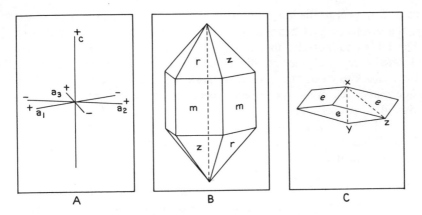

Fig. 20-6. Hexagonal minerals. (A) Conventional system of desig-
nating axes of hexagonal minerals. (B) Quartz crystal: *m*, unit
prism; *r*, positive unit rhombohedron; *z*, negative unit rhombohedron. Broken line is the *c*-crystallographic axis which equals
the optic axis (C) Calcite crystal: *e*, negative rhombohedron; *xy*,
is the *c*-crystallographic axis which equals the optic axis: *xz*,
short diagonal of rhombohedron.

Our knowledge of glide planes in rock-forming minerals is based on both
experimental data and the laboratory study of specimens collected in the
field. Experimental study has largely concentrated on calcite and quartz.

By the use of the Universal Stage—a device that can be attached to the
petrographic microscope and rotated on horizontal and vertical axes—it is
possible to prepare *petrofabric diagrams* on an equal-area net. In Fig. 20-6B
the quartz crystal is oriented so that the *c*-optic axis is vertical. Of course,
quartz grains in rocks are very irregular. But they all have a *c*-optic axis.
Figure 20-7 was prepared from a mica schist in New Hampshire. Figure
20-7A is a point diagram of the *c*-axes of 400 quartz grains. The contour
diagram, Fig. 20-7C, shows there is no strong preferred orientation, but the
perimenter of the circle is slightly favored. Since *aB* is the schistosity, there is
a weak girdle perpendicular to the schistosity and around *B*. Figure 20-7B
is based on the perpendiculars to the cleavage of 200 biotite flakes. The
contour diagram (Fig. 20-7D) shows, not unexpectedly, that the cleavage
of the biotite flakes is parallel to the schistosity of the rock.

Many thin sections of calcite show that a single grain is composed of
alternating sheets or layers with different optical behavior. When the nicols
of the microscope are crossed, one set may be dark when the other set is
light. These are twin lamellae, commonly parallel to the rhombohedron *e*
(01$\bar{1}$2) (see Fig. 20-6C). Many petrofabric diagrams of calcite represent the
attitude of the perpendiculars to these twin lamellae. Experimental data
show that calcite glides on the rhombohedron *e* (01$\bar{1}$2); the short diagonal

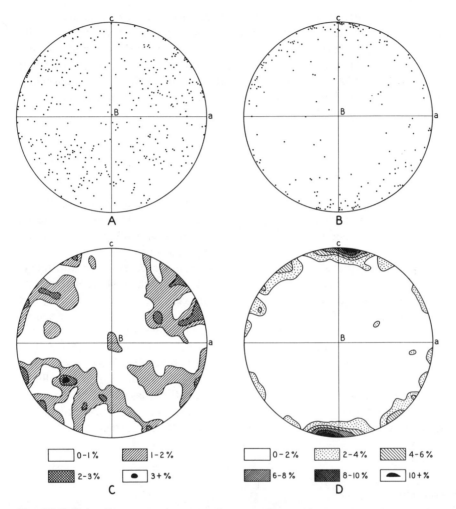

Fig. 20-7. Point diagrams and contour diagrams. Plots on lower hemisphere. (A) Point diagram. 400 optic axes of quartz from mica schist, Mt. Clough, New Hampshire. (B) Point diagram. 200 poles of cleavage of biotite from mica schist, Mt. Clough, New Hampshire. (C) Contour diagram of (A). (D) Contour diagram of (B).

(*xz* of Fig. 20-6C) of the rhombohedron is parallel to the translation direction *a* of the rock. Moreover, the acute angle between this short diagonal and the *c*-crystallographic axis (*yxz* Fig. 20-6C) generally opens in the glide direction. There is also twin gliding on *m* (10$\bar{1}$0) and translation gliding on *r* (10$\bar{1}$1) and *f*(02$\bar{2}$1).[8] On each of these glide planes there is a glide direction and glide sense.

Until recently quartz has been difficult to study experimentally, but apparatus to handle high temperatures and high pressures now permits this.[9] There is now evidence that quartz possesses the following glide planes: (a) base, c (0001), glide direction parallel to an a-axis; (b) prism, m (10$\bar{1}$0) parallel to the c [0001] axis. So-called deformation lamellae are a fraction of a millimeter thick and are recognized by differences in birefringence. They are inclined 10° to 30° to the base, c (0001).

RECRYSTALLIZATION

Recrystallization is another mechanism aiding plastic deformation. Rocks can crystallize without any change in shape, as is shown by limestone altered to marble near igneous intrusions. The number of crystals per unit volume decreases, but the size of the individual crystals increases.

Under conditions of differential pressure, however, solution and recrystallization may proceed in such a way that the rock is shortened in one direction and lengthened in another. The process may be explained by the *Riecke principle*. According to this principle the solutions in the pore spaces of the rocks dissolve that portion of the crystal under greatest stress. At the same time there is precipitation on that portion of the crystal subjected to the least stress. In this way the grain changes shape. If all the crystals in a body of rock are similarly affected, the mass as a whole changes shape.

It is obvious from observations in the field that plastic deformation and recrystallization have often been simultaneous, and we must accept the principle that recrystallization greatly facilitates plastic deformation.

Dynamic Analysis
of Petrofabric Diagrams

The orientation of the principal stress axis may be deduced from an analysis of the petrofabric diagrams.[8] A marble or limestone is investigated in the following way. The data are obtained from 100 or so grains of calcite in a thin section mounted on a Universal Stage. The orientation of the most prominent twin lamellae e (01$\bar{1}$2) is measured in a grain; the orientation of the c-optic axis in the same grain is measured. A restriction is that the plane of the thin section should be parallel to the glide direction in the mineral.

Such a grain is recognized by the fact that the angle between the optic axis and the perpendicular to the twin plane should be 24° to 28°.

The data are then plotted on an equal-area net (Fig. 20-8). The twin lamella, e', of grain 1 is represented by the line trending from the lower left to upper right. The perpendicular to this, $\perp e'$, is on the upper left part of the circumference and the pole of the optic axis, c', is at the top. The two points are on the circumference in this instance because the twin lamella is perpendicular to the thin section. The greatest principal stress axis, σ'_1, is then plotted on the plane containing $\perp e'$ and c', at a distance of 45° from $\perp e'$ in a direction away from c'. The least principal stress axis, σ'_3, is 45° from

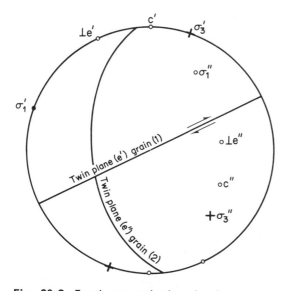

Fig. 20-8. Equal-area projection showing graphic method for constructing principal stress axes that are best oriented to produce twin-gliding in calcite. (Permission Geological Society of America, N. L. Carter and C. B. Raleigh.[8])

$\perp e'$, on the same plane, but toward c'. The more general case, in which the twin lamella is inclined to the thin section, is shown as plane 2. The points corresponding to those described for plane 1 are $\perp e''$, c'', σ''_1, and σ''_3. Thus for each grain the orientation of the principal stress axes can be determined. The basic assumption in this analysis is that the glide plane and glide direction make an angle of 45° to the greatest principal stress axis.

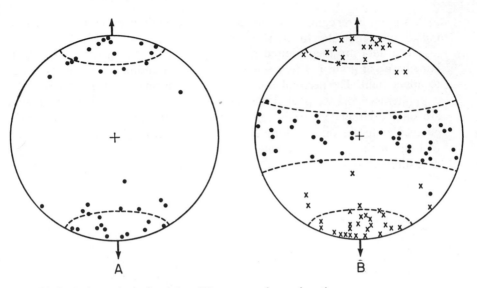

Fig. 20-9. Deformation of calcite. (A) c-axes of prominently twinned calcite grains in experimentally extended Yule Marble. (B) Least principal stress axis (σ_3', crosses) and greatest principal stress axis (σ_1', dots) constructed by analysis outlined in text. (σ_3' in (B) agrees with axis of tension in (A).) (After Carter and Raleigh. Permission Geological Society of America, N. L. Carter and C. B. Raleigh.[8])

A few examples may be cited. Figures 20-9A and 20-9B represent data obtained from twinned calcite grains in Yule Marble that had been extended experimentally.[8] The axis of tension extends from top to bottom. Figure 20-9A shows the orientation of the optic axes of calcite after deformation. Figure 20-9B shows the orientation of the greatest (solid dots) and least (crosses) principal stress axes as determined in specimens obtained from the rock after deformation. The least principal stress axis, σ_3 is deduced to be oriented from top to bottom in agreement with experimental conditions. The greatest and intermediate principal stress axes ($\sigma_2 = \sigma_1$) are equal and form a girdle on the equator of the diagram.

Figure 20-10 shows the dynamic analysis of calcite twin lamellae in a fold.[8] Figure 20-10A shows the fold. Figure 20-10B shows the orientation of σ_1 and σ_3, as deduced from the calcite twin lamellae, in the upper part of the hinge. Figure 20-10C shows the orientations in the lower part of the hinge. Although there is considerable scatter, in Fig. 20-10B the axes of greatest compression tend to be on the right and left sides of the diagram, whereas σ_3 forms a north-south girdle across the center of the diagram. Conversely, in Fig. 20-10C, σ_3 forms an east-west girdle across the center of the diagram, whereas σ_1 is at the top and bottom. This is what one would expect—compression parallel to the bedding on the inside of the fold, tension parallel to the bedding on the outside.

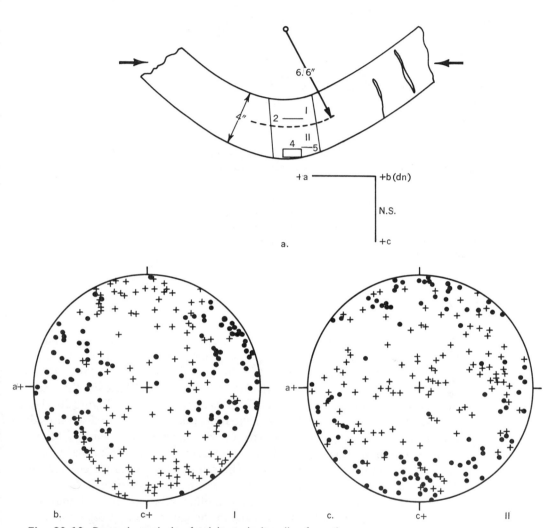

Fig. 20-10. Dynamic analysis of calcite twin lamellae from the axial region of a small fold in limestone. (A) Sketch of the fold showing orientation of thin sections in Regions I and II from which data in (B) and (C) were obtained. (B), (C). Compression axes (dots) and extension axes (crosses) derived from twins in Regions I (B) and II (C). The center of concentration of σ_1' axes in Region I is parallel to the a reference axis and the σ_3' axes are concentrated near the c reference axis spreading into a broad girdle normal to a. The reverse is true for Region II, but the data show considerably more scatter. (Permission Geological Society of America, N. L. Carter and C. B. Raleigh.[8])

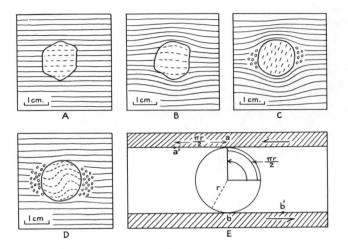

Fig. 20-11. Rotation of garnets. (A) Garnet that grew by replacement, no rotation. (B) Garnet the grew partially by replacement, partially by pushing aside the adjacent schist. (C) Garnet that rotated after it grew. (D) Garnet that rotated as it grew. (E) Calculation of differential movement (see text).

Rotated Minerals

Rotated minerals such as garnet, albite, and staurolite not only give striking evidence of significant differential movement in rocks, but also permit a rough quantitative estimate of the amount of the movements.[4] Figure 20-11 illustrates some of the various structural relations of garnet to schistosity that may be observed in thin sections of rocks from metamorphic areas. Figure 20-11A represents a garnet crystal that is euhedral—that is, has a well-developed crystal habit—because it grew quietly in the schist by replacement. Many of the atoms in the garnet entered from the adjacent schist while other atoms diffused out into the adjacent schist. The schistosity is undistorted. Figure 20-11B is a garnet that formed in a similar manner but as it grew it deformed the schistosity. But Fig. 20-11C shows another relationship that may be observed in some garnets. Small streaks in the garnet lie at a high angle to the schistosity. Similar impurities in the adjacent parts of the schist are parallel to the schistosity. These relations indicate that the garnet was rotated after it formed. This rotation is presumably the result of differential movements parallel to the schistosity as in the analogy of a sliding pack of cards. Still another relationship is shown in Fig. 20-11D. This shows that the garnet rotated as it grew. The interior of the garnet had rotated 90° by the time crystallization ceased, but shells progressively further from the center rotated less and less because they are younger.

At least one qualifying remark is necessary. In Figs. 20-11C and 20-11D it is apparent that the axis of rotation is perpendicular to the page of the paper. Moreover, it is clear that in a thin section cut parallel to this axis, the streaks of impurities would appear to be parallel to the schistosity. In fact, the relations in such a thin section would be like those in Figs. 20-11A and B. Therefore, in order to reach the conclusion made in the first paragraph concerning Figs. 20-11A and B, it is necessary to have two thin sections more or less at right angles to one another.

The axis of rotation of the garnets is *b*. The direction in the plane of schistosity at right angles to *b* is *a*. The perpendicular to the schistosity is *c*.

To determine the amount of differential movement let us consider a case such as that shown in Fig. 20-11C, but in which the streaks of impurities lie at a 90° angle to the schistosity. Furthermore, let us assume that the rotation of the garnets may be compared to the rolling of a sphere between two boards shearing over one another (Fig. 20-11E). If *r* is the radius of the garnet the outer edge of the garnet has rotated a distance $\pi r/2$ (one-quarter of the circumference). A point on the upper board that was originally at *a* has moved to *a'*, and a point on the lower board originally at *b* has moved to *b'*. The inferred movement of the upper board relative to the lower board is consequently πr. That is, in a layer with the thickness $2r$, the differential movement is πr if the garnets have rotated 90°. If *t* is the thickness of the bed, then the differential movement is $\pi t/2$. If all the garnets in beds 1000 feet thick show such rotation, it follows that the top of the beds has sheared 1570 feet relative to the bottom of the beds. Fairbairn states that the differential movement may often be three times the thickness and in one extreme case was as high as 5.6 times the thickness. Although the analogy to a sphere being rolled between two boards is undoubtedly too idealized, such calculations give clear evidence of the intense differential movements that have taken place.

The schistosity associated with such rotated minerals is obviously not a simple flow cleavage. It could be (1) bedding cleavage, (2) initially a flow cleavage that was later utilized for differential movements, or (3) a shear cleavage that has been transformed into a schistosity (page 398).

An example of the use of rotated garnets in unraveling tectonic events has been described from southeastern Vermont.[10] During the Mid-Devonian deformation there were two stages of garnet rotation. The axes of rotation were not parallel during the two stages but had a general north-south bearing and variable plunges. Large recumbent folds developed during stage I. Doming during stage II deformed the recumbent folds. The garnets were rotated as they grew due to differential movements on the limb of the folds. Some garnets that were rotated counterclockwise (looking north) during the stage of recumbent folding were rotated clockwise during the doming. The amount of rotation was not uncommonly 90° to 180°, in some instances as much as 270°, and in one case 560°.

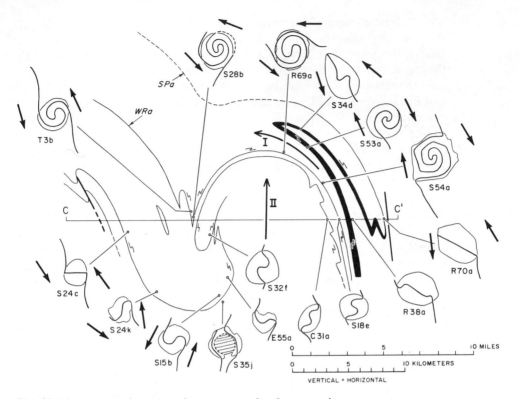

Fig. 20-12. Relation of rotation of garnets to regional structure in southeastern Vermont. East-west section; because of plunge of structure the data can be projected into one section. Є is near base of Cambrian; rocks beneath it are Precambrian. *D* is near base of Devonian. *Dsp* is volcanic unit in Devonian. Note inversion of structure in places: on west side of dome (II) the Precambrian is on top of Cambrian in a synform. On east side of dome the heavy dark band (*Dsp*) is a sharp antiform, with younger Devonian rocks in core; heavy arrows show direction of rotation of garnets. Heavy arrow I shows direction of early movements, heavy arrow II shows later doming. (Permission Interscience Publishers, division of John Wiley & Sons.)

Figure 20-12, an east-west cross section 20 miles long, shows the relations of the rotations to the regional structure. Because of the northerly plunge of the structure it is possible to show the structure above and below the line of the cross section. The cross section is approximately at right angles to the axes of rotation of the garnets. If a spiral turns toward the left going from the outside toward the center of the garnet, the rotation is counterclockwise. If the spiral turns right, the rotation is clockwise. Thus specimen S28b has rotated counterclockwise 560°. Most of the interior of specimen S54a has rotated counterclockwise, but the last rotation was clockwise.

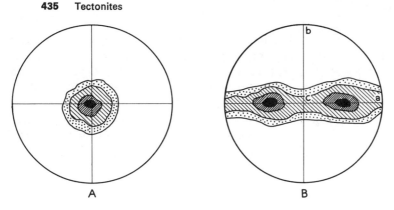

Fig. 20-13. Petrofabric diagram of tectonites. Contour diagrams of perpendiculars to cleavage of biotite. (A) *S*-tectonite. (B) *B*-tectonite.

Tectonites

A *tectonite* is a deformed rock, the fabric of which is due to the systematic movement of the individual units under a common external force. A *nontectonite* is a rock that results from the accumulation of many separate components, each of which moved into place independently of its neighbors; all undeformed sedimentary and igneous rocks belong in this category.

Two principal kinds of tectonites are frequently referred to in the literature: *S-tectonites and B-tectonites.*

In an *S*-tectonite the prominent structural feature is an *S*-plane, which may be a visible foliation or a theoretical plane deduced from a petrofabric diagram. Interpretation involves slip planes, which are real or theoretical surfaces. Each grain is believed to rotate until one of its glide planes becomes subparallel to a slip plane. When the shearing stress parallel to the glide direction exceeds a critical value, gliding takes place. The grain remains locked in this position and continues to deform by gliding. Since many grains do this, the minerals attain a preferred orientation. For example, visualize a quartz grain that rotates until a prism plane is essentially parallel to the slip plane and the *c*-crystallographic axis is essentially parallel to the direction of movement *a* in the slip plane. If the grain then becomes locked in this position, but now deforms by gliding, the *c*-axis will show a strong maximum in the *a* direction of the rock. Fig. (20-13A).

In a *B*-tectonite a lineation is prominent, and in some instances *S*-planes may be absent. The petrofabric diagram of a *B*-tectonite is usually characterized by a girdle, a more or less complete belt in which the points are concentrated; there may be one or more maxima within this girdle (Fig. 20-13B).

The girdle characteristic of *B*-tectonites may develop in one of several ways. (1) Several slip planes intersecting in *b* might be utilized for gliding. For example, if different calcite grains were to glide parallel to the short diagonal of the negative rhombohedron (xz of Fig. 20-6C) simultaneously along several such slip planes, the c-crystallographic axis (xy of Fig. 20-6C) would lie in a girdle about b. (2) If a single slip plane rotating about b were utilized for this same type of gliding, the c-crystallographic axis would lie in a girdle about b. (3) Individual grains might rotate until the plane containing the c-crystallographic axis (xy) and the short diagonal of the negative rhombohedron (xz) is more or less parallel to the ac plane of the fabric. The grain would then tend to stay in this position and deform by gliding.

References

[1] Cloos, Ernst, 1947, Oölite deformation in the South Mountain Fold, Maryland, *Geol. Soc. Amer. Bull.* 58: 843–918.

[2] Bridgman, P. W., 1952, *Studies in large plastic flow and fracture*, 362 pp., New York: McGraw-Hill Book Company, Inc.

[3] Nadai, A., 1950, *Theory of flow and fracture of solids*, 572 pp., New York: McGraw-Hill Book Company, Inc. See especially pp. 24–26.

[4] Fairbairn, H. W., 1949, *Structural petrology of deformed rocks*, 2d ed., Cambridge: Addison-Wesley Press.

[5] Knopf, E. B., and Ingerson, E., 1938, *Structural petrology*, Geol. Soc. Amer., Mem. 6.

[6] Turner, Francis J., and Weiss, Lionel E., 1963, *Structural analysis of metamorphic tectonites*, 545 pp., New York: McGraw-Hill Book Company, Inc.

[7] Moffatt, William G., Pearsall, George W., and Wulff, John, 1964, *The structure and properties of material, Vol. 1, Structure*, 236 pp., New York: John Wiley & Sons.

[8] Carter, Neville L., and Raleigh, C. Barry, 1969, Principal stress directions from plastic flow in crystals, *Geol. Soc. Amer. Bull.* 80: 1231–64.

[9] Baëta, R. D., and Ashbee, K. H. G., 1969, Slip systems in quartz, *Amer. Mineral.* 54: 1551–82.

[10] Rosenfeld, John L., *Garnet rotations due to major Paleozoic deformations in southeast Vermont*, in E-an Zen et al., 1968, *Studies in Appalachian geology, northern and maritime*, 475 pp., New York: John Wiley & Sons.

21

IMPACT STRUCTURES

Introduction

Impact structures are large circular features that are the result of a sudden shock; they have also been called *cryptoexplosion structures*. Both these terms wisely avoid the issue of the ultimate cause of the shock. The term *cryptovolcanic structure* obviously ascribed the origin to volcanic processes. The term *astroblem* ascribes the origin to impact by meteors. Such structural features have been known on earth and moon for many decades, and their origin has been debated. But only since lunar exploration began in earnest has the subject been extensively considered.[1,2,3] The first part of this chapter will be confined to these features on earth. A later section will consider such features on the moon.

Physical Features

Impact structures are circular to oval in plan, and range in maximum diameter from less than a mile to 50 miles. They may be expressed as craters or as circular areas in which a central block has been uplifted.

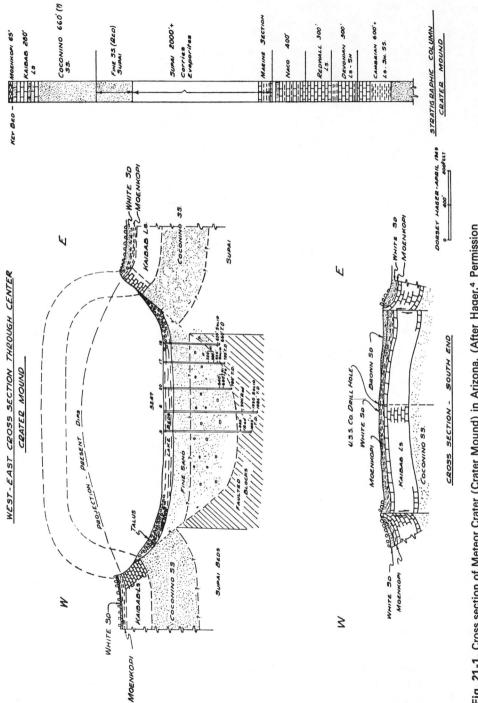

Fig. 21-1. Cross section of Meteor Crater (Crater Mound) in Arizona, (After Hager.[4] Permission American Association of Petroleum Geologists.)

The craters are conspicuous depressions below the general surface. Meteor Crater in Arizona[4,5] is a slightly oval depression about 3500 feet in diameter. The country rock consists of flat-lying Upper Paleozoic and Lower Mesozoic strata. The dry flat floor, underlain by lake sediments, is 570 feet

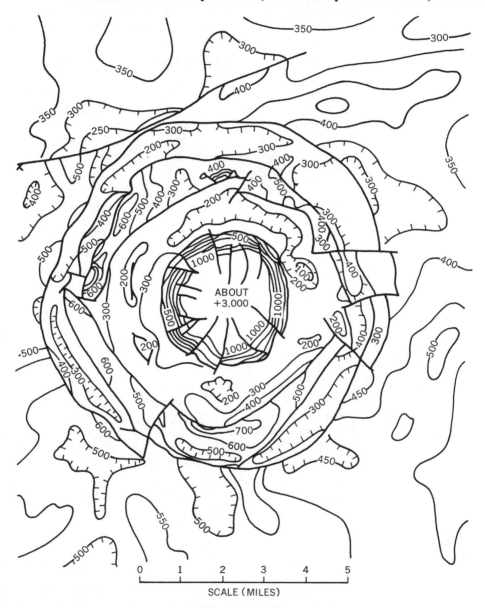

Fig. 21-2. Structure contour map of Wells Creek Structure, Tennessee. (After Richard G. Stearn.[7] Permission Mono Book Corporation.)

below the crater rim (Fig. 21-1). The crater is in the middle of a dome, several miles across, that rises 200 feet above the surrounding plain. A concentric syncline, structurally 20 feet deep, lies 2500 to 4000 feet from the center of the crater. About 20 drill holes help determine the geology on the walls and in the center of the crater. Silica glass is undoubtedly genetically associated with the formation of the crater, but meteoric material in the vicinity may be a coincidence.

The New Quebec Crater in Northern Canada is 2 miles in diameter; 1300 feet deep, it is half filled with water. The rim rises 300 to 500 feet above the surrounding plain; the country rock is Archean gneiss. West Hawk Lake in Manitoba is nearly circular in diameter and 365 feet deep; it is believed to occupy an impact crater.[6] The Ries Basin in Germany is 14 miles across.

Impact structures are also expressed as circular disturbed areas. Although details vary considerably, there is a central area of uplift, surrounded by one or more concentric synclines. Faulting and brecciation are common. The disturbed area in the Wells Creek structure of Tennessee[7] is nearly 10 miles in diameter (Fig. 21-2) in an area of essentially flat Paleozoic strata. A central block 2 miles in diameter has been uplifted 2500 feet relative to the flat strata outside the disturbance, but has been eroded to a lowland. Two concentric synclines, separated by an anticline, are depressed 200 feet relative to the surrounding plateau. Concentric, radial, and diversely oriented faults are common.

The Manicouagan structure in Quebec[8] is outlined by Ordovician rocks in a circular syncline that is 38 miles in diameter; but the entire disturbed zone may be 65 miles in diameter. The rocks outside the syncline are Precambrian. Those inside are also Precambrian, but many of them have been reworked by shock metamorphism; igneous rocks of Triassic age are also present. The authors suggest that the structure may be due either to meteor impact or igneous intrusion.

Some of the impact structures are very old. The Flynn Creek Crater in Tennessee[9] formed in the Devonian in Paleozoic rocks as a crater 3.6 kilometers in diameter and 150 meters deep. Late Devonian seas invaded the region and filled the crater with mud.

The topographic expression of many impact structures is due to differential erosion and not due directly to cratering.

Shatter Cones

Many impact structures are characterized by *shatter cones*.[10] These cones are similar in appearance to cone-in-cone structure found in sedimentary rocks.[11] Shatter cones are bounded by discontinuous, overlapping, striated conical fractures. The base of the cones may range in diameter from an inch to a foot. At Vredefort dome in South Africa[12] the angle of the apex of the cones ranges from 90° to 122°. Studies at several areas have shown that the cones are systematically oriented relative to the impact structure. At Vrede-fort, for example, the cones open downward toward the center of the dome, but if the strata are tilted back to horizontal, the axes plunge outward at low angles and the apices point away from the dome.

Shatter cones at Sudbury, Ontario[13,14] are developed in a belt 11 miles wide around the outer contact of the eruptive and are confined to preintrusive rocks. In general the axes dip toward the Sudbury basin.

Mineralogy

Special mineralogical features[15] are characteristic of some of the rocks associated with impact structures; these rocks are sometimes called impactites. High-pressure effects include the development of high-pressure polymorphs of silica such as coesite and stishovite. High-temperature effects include the conversion of quartz to silica glass (lechateliérite). Evidence of high strain rates include the shattering of crystals, the formation of shock lamellae in quartz, and conversion of quartz and feldspar to isotropic phases.

Mechanics of Impacts

The term impact structure or cryptoexplosion structure implies a sudden blow or shock. This blow may be due to meteor impact or volcanic explosion. The presumed sequence of events in the formation of a meteor crater is shown in Fig. 21-3. It has been calculated[5,16,17] that the meteor forming Meteor Crater in Arizona weighed 63,000 tons, had a velocity of 15 km/sec, and the total shock energy was 1.7 megatons. If the density of the meteorite was 7.85 g/cm^3, the diameter was 24.8 m (81 feet).

Craters, of course, are a consequence of normal eruptive processes in central eruptions (Chap. 15). Craters have also resulted from collapse over underground nuclear explosions.[18,19]

On the theory of meteorite impact, the structures produced at depth is the result of elastic rebound.[20] The rock is suddenly compressed by the impact and almost instantly springs back into its original position. But the uplift is concentrated in the center of the deformed area, whereas the outer regions remain depressed. Theoretically the volume of the uplift should be about equal to the volume of the depressed portion.

Lunar Geology

A vast literature has accumulated in recent years on the geology of the moon.[21,22,23] Many of the investigations are concerned, quite properly, with the mineralogy, petrology, geochemistry, and seismology of the moon. It will suffice here to say that the moon is composed largely of basaltic rocks

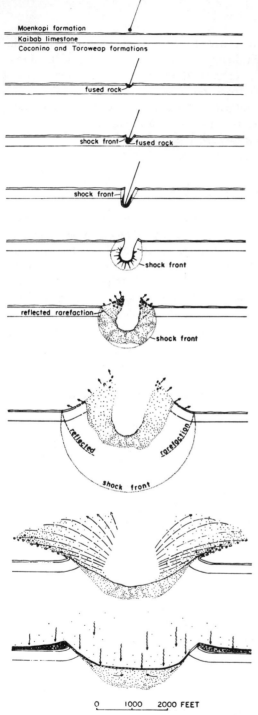

1. Meteorite approaches ground at 15 km/sec.

2. Meteorite enters ground, compressing and fusing rocks ahead and flattening by compression and by lateral flow. Shock into meteorite reaches back side of meteorite.

3. Rarefaction wave is reflected back through meteorite, and meteorite is decompressed, but still moves at about 5 km/sec into ground. Most of energy has been transferred to compressed fused rock ahead of meteorite.

4. Compressed slug of fused rock and trailing meteorite are deflected laterally along the path of penetration. Meteorite becomes liner of transient cavity.

5. Shock propagates away from cavity, cavity expands, and fused and strongly shocked rock and meteoritic material are shot out in the moving mass behind the shock front.

6. Shell of breccia with mixed fragments and dispersed fused rock and meteoritic material is formed around cavity. Shock is reflected as rarefaction wave from surface of ground and momentum is trapped in material above cavity.

7. Shock and reflected rarefaction reach limit at which beds will be overturned. Material behind rarefaction is thrown out along ballistic trajectories.

8. Fragments thrown out of crater maintain approximate relative positions except for material thrown to great height. Shell of breccia with mixed meteoritic material and fused rock is sheared out along walls of crater; upper part of mixed breccia is ejected.

9. Fragments thrown out along low trajectories land and become stacked in an order inverted from the order in which they were ejected. Mixed breccia along walls of crater slumps back toward center of crater. Fragments thrown to great height shower down to form layer of mixed debris.

Fig. 21-3. Evolution of a meteor crater. Based on Meteor Crater, Arizona. (From Shoemaker.[16])

and presumably of meteorite fragments. The surficial rocks consist of material ranging in size from dust to blocks many feet across. These rocks are both crystalline and glassy.

The broader topographic features of the moon have been known for a long time, but in recent years photographs obtained from orbiting satellites and from actual landings have greatly increased our knowledge of the details. The geological structure must be deduced largely from the topographic features. The largest morphological units are the mare and highlands. The most conspicuous and intriguing features are the craters. They range in size from those that are a few feet across to those that are several tens of miles in diameter, such as Eratothenes, 38 miles across, and Copernicus, 58 miles across. A few huge craters are: Sinus Iridum, 160 miles; Mare Crisium, 305 miles; and Mare Imbrium, 1630 miles. The crater walls range in height from a few feet to many thousands of feet. The walls of many of the larger craters are terraced, with the treads separated by steep scarps. Such features are very similar to those at Kilauea, and are due to the collapse of the walls of an excavation, whatever its origin. Rays, due to differences in the albedo of the rocks, radiate out for miles from some of the younger craters. They represent lines of material ejected from the crater.

The relative ages of the craters can be deduced from several lines of evidence: (1) a younger crater will truncate the rim of an older crater; (2) younger craters are sharper than older craters; (3) rays of younger craters are better preserved than the rays of older craters; (4) a younger crater will truncate the rays of an older crater; and (5) the mare material, whatever its origin, penetrates some craters, whereas it is itself cut by other craters.

The random distribution of the craters suggests that most, if not all, are the result of meteorite impact. Alignment of craters would suggest that a fracture in the lunar crust controls the outpouring of lava. But there is no striking alignment to suggest that any of the craters are volcanic. Moreover, large caldera on earth are the result of the eruption of highly gaseous siliceous magma, a type that probably does not exist on the moon.

Rilles are long channels, many miles long, thousands of feet wide, and hundreds of feet deep. Some are straight, others are winding, and some are distinctly meandering. Their origin is still enigmatic.

References

[1] Green, J., chairman, 1969, *Geological problems in lunar research*, Annals New York Academy of Sciences, vol. 123, art. 2, pp. 367–1257.

[2] French, Bevan M., and Short, Nicholas M., 1968, *Shock metamorphism of natural materials*, 644 pp., Baltimore: Mono Press.

[3] Freeberg, Jacquelyn H., 1969, *Terrestrial impact structures—A bibliography, 1965–68*, U. S. Geol. Surv., Bull. 1320, 39 pp.

[4] Hager, Dorsey, 1953, *Crater mound (Meteor Crater), Arizona, a geologic feature,* Amer. Assoc. Petrol. Geol. Bull. 37: 821–57.

[5] Shoemaker, E. M., *Impact mechanics at Meteor Crater, Arizona, in* Middlehurst, B. M., and Kuiper, G. P., eds., 1963, *The solar system, vol. 4, The moon, meteorites, and comets,* Chicago: The University of Chicago Press, pp. 301–36.

[6] Short, Nicholas M., 1970, Anatomy of a meteorite impact crater: West Hawk Lake, Manitoba, Canada, *Geol. Soc. Amer. Bull.* 81: 609–48.

[7] Stearn, Richard G., *The Wells Creek structure, Tennessee, in* French, Bevan M., and Short, Nicholas M., 1968, *Shock metamorphism of natural materials,* pp. 323–37.

[8] Murtough, J. C., and Currie, K. L., 1969, *Preliminary study of the Manicouagan structure,* Quebec Department of Natural Resources, Preliminary Report 583, 9 pp.

[9] Roddy, David J., The Flynn Creek Crater, Tennessee, *in* French, Bevan M., and Short, Nicholas M., 1968, *Shock metamorphism of natural materials,* pp. 291–322.

[10] Dietz, Robert S., Shatter cones in cryptoexplosion structures, *in* French, Bevan M., and Short, Nicholas M., 1968, *Shock metamorphism of natural materials,* pp. 267–85.

[11] Amstutz, G. C., 1965, *A morphological comparison of diagnostic cone-in-cone structure and shatter cones,* Annals New York Academy of Sciences, vol. 123: 1050–56.

[12] Manton, W. I., 1965, *The orientation and origin of shatter cones in the Vredefort Rings,* Annals New York Academy of Sciences, vol. 123 art. 2, pp. 1017–49.

[13] Bray, J. G., et al., 1966, *Shatter cones at Sudbury,* J. Geol. 74: 243–45.

[14] French, Bevan M., Sudbury Structure, Ontario, some petrographic evidence for origin by meteor impact, *in* French, Bevan M., and Short, Nicholas M., 1968, *Shock metamorphism of natural materials,* pp. 383–412.

[15] French, Bevan M., Shock metamorphism as a geological process, *in* French, Bevan M., and Short, Nicholas M., 1968, *Shock metamorphism of natural materials,* pp. 1–17.

[16] Shoemaker, Eugene M., 1960, *Penetration mechanics of high velocity meteorities, illustrated by Meteor Crater, Arizona,* Report 21st Int. Geol. Congr., Session Norden, part XVIII, pp. 418–34.

[17] Gault, Donald E., et al., Impact cratering mechanics and structures, *in* French, Bevan M., and Short, Nicholas M., 1968, *Shock metamorphism of natural materials,* pp. 87–99.

[18] Houser, F. M., *Application of geology to underground nuclear testing, Nevada Test Site,* pp. 21–33; Edwin B. Eckel, ed., 1968, *Nevada Test Site,* Mem. 110, Geol. Soc. Amer. 290 pp.

[19] Barosh, Patrick, J., 1968, Relationships of explosion-produced fracture patterns to geologic structure in Yucca Flat, pp. 199–217, *Nevada Test Site,* Mem. 110, Geol. Soc. Amer., 290 pp.

[20] Boon, John D., and Albritton, Claude C., Jr., 1936, Meteor craters and their

possible relationship to "cryptovolcanic structures," *Field and laboratory* 4: 1–9, Southern Methodist University.

[21] Baldwin, R. B., 1949, *The face of the moon*, 219 pp., Chicago: The University of Chicago Press.

[22] Abelson, Philip H., 1970, The moon issue, *Science* 167: 449–784.

[23] U. S. Geological Survey, *Geological atlas of the moon.*

22

GEOPHYSICAL METHODS IN STRUCTURAL GEOLOGY: GRAVITATIONAL AND MAGNETIC

Introduction

For many years structural geologists had to rely exclusively on data obtained by direct observation. Most of the information came from natural exposures, although some facts were revealed by artificial openings such as mines, tunnels, railroad cuts, highway cuts, and dam sites. Data obtained from bore holes may also be considered in this category because the cores or cuttings were examined megascopically and sometimes microscopically. But in the last 50 years geophysical methods have become increasingly important in deducing the location and shapes of subsurface rock bodies. At first this was largely due to the economic importance of geophysical methods in mining and petroleum geology. But in recent decades geophysical methods have been used extensively in pure research, including investigations of the oceans.

Geophysical data are used most successfully when combined with geological information. The geophysical methods and geological methods are so specialized that one individual cannot hope to be fully trained in both fields. Nevertheless, the geologist should be well versed in the principles of geophysical exploration, and, conversely, the geophysicist should be cognizant of geological principles. Many problems will demand the cooperation of men trained in both fields.

It is beyond the scope of this book to discuss the techniques involved in geophysical studies, and only a rather general discussion of the subject can be presented. For a complete treatment, the student is referred to textbooks in geophysical prospecting.[1,2,3] But the geologist should be thoroughly familiar not only with the methods that can be used, but also with the extent of their usefulness and their limitations. The broad physical principles underlying the various methods are discussed briefly here, and some examples of geophysical studies of structural problems are presented.

The results of geophysical investigations may range from crudely qualitative to very quantitative. The data may merely indicate the presence of a rock body that differs in composition from its surroundings. In some cases the data may indicate the depth of the rock body. In still other instances the lithology and shape of the body may be deduced, especially when combined with local geological information. In recent years geophysical methods have contributed very important information on the character of the movement along faults associated with earthquakes.

In many cases the geophysical methods are employed in areas where surface exposures are rare or absent. In countries that have not been adequately studied geologically, the interpretation of the data may be difficult. In geological provinces that are reasonably well known, however, the correct solution might be readily obtained. If, for example, in the Gulf Coast of the United States a circular area a mile or so in diameter is occupied by a rock that transmits elastic waves with much greater speed than do its surroundings, a salt dome is probably present. In the Lake Superior region, strong positive magnetic anomalies suggest a deposit of iron ore.

Geophysical Methods

There are six principal geophysical methods: gravitational, magnetic, seismic, electrical, radioactive, and thermal. The gravitational, magnetic, radioactive, and thermal methods involve the measurement of properties that are inherent to the rocks of the crust. The seismic method is based on the effects of energy that is introduced artificially or is released during earthquakes. The electrical methods involve the effects of both natural currents and artificially introduced currents.

The coverage in the ensuing pages is restricted for several reasons. We are primarily concerned with geological structure. Consequently, methods that are concerned with stratigraphic sequence, such as electrical logging, radioactivity logging, and continuous seismic velocity logging, will not be considered. Moreover, we shall not undertake a study of the interior of the earth, for which much of the information comes from seismology. Moreover, since regional geology is not an objective of this book, geophysical studies of large areas will not be undertaken. We are primarily concerned with the application of geophysical methods to those structures that range in size from a few hundred feet to a few tens of miles across.

Gravitational Methods

PRINCIPLES

Gravitation is the force by which, due to mass, all bodies attract each other. This force is directly proportional to the product of the masses of the two bodies concerned and is inversely proportional to the square of the distance separating them. This principle may be expressed as

$$F \propto \frac{m_1 m_2}{r^2} \tag{1}$$

Here F is the force, m_1 and m_2 are the masses of the two bodies, and r is the distance between them. In the form of an equation,

$$F = \frac{k \cdot m_1 \cdot m_2}{r^2} \tag{2}$$

In this equation m_1 and m_2 are measured in grams, r is in centimeters, and F is in dynes. A *dyne* is the force necessary to give a mass of one gram an acceleration of one centimeter per second per second. If two masses, each weighing a gram, are one centimeter apart, equation (2) becomes $F = k$. That is, k, known as the *gravitational constant*, is the force of attraction of two equal masses of one gram each at a distance of one centimeter; $k = 6.673 \times 10^{-8}$ c.g.s. units.

Gravity is the force by which the earth attracts bodies toward it. If a stationary body suspended in air a short distance above sea level is released, it will move toward the earth with a constantly increasing speed; that is, it accelerates. The acceleration is independent of the mass of the released body. It is thus more satisfactory to express the force of gravity in terms of the acceleration it causes. A *gal* is an acceleration of one centimeter per second per second; a *milligal* is one-thousandth of a gal. At sea level the acceleration of gravity is approximately 980 gals (approximately 32 feet per second per second).

Because the earth is a rotating oblate spheroid, flattened at the poles, the normal acceleration of gravity differs with latitude; at the equator it is 978.049 gals, at the poles it is 983.221 gals. The general equation is

$$g = 978.049(1 + 0.00052884 \sin^2 \phi - 0.0000059 \sin^2 2\phi) \tag{3}$$

In this equation g is the normal gravity at sea level, and ϕ is the latitude.

Equation (3) is based on data for the whole earth. In an ideal, homogeneous earth, the value of gravity could be calculated if the latitude were known. Actually, however, gravity usually differs from the theoretical value. This is due in part to the fact that the earth is heterogeneous; even at the surface, the rocks are by no means uniform. If an unusually heavy rock is near the surface, gravity will exceed the theoretical value; if an unusually light rock is near the surface, gravity will be less than the theoretical value.

Some of the gravitational methods, such as those utilizing pendulums and gravimeters, measure the value of gravity directly. Other methods, such as those using torsion balances, measure the rate of change of gravity and the deviation of equipotential surfaces from a spherical shape.

METHODS

Pendulum method

By swinging a simple pendulum, we may determine the value of gravity in gals according to the equation

$$g = 4\pi^2 n^2 L \tag{4}$$

In this equation g is the value of gravity in gals, L is the length of the pendulum, and n is the number of vibrations the pendulum makes in unit time; in the C.G.S. system L is measured in centimeters and n in seconds; g is in gals.

This method has been used for many years by the U.S. Coast and Geodetic Survey and similar organizations to obtain precise values of gravity. A single observation, however, consumes considerable time—several hours—and for this reason the method is not practical for ordinary field mapping.

Gravimeter method

The great advantage of gravity meters is that the force of gravity, relative to some standard station, can be measured within a few minutes. Moreover, light gravimeters, weighing as little as five pounds, are available for use in regions accessible only on foot. The mechanical gravimeters use the elastic force of springs and the torsion of wires. The deformation is magnified by optical, mechanical, or electrical means to give an accuracy of 1 in 10,000,000. The instruments must be calibrated before use. Moreover, there is some drift during field work, so that some stations must be reoccupied to determine the amount of the drift.

Both the pendulum apparatus and gravimeter were devised originally to measure gravity on land. Measurements at sea presented a special problem, as the constant vertical oscillation of the ship involved an accceleration that could not be evaluated. In the 1920s a pendulum was devised for use in submarines, because they can submerge to depths below the oscillation of the waves.[4] But in recent years an apparatus has been devised for use on shipboard.[5,6]

Corrections applied in pendulum and
gravimeter methods

In order to compare gravity measurements at adjacent stations, it is necessary to make a number of corrections. After the value of gravity is determined by the instruments, the following corrections are made: (1) free air correction; (2) Bouguer reduction; (3) terrain (topographic) correction; (4) isostatic correction.

Free air correction. Most stations are above sea level, and the force of gravity becomes progressively less as the altitude becomes greater. For example, if gravity at some station at sea level were 980.000 gals, gravity measured from a captive balloon 300 meters above the station would be only 979.907 gals. For all stations above sea level, therefore, a correction must be added proportional to the altitude. The correction is +30.86 milligals per hundred meters, or +9.406 milligals per hundred feet.

Bouguer reduction. The rock between the station and sea level increases the value of gravity; the amount is proportional to: (1) the altitude of the station and (2) the density of the rocks between the station and sea level. For example, if the gravity station is at an altitude of 300 meters on a flat tableland underlain by granite, the rock between the station and sea level exerts a gravitational attraction of 33 milligals. This correction, therefore, must be subtracted when one calculates the value of gravity at sea level.

Thus, whereas the free air correction is positive, the Bouguer correction is negative; but the free air correction is the greater—approximately three times as large as the Bouguer correction. Morevoer, it is obvious that the free air and Bouguer corrections can be combined into one correction according to the equation

$$g_0'' = g + H(0.0003086 - 0.0000421\delta)\text{gals} \tag{5}$$

in this equation g_0'' = gravity at sea level with free air and Bouguer corrections; g = measured value of gravity; H = altitude in meters; δ = density of rocks between station and sea level.

Topographic (terrain) correction. Hills that rise above the instrument, or depressions that descend below it, exert an influence on the measured value of gravity. If good contour maps are not available, it may be necessary for the field party to obtain sufficient data for calculating the topographic data. This is particularly true if considerable relief exists near the station.

One procedure is to divide the country into compartments by radii extending out from the station and by circles concentric about it, as illustrated in Fig. 22-1, where S is the station. The gravitational effect of each compartment, such as *abcd*, is

$$\Delta g = k \cdot \delta \cdot \phi(r_1 + \sqrt{r_2^2 + h^2} - \sqrt{r_1^2 + h^2} - r_2) \tag{6}$$

In this equation Δg is the gravitational effect of compartment in gals; k = gravitational constant (see p. 448); δ = density of rocks; ϕ = angle between

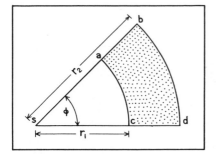

Fig. 22-1. Calculation of terrain correction in gravity measurements. See text. (After C. A. Heiland.)

radii, r_1 = radius in centimeters of inner circle bounding compartment; r_2 = radius in centimeters of outer circle bounding compartment; and h = average elevation in centimeters of the compartment above or below the instrument.

Charts and tables for making these corrections are available.[7]

Regardless of whether the average altitude of the compartment is greater or less than that of the instrument, the observed gravity is too low. In other words, the correction Δg should be added to the observed gravity.

Isostatic correction. A correction for isostatic compensation is unnecessary in the study of local structure. On the other hand, such corrections are of great importance in geodetic work and structure of continental proportions.

Calculation of gravity anomaly

The theoretical or normal value of gravity for a station is calculated by equation (3). The corrected value of gravity is obtained from the observed value by making the free air, Bouguer, and terrain corrections. This difference between the corrected value and the normal value is the *gravity anomaly*. That is,

$$\Delta g_0'' = g_0 - g_0'' \tag{7}$$

where $\Delta g_0''$ = gravity anomaly, g_0'' = normal gravity, and g_0 = observed gravity, with free air, Bouguer, and terrain corrections. The gravity anomaly is usually expressed in milligals. The gravity anomalies calculated by geodesists include an isostatic correction.

The gravity anomalies for each station are plotted on a map, and then isogals are drawn. *Isogals* are lines of equal gravity anomaly, and they are drawn in the same way as contour lines, interpolating wherever necessary. Figure 22-2A is an isogal map of the Wellington field in Colorado; the isogal interval is 0.2 milligals.[8]

In some areas, the effects of the local structure may be concealed by a strong regional change in the gravity anomalies, and some method of adjust-

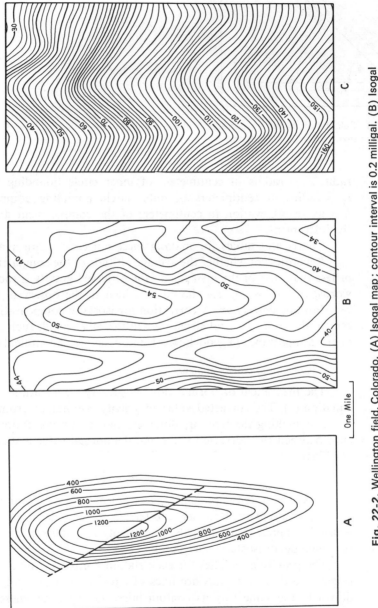

Fig. 22-2. Wellington field, Colorado. (A) Isogal map; contour interval is 0.2 milligal. (B) Isogal map, adjusted for regional gradient of 2.2 milligals per mile south; contour interval is 0.2 milligal. (C) Structure contour map on the Muddy Sand; contour interval is 100 feet. (After J. H. Wilson,[8] *Geophysics.*)

ment must be used. In Fig. 22-2A, the value of gravity increases toward the south at the rate of 2.2 milligals per mile; the average east-west change is zero. Figure 22-2B is the adjusted map after the regional effect has been eliminated. Figure 22-2C is the structure contour map of the region; it is clearly expressed by the gravity map.

Torsion balance method

For 20 years before gravimeters were introduced (they appeared about 1935), torsion balances were used for gravitational studies. These instruments do not measure the value of gravity, but instead measure the gravity gradient and the curvature of equipotential surfaces. The horizontal gravity gradient is measured in the Eotvös unit, which is 1×10^{-9} dyne per horizontal centimeter.

RELATION BETWEEN GRAVITY AND STRUCTURE

The use of gravitational methods is based on the simple principle that the gravitational effect of a rock body is a function of its density, size, shape, and distance from the observer. The presence of folds, buried ridges, salt domes, faults, and igneous intrusions may be deduced from gravity data. Moreover, the attitude of faults and the shape, size, and depth of rock bodies may be inferred with various degrees of precision.

In deducing the depth, size, and shape of rock bodies that cause gravity anomalies, a series of hypothetical models are tested. The parameters of these bodies—density, depth, shape, etc.—are varied. The gravitational effect of each of these models is calculated. The model that best explains the observed anomalies is considered to be correct. A few examples of some possible geometric forms involved are given here.

If the body is a horizontal cylinder

$$Z = \frac{W}{2} \tag{8}$$

where Z is depth of center of gravity of mass causing anomaly, and W is width of anomaly in a profile perpendicular to long axis of the cylinder.

$$R = \frac{\gamma^2}{12.77\rho_c} \tag{9}$$

where R = radius of cylinder, γ = difference between highest and lowest anomalies, ρ_c = density contrast (difference in density of cylinder and surrounding rocks).

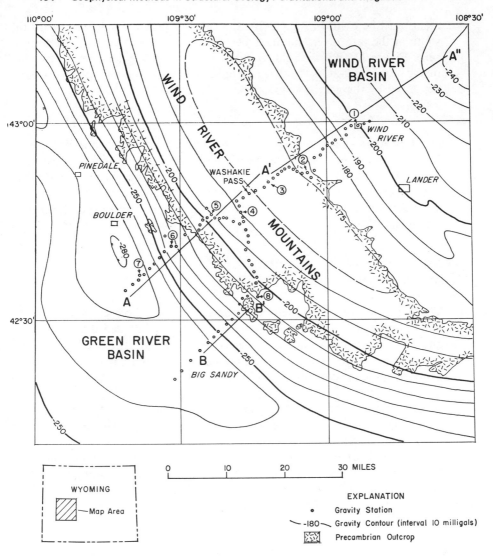

Fig. 22-3. Bouguer anomaly map of Wind River Mountains, Wyoming. (Permission Geological Society of America, R. R. Berg and F. E. Romberg.[9])

For a sphere:

$$Z = 0.652 \, W \qquad (10)$$

and radius is as in equation (9).

For a vertical cylinder, if the distance to the top of the cylinder is zero:

$$\gamma = 2\pi k \rho_c [1 + R - \sqrt{1^2 + R^2}] \qquad (11)$$

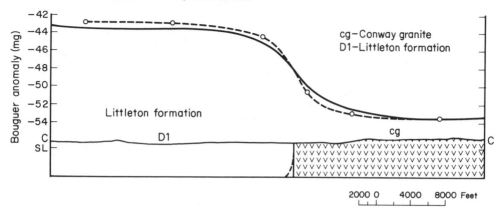

Fig. 22-4. Relation of gravity anomaly to a granitic stock. Solid line is observed Bouguer anomaly profile. Broken line is computed curve on basis of distribution of rock masses as shown. Density contrast between formations is assumed to be 0.25 g/cm³. Mad River stock, New Hampshire. (After Bean.[10])

Anticlines, especially those in which denser basement rocks underlie a less dense cover, are indicated by gravity maps. Figure 22-3 is a Bouguer anomaly map of the Wind River Mountains and adjacent basins in Wyoming.[9] Over the sedimentary basins the anomaly is as low as −280 milligals, whereas over the Precambrian rocks of the Wind River Range it reaches a high value (relatively) of −175 milligals. Obviously the density of both groups of rocks ranges considerably in value, but the Precambrian averages about 2.68 g/cm³, whereas the basin rocks average about 2.36 g/cm³. The average density contrast is 0.32 g/cm³. A gravity high thus corresponds to an anticline. Normally a salt dome is indicated by a negative anomaly, but a thick cap rock will cause a positive anomaly.

This reduces to

$$\gamma = 4.193 \times 10^{-7} \rho_c [1 + R - \sqrt{1^2 + R^2}] \qquad (12)$$

where γ = difference in highest and lowest anomalies, k is gravitational constant, l is length of cylinder, R is radius of cylinder.

An additional term is introduced into the equation if the top of the cylinder is below the surface.

Figure 22-4 is the geological profile and Bouguer anomaly profile for the southwestern part of a stock of Conway Granite in New Hampshire.[10] The basic data are: γ = −11 milligals (0.011 gals), ρ_c = 0.25 g/cm³, l = length of cylinder (in centimeters), to be determined; R = 3 miles

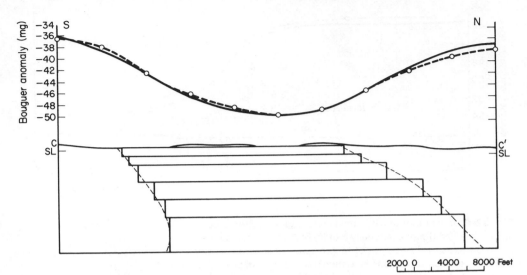

Fig. 22-5. Relation of gravity anomaly to a granitic stock. North-south profile. Solid line is observed Bouguer anomaly profile, broken line is calculated anomaly profile on basis of assumptions shown. Stock composed of Lebanon granite (*lg*); surrounding rocks are metamorphosed sedimentary and volcanic rocks. Density contrast 0.13 g/cm³. Near Hanover, New Hampshire. (After Bean.[10])

(= 3 × 1609.3 × 10² cm = 4.83 × 10⁵ cm). Using equation (12), we calculate *l* to be 1200 meters (4000 feet). This means that the Conway Granite extends to a depth of at least 4000 feet. But it tells us nothing about the rock beneath the depth of 4000 feet—whether it is Conway Granite, schist, or some other rock.

For more complicated shapes the model should consist of successive discs stacked on top of one another. The upper part of Fig. 22-5 shows a gravity profile through the Lebanon granite in western New Hamphsire.[10] Geological evidence suggested this is an elliptical cylinder plunging steeply to the north. The lower part of the figure shows a model consisting of a series of discs, the centers of which are staggered. In this case the density contrast extends to 12,000 feet.

Figure 22-6 shows two geological and gravity profiles through the Boulder batholith of Montana.[11] The profiles along a N.35°E. line are in the upper half of the figure, whereas the profiles along a N.55°W. line are on the lower half. Various models are possible, but generally agree that the density contrast extends to 10 or 15 km.

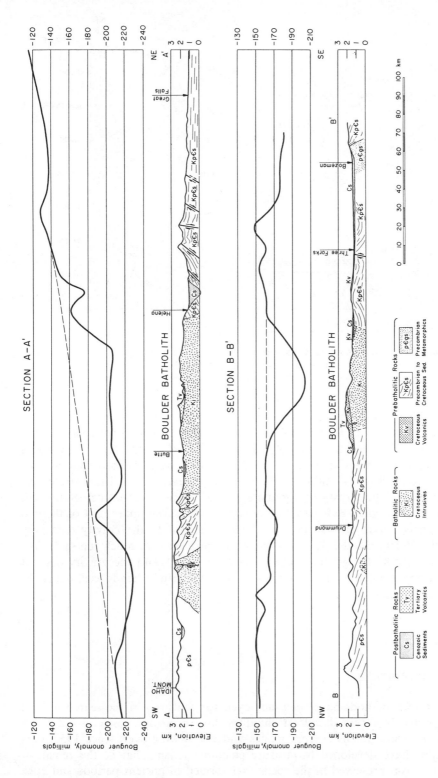

Fig. 22-6. Relation of gravity to a granite sheet. Boulder batholith, Montana. Two profiles shown; upper profile is parallel to long dimension of batholith, lower profile is at right angles to long dimension of batholith. Bouguer anomaly profiles shown as solid lines. (After Biehler and Bonini. Permission Geological Society of America, S. Biehler and W. E. Bonini.[11])

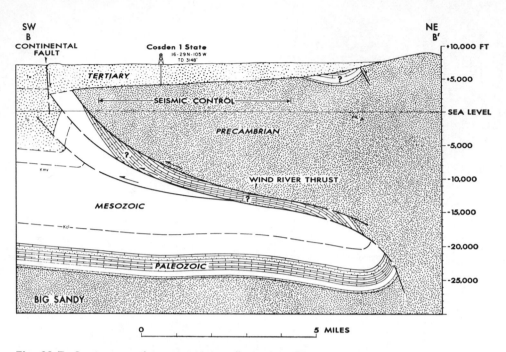

Fig. 22-7. Structure section on southwest flank of the Wind River Range, Wyoming, based on seismic data; see Fig. 22-8 for modification. (Permission Geological Society of America, R. R. Berg and F. E. Romberg.)

Figures 22-7 and 22-8 show how gravity data may be applied to determine the attitude of a thrust fault on the southwest flank of the Wind River Range of Wyoming. Figure 22-7 is a somewhat diagrammatic cross section, showing Precambrian crystalline rocks thrust over Paleozoic, Mesozoic, and Tertiary rocks. Although the surface geological data demonstrates the presence of the thrust, the dip and extent of the fault cannot be determined. Model A (Fig. 22-8) is based on seismic studies. The upper part of the two anomaly curves shows that this model does not fit the gravity data very satisfactorily. Model B, prepared by trial and error in much the same way as the model for granite bodies, agrees very well with the gravity data. It thus appears that the thrust plane dips about 15° and extends back under the mountains for 12 miles.

Magnetic Methods

INTRODUCTION

Magnetic methods have been employed for centuries in the search for ore deposits. In recent decades techniques have been developed to determine the magnetic intensity from planes and ships. Moreover, since 1950 techniques have developed to measure paleomagnetism, that is, the remanent magnetism preserved in the rocks that formed in ancient periods and eras.

458

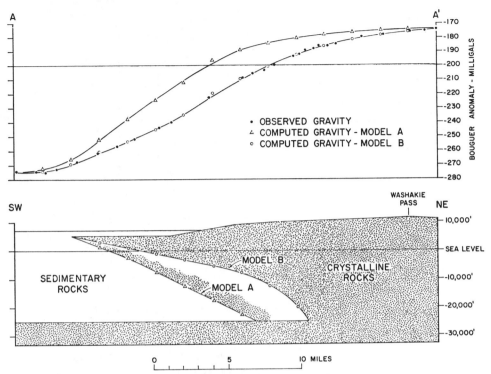

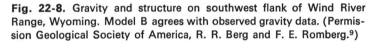

Fig. 22-8. Gravity and structure on southwest flank of Wind River Range, Wyoming. Model B agrees with observed gravity data. (Permission Geological Society of America, R. R. Berg and F. E. Romberg.[9])

PRINCIPLES

Magnets are characterized by two *poles*, one of which is known as the *north-seeking* or *north pole*, the other as the *south-seeking* or *south pole*. Like poles repel each other, unlike poles attract. The force between two magnetic poles is proportional to the strength of each pole and inversely proportional to the square of the distance between the poles. That is,

$$F = \frac{SS'}{d^2} K \tag{13}$$

In this equation F is the force; S and S' are the strengths of the two poles; d is the distance between the two poles; and K is a constant of proportionality that depends upon the units chosen. A magnetic pole, placed in a field that has a strength of one *gauss* or *oersted* is acted upon by a force of one dyne.

In magnetic studies related to geological problems, it is more convenient to use the *gamma*, which is 1/100,000 part of a gauss. That is, 1 gamma = 1 gauss $\times$ 10^{-5}.

The *magnetic field* of a magnet is that region in space into which the influence of the magnet extends. The *lines of force* of a magnetic field are

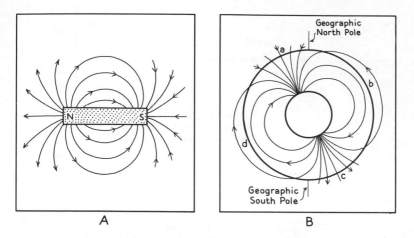

Fig. 22-9. Magnetic fields. (A) Magnetic field around a bar magnet. (B) Magnetic field of the earth.

the imaginary lines drawn through the field in such a way that they extend in the direction in which small magnetic poles would tend to move. Figure 22-9A illustrates the field surrounding a bar magnet. The orientation of lines of force about a magnet is shown very clearly by scattering iron filings in the field because the filings align themselves parallel to the lines of force.

A suitable substance immersed in a magnetic field will be magnetized. This is *induced magnetization*. The susceptibility of a substance is the ratio of the strength of the induced magnetization to the strength of the field. That is, for a field that is normal to the surface of a magnetic material

$$k = \frac{I}{H} \tag{14}$$

where k = susceptibility, I = strength of induced magnetization in gauss, H = strength of field. Some values of susceptibility are: magnetite, 0.3 to 0.8; pyrrhotite, 0.125; and ilmenite, 0.135.

The earth is a gigantic magnet; in a cross section through the earth, the magnetic lines of force are distributed as shown in Fig. 22-9B. The magnetic poles are displaced relative to the geographical poles. Point a is on the west side of Hudson Bay at about 70°N. latitude.

A magnetic needle suspended in such a way that it is free to turn in all directions tends to align itself parallel to the lines of force. At points b and d of Fig. 22-9B, the needle would be horizontal; at points a and c, it would be perpendicular to the surface of the earth. At any intermediate point it would be inclined, and the *magnetic dip* is the angle between the lines of force and the horizontal. The *magnetic declination* at a point is the angular difference between geographic north and the vertical plane that contains the inclined needle.

In Fig. 22-10, which applies to the northern hemisphere, X points north

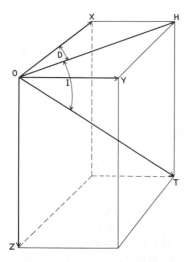

Fig. 22-10. Vector diagram of the earth's magnetic field in the Northern Hemisphere. (After G. A. Heiland.)

and Y points east. If T is the total intensity of the magnetic force, H is the horizontal component, Z is the vertical component, X is the north horizontal component, and Y is the east horizontal component. The magnetic dip I is measured in the plane containing H and T; the magnetic declination D is measured in the horizontal plane that contains H, X, and Y.

In the United States, the magnetic dip ranges from 57 to 80 degrees, and the horizontal intensity ranges from 0.15 to 0.26 gauss.

The lithosphere lacks homogeneity in its magnetic properties. Some rocks are relatively magnetic, others are weakly magnetic. Consequently, the intensity of the magnetism and the attitude of the lines of force differ from place to place.

TECHNIQUE

For ground surveys, various types of magnetometers are used. Dip needles have been employed for many years in prospecting for ores. The Hotchkiss superdip is a refined version. These instruments are used primarily in the search for magnetite and pyrrhotite bodies. One type of Schmidt balance measures the relative strength of the vertical component of the magnetic field of the earth. In this instrument the magnetic system is suspended on a knife-edge that is oriented at right angles to the magnetic meridian (magnetic north). Another type of Schmidt balance measures the relative strength of the horizontal component of the magnetic field of the earth. In this instrument the magnetic system is oriented parallel to the magnetic meridian. Nuclear magnetometers based on the principle of nuclear resonance measure the total intensity of the magnetic field of the earth. They may be used in planes or at sea.

By the use of airborne magnetometers, large areas can be surveyed very rapidly. One type of instrument is a self-orienting detector element which hangs down below and back of the plane (the "bird") or is in the tail section. These instruments measure the total intensity of the magnetic field of the earth. Surveys are generally flown at an altitude of a few hundred or thousand feet above the ground. The flight paths are generally one-half to one mile apart. It is necessary to know the course, altitude, and groundspeed of the plane, because the data are recorded at a constant rate. The basic data are profiles that are converted into contour maps, showing the total magnetic intensity in gammas, less an arbitrarily chosen figure, as only relative values are important. From equation (13) (p. 459) it is apparent that the complexity of the profile is inversely proportional to the altitude of the plane relative to the land surface; that is, the higher the plane the smoother the profiles.

In the more refined ground surveys, several corrections may be applied. These include corrections for temperature, diurnal variation, magnetic storms, and long-period changes in the field of the earth if surveys made at different times are to be compared.

The *magnetic anomaly* is determined algebraically by subtracting the theoretical value of the magnetic intensity from the observed values as corrected. The theoretical value can be determined from charts issued by the U.S. Coast and Geodetic Survey. But generally an arbitrary base is chosen and the anomaly is determined by subtracting this base from the observed values. Lines of equal anomaly can be drawn as in any contour map.

RELATION OF MAGNETIC INTENSITY TO GEOLOGIC STRUCTURE

The utilization of magnetic data to interpret geologic structure is analogous to the use of gravitational data.[12] But there is one basic difference that makes the interpretation of magnetic data more difficult. The intensity of magnetization is a vector quantity, having both direction and magnitude, whereas gravitational intensity has only magnitude. Remanent magnetism and reversed magnetization, subjects that are discussed below, also complicate the analysis.

Magnetic data may be used both qualitatively and semiquantitatively. In the qualitative methods a magnetic body is inferred to be present, but no effort is made to deduce its distance, size, and shape. The semiquantitative methods are similar to those used in gravitational studies, but more discretion is necessary in using them. Various models are analyzed and the one that best agrees with the observed data is assumed to be correct. A few equations are given blow. These equations assume that the magnetic polarization is vertical.

The vertical intensity of magnetization directly above the center of the body is given in the following equations:

(1) Sphere

$$V = 8.30\frac{R^3 I}{Z^3} \tag{15}$$

(2) Horizontal cylinder

$$V = 6.28\frac{R^2 I}{Z^2} \tag{16}$$

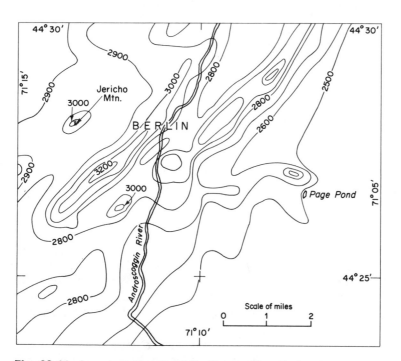

Fig. 22-11. Aeromagnetic map of Berlin area, New Hampshire. Flown approximately 1000 feet above ground. Total vertical intensity, contour interval 100 gammas, related to an assumed base (After U. S. Geological Survey, Bromery, et al.[14]) For geologic map of same area see Fig. 22-12.

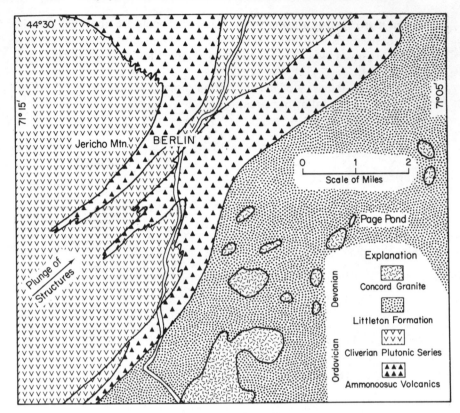

Fig. 22-12. Geologic map of Berlin, New Hampshire. (After M. P. Billings and K. Fowler-Billings.[15]) Compare with Fig. 22-11.

(3) Vertical sheet

$$V = 2It\left[\frac{1}{Z_1} - \frac{1}{Z_2}\right] \tag{17}$$

where V = vertical intensity of magnetization in gauss, R = radius of sphere or cylinder, t = thickness of sheet, I = intensity of induced magnetization in gauss, Z = depth to center of sphere or cylinder, Z_1 = depth to top of sheet, Z_2 = depth to bottom of sheet.

I is calculated from equation (14), in which H is taken as 0.6 gauss. The susceptibility of rocks can be obtained from tables.[3,13] A few average values are: granite, 0.0027; gabbro, 0.0072; and basalt, 0.0143.

Lengths are theoretically given in centimeters, but any unit can be used, because the equations involve only ratios of lengths.

Magnetic maps may be used in a qualitative way to trace formations. Some formations are more magnetic than others. Aeromagnetic maps are especially valuable in doing this. The correlation between a geologic map and an aeromagnetic map is shown in Figs. 22-11 and 22-12. In general the magnetic highs correspond to the Ammonoosuc Volcanics, due to magnetite in this formation.[14,15] There are, however, two unexplained highs, one at Jericho Mountain and the other northwest of Page Pond.

Figure 22-13 represents a hypothetical case. An aeromagnetic map shows a residual positive anomaly of 100 gammas trending north-south and about 2 miles wide. In the east-west profile the maximum is at a. The rocks at the surface are limestone. A bore hole at d penetrates 10,000 feet of flat limestone, and continues downward through an unconformity into 500 feet of gabbro and bottoms at a depth of 10,500 feet. What is the cause of the anomaly?

It is assumed that the magnetic high is due to an anticline trending north-south. As shown in Fig. 22-13, the geometry may be approximated by using a horizontal cylinder. Using equation (14), k for gabbro may be taken as 1000×10^{-6}; a reasonable figure for H is 0.6. We can then calculate I as 6×10^{-4}. Using equation (16), we take Z as 10,000 feet (the depth to the axis of the cylinder, Fig. 22-13). If V is 100 gammas, R may then be calculated to be about 5000 feet; this is the amplitude of the anticline.

The relation between geology and an aeromagnetic map of a large area in Alaska has been described.[16]

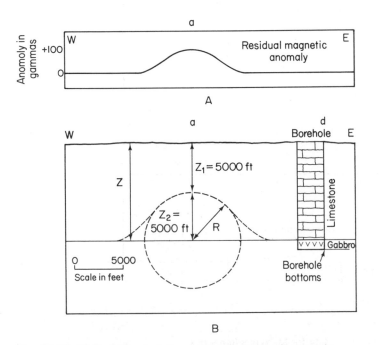

Fig. 22-13. Method of calculating cause of a magnetic anomaly. (A) Magnetic anomaly. (B) Shape of anticlinal uplift of basement is approximated by a horizontal cylinder, the axis of which bears north-south.

The formulas for dipping contacts are rather complicated. The anomaly is always higher on that side of the dike toward which it dips. Moreover, a trough is present on the side of the dike away from which the dike dips.[3]

A rather unusual case, in which both the gravity and magnetic data seem contrary to expectations, is illustrated by the Merrymeeting Lake stock in New Hampshire.[17] This stock, $6\frac{1}{2}$ miles in diameter, is composed chiefly of Conway granite, in which the only dark mineral is biotite, constituting about 3 percent of the rock. A body of gabbro, over a mile in diameter, lies directly on the northeast contact of the stock. The country rock consists of older schists and quartz diorite.

The Conway granite should give a gravity low because the specific gravity is less than that of the surrounding schists. Instead, as shown in Fig. 22-14, it is associated with a gravity high, which reaches a maximum

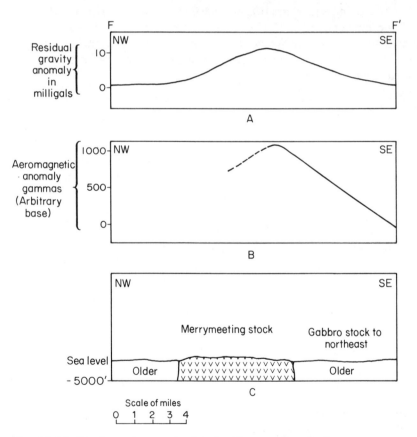

Fig. 22-14. Gravity and magnetic data for Merrymeeting stock, New Hampshire. (A) Residual Bouguer gravity anomaly. (B) Aeromagnetic anomaly. (C) Assumed shape of stock. (After Joyner.[17])

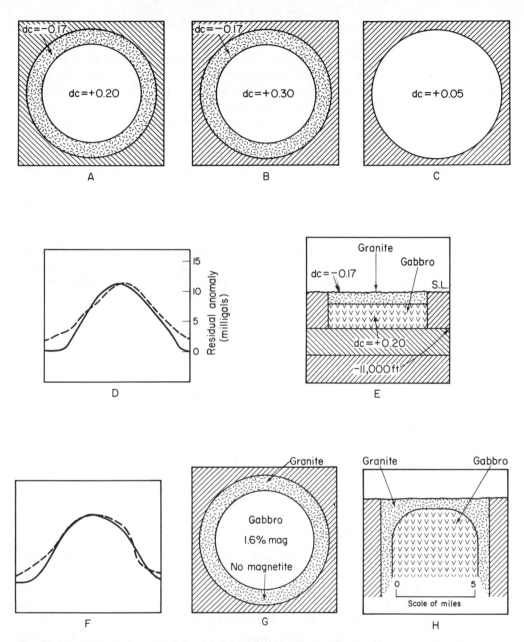

Fig. 22-15. Interpretation of gravity and magnetic anomalies in Merrymeeting stock, New Hampshire. (A), (B), and (C) assume gravity anomaly due to spherical mass, with density contrasts (compared with country rock) as shown. (D) Solid line is observed residual Bouguer gravity anomaly; dotted line is calculated anomaly curve for models in (A), (B), and (C); all three are the same. (E) Assumes gravity anomalies are due to cylindrical bodies with vertical axes. (F) Solid line is observed residual Bouguer gravity anomaly. Dotted line is curve calculated for model shown in (D). (G) Model used to explain magnetic anomaly; vertical cylinders, as shown in (E) would give a similar result. (H) Final interpretation, based on gravity and magnetic anomalies and field geology. (Permission Geological Society of America and W. B. Joyner.[17])

467

of 11 milligals on the residual curve, that is, the curve obtained by removing the regional gradient from the original curve. The magnetic airborne survey also shows an unexpected positive anomaly amounting to 1100 gammas (Fig. 22-14B). Unfortunately the aeromagnetic survey did not cover the northern half of the stock. The gravity and magnetic anomalies suggest that a denser, magnetite-rich rock may underlie the Conway granite. Two geological facts also suggest this: (1) the presence of the gabbro stock on the northeast contact and (2) the abundance of gabbro in the Belknap Mountain complex ten miles to the west.

Figure 22-15 shows some of the gravity models that have been analyzed. In one model (Fig. 22-15A) an outer shell with a density contrast (compared with the country rock) of -0.17 g/cm^3 encloses a spherical body with a density contrast of $+0.20$ g/cm^3. Figures 22-15B and 22-15C are other possible models. In Fig. 22-15D the observed residual gravity anomaly is shown as a solid line. The models in Figs. 22-15A, B, C all give the same anomalies, as shown by the broken line in Fig. 22-15D. Still another model is shown in Fig. 22-15E—a disc of Conway granite, 2000 feet thick, overlies a disc of gabbro. The calculated gravity anomalies for this model are shown as a broken line in Fig. 22-15F. The agreement with the observed curve is very satisfactory.

The model shown in Fig. 22-15G was one of many constructed to try to explain the magnetic anomaly. A sphere of gabbro with 1.6 percent magnetite is surrounded by a shell of Conway granite with no magnetite. Outside of this is the country rock with no magnetite. The anomalies calculated for this model agree very satisfactorily with the observed data. The inferred structure, based on gravity and magnetic data, as well as familiarity with the field geology, is shown in Fig. 22-15H. The Belknap Mountains ten miles to the west show a structure consistent with this interpretation.

PALEOMAGNETISM

During the past two decades many investigations in paleomagnetism have been made.[18] Some ancient rocks possess a remanent magnetism that is preserved from some earlier stage in their history. The orientation of this paleomagnetism—that is, its bearing and plunge—is of great significance. It usually differs from that of the present field. Of course, in many rocks all paleomagnetism has been wiped out by the induced magnetism related to later magnetic fields of the earth.

The best rocks for paleomagnetic studies are ferruginous sedimentary rocks and mafic igneous rocks. At each site (locality) the attitude of the bedding must be recorded, so that the bed can be figuratively rotated back into the horizontal position if the remanent magnetism was obtained prior to folding. Several samples are collected at the site, preferably by shallow drilling, to obtain a core of fresh rock an inch or more in diameter. The sample must be oriented, that is, marked in such a way as to identify north, horizontal, and up and down. From each sample several specimens are cut. Thus as many as ten specimens may be available from one site. The specimens are placed in a magnetometer, of which there are several types, to determine the bearing and plunge of the remanent magnetism. During the analysis the specimens are "cleansed" to get rid of the "soft" magnetism; the "hard" magnetism that is left is considered to be the remanent magnetism. The cleansing may be done magnetically or thermally. The data for all the specimens from a site are averaged. From this average it is possible to determine the location of the paleomagnetic pole at the time the rock acquired the remanent magnetism. Although this pole is strictly the magnetic pole, it is assumed to be close to the pole of rotation of the earth. The pole is in the direction of the bearing of the remanent magnetism. The distance in degrees is

$$\cot p = \tfrac{1}{2} \tan I \tag{18}$$

where p is the angular distance along a great circle between the site and the pole, I is the angle of inclination (plunge) of the remanent magnetism. If $I = 20°$, then $p = 80°$. The polarization must also be determined, that is, whether the down direction is north- or south-seeking. Statistical tests and criteria of reliability are applied. If the sites are in one area, the data for the sites may be averaged before calculating the position of the paleomagnetic pole.

Paleomagnetism has extensive application in the problem of polar wandering, continental drift, and pole reversals. The last, referring to the switching of the north and south poles, may be used as a means of correlation and in calculating the rate of sea-floor spreading. But all these fascinating subjects are beyond the scope of this book.

A few examples of the application of paleomagnetic data to local structural problems may be described.

The Lewis overthrust is a large overthrust in Montana and Alberta.[19] Precambrian and Paleozoic sedimentary rocks have been thrust several tens of miles northeasterly over Cretaceous rocks. The thrusting is early Tertiary. In Alberta the trace trends northwesterly for 30 miles north of the United States border, then swings westward for 22 miles to Kootenay Pass, whence it trends north. If the thrust sheet moved about N.80°E., then all irregularities in the trace of the fault are due to differential erosion of a warped thrust plane. On the other hand, the thrust block may have moved east in the Flathead Range (Fig. 22-16), but may have been rotated to move northeast in the Clark Range. If a remanent magnetism had been imposed on the rocks in Precambrian time, it should have been affected by a rotation. As shown in Fig. 22-17, the directions of magnetization are the same throughout the area. One would conclude there has been no rotation. The paleomagnetic pole deduced from this diagram is in the mid-Pacific 7° south of

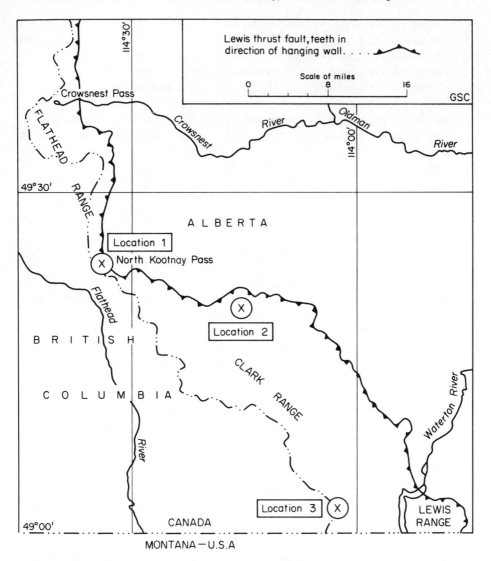

Fig. 22-16. Location of paleomagnetic studies in the Lewis overthrust sheet. (After Norris and Black. Permission Alberta Geological Society, D. K. Norris and R. F. Black.[19])

the equator, similar to the location deduced from other Late Precambrian rocks.

The folds of the Valley and Ridge Province of the Appalachian Highlands of Pennsylvania are disposed in a great arc, the axes trending N.15°E. in the south-central part of the state, N.45°E. in the central part, and N.70°E. in the eastern part. Were the folds pushed into the arc or does the arc predate

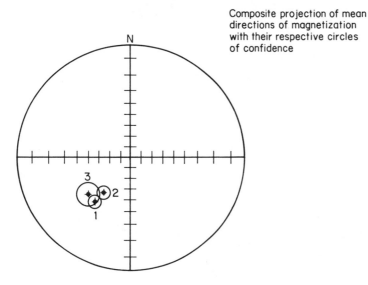

Composite projection of mean
directions of magnetization
with their respective circles
of confidence

Fig. 22-17. Remanent magnetism in three areas on the Lewis overthrust sheet shown in Fig. 22-16. (Permission Alberta Geological Society, D. K. Norris and R. F. Black.[19])

the folds? If the folds were pushed into the arcuate pattern, the direction of remanent magnetization, if prefolding, should have been rotated about a vertical axis.[20] A special study to test this hypothesis was made of the red siltstones of the Mississippian Mauch Chunk Formation. The paleomagnetic directions were determined and the beds figuratively tilted back into a horizontal position. The direction of magnetization is uniform regardless of the position in the state of Pennsylvania (Fig. 22-18), averaging 10° in a direction S.18°E. The uniformity of attitude indicates that the arc is the shape of the original geosyncline and not the result of a great push. The north pole at the time of sedimentation was at 43°N., 127°E., which is near Japan.

OCEANIC ANOMALY BELTS

During the last 15 years magnetic anomaly maps have been prepared for large areas in the oceans. Surverys are made by ship or plane. In many places alternating positive and negative anomaly belts, which may be 10 to 20 miles wide, trend north-south. Along certain lines, such as the fracture zones of the Pacific (Fig. 22-19) the anomaly belts are offset, presumably due to faults. If the belts on opposite sides can be successfully correlated, the strike-slip component of the fault can be calculated.

These anomaly belts are also being used to calculate the rate of sea-floor spreading (Fig. 23-14). Although there are several variants of this hypothesis,

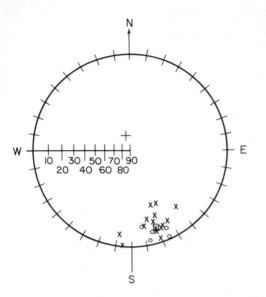

Fig. 22-18. Remanent magnetism in Mauch Chunk (Mississippian) Formation of Pennsylvania. Specimens thermally cleansed and corrected for dip. Crosses are plots on lower hemisphere, circles are plots on upper hemisphere. Mean plunge of magnetic vector is 10°S. 18°E. (After Knowles and Opdyke. Permission American Geophysical Union, R. R. Knowles and N. D. Opdyke.[20])

one assumes that convection currents are rising beneath the oceanic ridges. These subcrustal currents move outward in opposite directions from the crest of the ridge, tearing it apart. New basaltic lava moves in to fill the void and becomes magnetized in accordance with the magnetic field of the earth at that time (page 460). The convection currents are such that the opposite sides of the ridge move away from the ridge at a rate of one to four centimeters a year. The rate is deduced on the assumption that the alternating magnetic anomaly belts represent normal and reverse magnetic earth fields (page 469) and that these can be dated by integration into a time scale based on the potassium-argon method.

With the passage of time the new crust gets further and further from the ridge. Thus, crust that formed directly over the apex of the convection current 1,000,000 years ago would now be 10 kilometers away from the crest of the ridge. Transform faults develop on the crest of the ridge. The oceanic crust underthrusts the continental areas along the Benioff Zone (page 489).

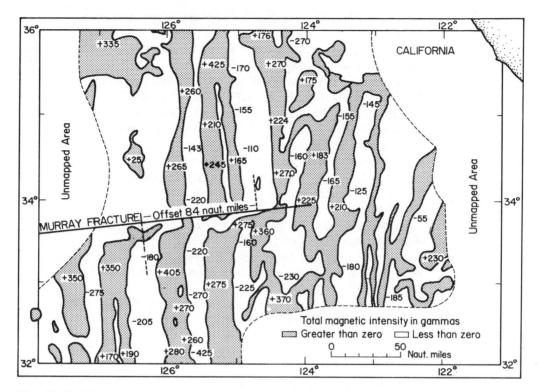

Fig. 22-19. Magnetic anomalies associated with Murray fracture zone, Eastern Pacific Ocean. Positive numbers show the greatest positive anomalies; negative numbers show the greatest negative anomalies. (After Mason.)

References

[1] Griffiths, D. H., and King, R. F., 1965, *Applied geophysics for engineers and geologists*, 223 pp., Oxford: Pergamon Press.

[2] Grant, F. S., and West, G. F., 1965, *Interpretation theory in applied geophysics*, 583 pp., New York: McGraw-Hill Book Company, Inc.

[3] Dobrin, Milton B., 1960, *Introduction to geophysical prospecting*, 2d ed., New York: McGraw-Hill Book Co., p. 446.

[4] Vening Meinez, F. A., *Gravity expeditions at sea*, Vol. 1 (1932), Vol. 2 (1934), Delft: Netherlands Geodetic Commission.

[5] La Coste, L., et al., 1967, La Coste and Romberg stabilized platform shipboard gravity meters, *Geophysics* 32: 99–109.

[6] Wing, Charles G., 1969, MIT vibrating string surface-ship gravimeter, *J. Geophys. Res.* 74: 5882–94.

[7] Hammer, S., 1939, Terrain corrections for gravimeter stations, *Geophysics* 4: 184–94.

[8] Wilson, J. H., 1941, Gravity meter survey of the Wellington Field, Larimer County, Colorado, *Geophysics* 6: 264–69.

[9] Berg, Robert R., and Romberg, Frederick E., 1966, Gravity profile across the Wind River Mountains, Wyoming, *Geol. Soc. Amer. Bull.* 77: 647–56.

[10] Bean, R. J., 1953, Relation of gravity anomalies to the geology of central Vermont and New Hampshire, *Geol. Soc. Amer. Bull.* 64: 509–38.

[11] Biehler, Shawn, and Bonini, William E., 1969, *A regional gravity study of the Boulder Batholith, Montana*, Geol. Soc. Amer., Memoir 115, 401–22.

[12] Vacquier, V., Steinland, Nelson Clarence, Henderson, Roland G., and Zietz, Isidore, 1951, *Interpretation of aeromagnetic maps*, Geol. Soc. Amer., Memoir 47, 151 pp.

[13] Clark, Sydney P., ed., 1966, *Handbook of physical constants*, rev. ed., Geol. Soc. Amer., Memoir 97, 587 pp.

[14] Bromery, Randolph W., Kirby, John R., Vargo, Joseph, et al., 1957, *Aeromagnetic map of Berlin and vicinity, New Hampshire*, U. S. Geol. Surv., Geophys. Invest. Map 9P 139.

[15] Billings, Marland P., and Fowler-Billings, Katharine, *Geology of the Gorham Quadrangle, New Hampshire*, in preparation.

[16] Brosgé, W. P., Brabb, E. E., and King, E. R., 1970, *Geologic interpretation of reconnaissance aeromagnetic survey of northeastern Alaska*, U. S. Geol. Surv., Bull. 1271F, 140 pp.

[17] Joyner, William B., 1963, Gravity in north-central New England, *Geol. Soc. Amer. Bull.* 74: 831–58.

[18] Irving, E., 1964, *Paleomagnetism and its application to geological and geophysical problems*, 399 pp., New York: John Wiley & Sons, Inc.

[19] Norris, D. K. and Black, R. F., 1962, Paleomagnetism and differential rotation in the Lewis Thrust Plate, *J. Alberta Soc. Petroleum Geol.* 10: 13–21.

[20] Knowles, Raymond R. and Opdyke, Neil D., 1968, Paleomagnetic results from the Mauch Chunk Formation: A test of the origin of curvature in the folded Appalachians of Pennsylvania, *J. Geophys. Res.* 73: 6515–28.

23

GEOPHYSICAL METHODS IN STRUCTURAL GEOLOGY: SEISMIC AND THERMAL

Seismic Methods

INTRODUCTION

The essence of the seismic methods is the accurate observation of elastic waves, generated naturally or artificially, that are transmitted through the rocks. The artificial waves are generally induced by expolsives, but other devices are also used. The energy released during earthquakes generates elastic waves that are indispensible in investigating the interior of the earth and larger structural features. Atomic explosions have been important. The velocity of elastic waves in rocks depends on numerous factors. Here we are concerned only with the longitudinal or P waves, in which the particles vibrate parallel to the direction of transmission, as in sound. The kind of rock is important. In alluvium and glacial drift, the velocities range from 1900 to 6400 feet per second; in shales and sandstones the velocities are from 3000 feet to 14,000 feet per second; in granite and gneiss the velocities are from 13,000 feet to 25,000 feet per second; and in salt the velocity is 15,000 to 25,000 feet per second.

All seismic methods are similar in the there is a *shot point* and one or more *receiving points* (Fig. 23-1). If dynamite is used, a hole a few feet to 100 feet deep is drilled at the shot point. After the charge, which varies from a fraction of a pound to several pounds of dynamite, has been placed, the hole is tamped with water or dirt. The instant of the explosion may be transmitted to the recording apparatus in various ways. Where the distance between the shot point and the recording apparatus is not great, the instant of shot is transmitted by wire. A contact in an electrical circuit is broken or made at the instant of explosion. If the distance is large, the instant of shot is transmitted by radio.

At each receiving point is a *detector*, known also as a *geophone, phone,* or *pickup*. In some kinds of work, as many as 36 detectors, set up along a profile line a few hundred feet long, record the energy sent out by one explosion. The detectors are connected by wire to amplifiers and a recording camera, often placed in a specially designed truck. The vibrations received by the detectors are recorded on rapidly-moving photographic paper; a timing device, usually a tuning fork, marks every hundredth of a second.

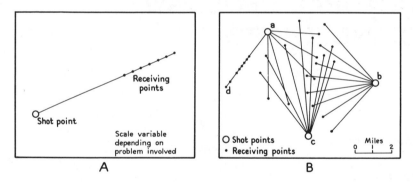

Fig. 23-1. Seismic surveying. (A) Profile shooting. (B) Fan shooting.

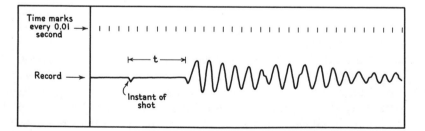

Fig. 23-2. Seismogram to show interval of time between instant of shot and arrival of first waves at receiving instruments.

The instant of explosion is also recorded on this same paper. Figure 23-2 is an example of such a record, which shows that the first impulse arrived 0.06 second after the instant of shot. Knowing the distance, it is possible to calculate the velocity of the elastic waves through the rocks.

$$v = \frac{x}{t} \tag{1}$$

Here v = velocity, x = distance, and t = time.

The data are plotted on a travel-time curve, an example of which is shown in the upper part of Fig. 23-3. The horizontal scale is the distance from the shot point to the detector; the vertical scale is the time that elapses between the instant of shot and the arrival of the first elastic wave. The less the slope of the curve, the higher the velocity. The velocity can be determined directly from the curve by the equation

$$v = \frac{x_2 - x_1}{t_2 - t_1} \tag{2}$$

In this equation v = velocity, x_2 and x_1 are distances represented by two points on the curve, and t_2 and t_1 are the times represented by the same two points.

In the upper part of Fig. 23-3, it is apparent that the part of the curve to the left of b indicates a velocity of 6000 feet per second, whereas the part of the curve to the right of b represents a velocity of 12,000 feet per second.

In recent years the processing of geophysical data has become highly sophisticated involving the use of computers. But the basic principles used in interpreting the data have not changed.

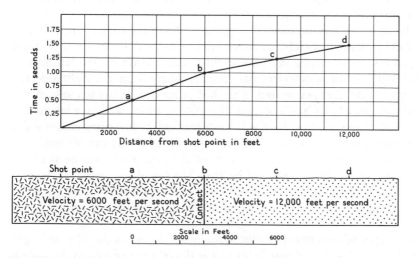

Fig. 23-3. Travel-time curve (upper diagram) obtained if relations are those shown in cross-section below. (In part after C. A Heiland.)

REFRACTION METHODS

In *fan shooting*, which has been particularly successful in locating salt domes, the receiving points are placed at similar distances near the circumference of a circle, at the center of which the shot point is located (Fig. 23-1B). For each shot point there are ordinarily 5 to 20 receiving points, which are 4 to 8 miles from the shot point. Several overlapping fans are usually employed.

In a region that has not been previously studied, the normal travel-time curve is determined by "shooting a profile"; that is, the travel-time from a single shot point to 5 to 10 receiving points on the same straight line is determined. In Fig. 23-1B, the line *ad* is such a profile. If the intervening rocks are relatively homogeneous, the curve should be a straight line.

If a body of unusual composition is present in the area, the travel times for some of the receiving points on the fan will lie off the normal curve. If a salt dome is present, the travel times are less, and the points lie below the normal curve—this is because the waves travel faster in salt than in the surrounding sediments. By noting on the map which receiving points are characterized by a lowered travel time, it is possible to locate the salt dome approximately. Other methods are usually utilized to outline the dome more accurately.

In *refraction profile shooting*, the shot point and the detectors are placed on a straight line. The same shot point may be used for successive locations of the detectors.

By way of introduction, the case of a vertical geological boundary may be considered, although the principles of refraction are not involved. This contact may be a fault, a sedimentary contact, or an intrusive contact. If the velocity of the seismic waves on opposite sides of the contact are different, a distinct break in the travel-time curve will be observed. In the lower diagram of Fig. 23-3, for example, the velocity in the formation on the left side of the vertical contact is 6000 feet per second, whereas on the right side the velocity is 12,000 feet per second. The shot point is near the left side of the diagram; at the end of half a second, the seismic waves will have reached *a*. At the end of one second, they will have reached *b*. But after crossing the vertical contact, they will travel at the rate of 12,000 feet per second. One and one-quarter seconds after the instant of shot, the waves will be at *c*, and at the end of 1.5 seconds, they will be at *d*. The travel-time curve is given in the graph above the structure section. No matter where the shot point or the receiving points, the break in the travel-time curve will always be at locality *b*.

Refracted seismic rays obey the same general laws as do refracted light waves. In Fig. 23-4, the velocity in layer *A* is 10,000 feet per second; in layer *B* the velocity is 20,000 feet per second. If an explosion occurs at *S*, the energy moves outward in all directions. Some of it, which follows the path *SR*

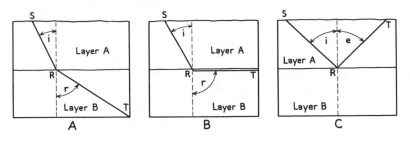

Fig. 23-4. Principles of refraction. (See text.)

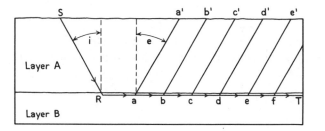

Fig. 23-5. Paths followed by refracted ray. (See text.)

(Fig. 23-4A), is, upon meeting layer B, refracted to follow RT; as in light

$$q = \frac{\sin i}{\sin r} = \frac{v_1}{v_2} \tag{3}$$

In this equation q = index of refraction; i = angle of incidence; r = angle of refraction; v_1 = velocity in upper layer; v_2 = velocity of lower layer.

The greater the angle of incidence, the greater the angle of refraction. There is a critical angle of incidence for which r is 90°; that is, RT is parallel to the contact; in the case illustrated (Fig. 23-4B), the critical angle is 30°. In the general case

$$\sin c = \frac{v_1}{v_2} = q \tag{4}$$

Here c = critical angle, and the other letters have the same meaning as in equation (3). If the angle of incidence exceeds the critical value, as in Fig. 23-4C, the energy is reflected; the angle of reflection e equals the angle of incidence i.

The wave that follows the path RT (Fig. 23-5) travels with the velocity of the lower layer, but it sends energy into the overlying layer. This energy is transmitted upward with the velocity of the upper layer. The angle of emergence is the same as is the angle of incidence. There are an infinite number of such rays; in Fig. 23-5, rays aa', bb', cc', dd', and ee' are only examples. Of the total energy expended at S only a small percentage behaves in this way.

The lower part of Fig. 23-6 illustrates the character of the wave fronts where the upper layer, 3000 feet thick, has a velocity of 10,000 feet per second; in the lower layer the velocity is 20,000 feet per second. The shot point is S; the position of the wave front at every tenth of a second is shown in Arabic numerals. At the end of 0.5 second, the waves have reached locality a; the energy has traveled directly along the surface of the earth at the rate of 10,000 feet per second. At the end of 1.0 second the energy has reached locality b, also having traveled along the surface of the earth at a speed of 10,000 feet per second. At the end of 1.1 seconds, energy that has followed the more circuitous route $Smnc$ reaches c before energy that has traveled directly from S to c. At all points to the right of b, the energy that travels in the lower layer arrives first. The travel-time curve for this case is given in the upper part of Fig. 23-6. The break in this curve occurs directly above b.

A *reversed profile* must be shot to determine the dip of the contact

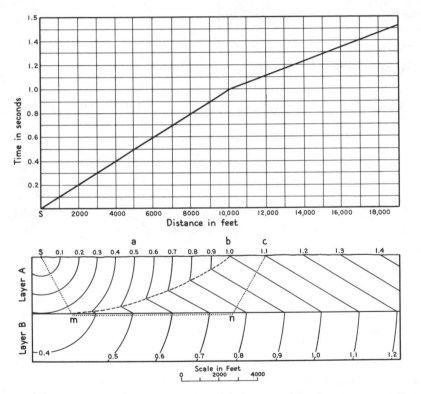

Fig. 23-6. Travel-time curve for horizontal layers. Lower diagram shows wave front each tenth of a second if S is shot point, if velocity in layer A is 10,000 feet per second, and if velocity in layer B is 20,000 feet per second. Upper diagram is travel-time curve for this case. (After H. R. Thornburgh, *Bulletin American Association Petroleum Geologists.*)

between the two layers. That is, after the original profile was shot with the shot point at the left (Fig. 23-6), a new profile should be shot with the shot point at the right. If the reversed profile is the mirror image of the original profile, the layers are horizontal. Otherwise the beds are dipping and the angle of dip can be determined by comparing the two profiles.

In actual field practice the travel-time curve is used for determining the velocities in the different layers. The slope of the curve is inversely proportional to the velocity (see p. 477). In Fig. 23-6, it is apparent from the left end of the travel-time curve that in one second the energy has traveled 10,000 feet; this is the velocity in the upper layer. The right-hand part of the curve, between 1.0 and 1.45 seconds, indicates a velocity of 20,000 feet per second. This is the velocity in the lower layer.

One would suppose that the travel-time curve for the uppermost layer would always intersect the abscissa and ordinate at zero as in Fig. 23-6. Often, however, the curve strikes the time scale above zero. This is due to a "weathered zone" in which the rocks, because of fracturing and weathering, are characterized by abnormally low velocities.

The thickness of the upper layer may be calculated in various ways. One equation is

$$d = \frac{x}{2\sqrt[2]{\dfrac{1+q}{1-q}}} \tag{5}$$

In this equation d = depth of the contact; x = "critical" distance on travel-time curve—that is, the distance from the origin to the break in the curve; and q is the index of refraction (see equation 4).

Travel-time curves may be used to obtain velocities and depths of more than two layers.

The problem is more complicated if the contact between two layers is inclined. The reversed profile does not give the same travel-time curve as does the first profile. Depths and angles of dip may be calculated, but space does not permit consideration of the methods employed in such cases.

REFLECTION METHODS

Reflections can be obtained from horizons as deep as 20 miles, but normally the investigator is concerned with shallower depths. If two layers of different velocities are in contact, some of the energy released by the explosion is reflected back toward the surface of the earth. In Fig. 23-7, S is the shot point, R is a receiving point. Of all the possible rays radiating from S, one is SE; the energy following this path is reflected at E and follows the path ER. The angle of reflection e equals the angle of incidence i.

If the beds are essentially horizontal, *correlations shooting* is employed. The shot point and the receiving points are on the same straight line (Fig.

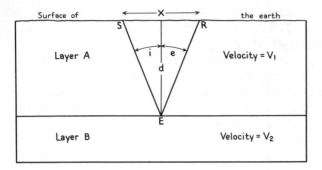

Fig. 23-7. Path followed by a reflected wave. (After C. A. Heiland.)

23-1A). Six or more geophones operate simultaneously at receiving points that are 20 to 100 feet apart. The distance from the shot point to the middle geophone is 100 to 2000 feet.

The waves from a reflecting horizon produce in the seismogram a striking change in amplitude, which is followed by numerous oscillations. Figure 23-8 shows the seismograms for six geophones that are placed 100, 200, 300, 400, 500, and 600 feet from the shot point. Such a record could be obtained from one shot, with the six geophones recording simultaneously. The time is given at the bottom of the figure, each mark representing one-hundredth of a second. The instant of shot is marked zero. The greater the distance between the shot point and the receiver, the greater is the time interval between the instant of shot and the arrival of the direct waves. The record of the direct waves is often blurred or lost because of the magnification by the instrument. In the uppermost record, in which the shot point and geophone are 100 feet apart, the reflected waves begin to arrive at 0.2 second. In all the seismograms the reflected waves arrive at approximately the same time, although, as is to be expected, they are a little later where the distance between the shot point and geophone is large.

The depth of a horizontal reflecting horizon is readily calculated:

$$d = \tfrac{1}{2} \sqrt{v_1^2 t^2 - x^2} \tag{6}$$

In this equation d = depth; v_1 = velocity in upper layer; t = time interval between instant of shot and arrival of reflected wave; and x is the distance between shot point and receiving point. The value of v_1 is determined for the surface waves by dividing the distance x by the time interval t.

Applying these equations to Fig. 23-8, it is first necessary to calculate the velocity in the surface layer. The lowest seismogram shows that the waves travel 600 feet in 0.06 seconds, indicating a velocity of 10,000 feet per second. This same seismogram shows that the reflected waves begin to arrive at approximately 0.21 second. Calculating the depth by equation (6), we get a depth of 1006 feet. Using equation (7), we get a depth of 1050 feet.

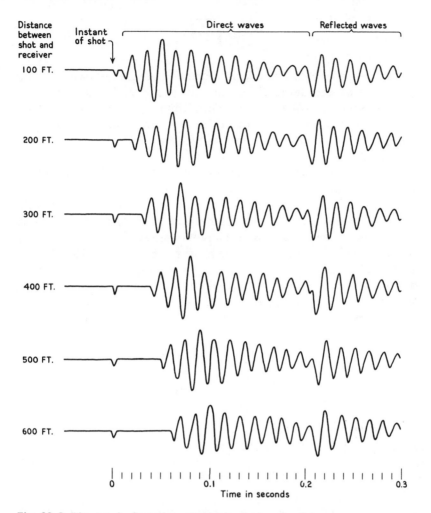

Fig. 23-8. Direct and reflected waves obtained when the distance between the shot point and the receiving point ranges from 100 to 600 feet. Velocity in layer above reflecting horizon is 10,000 feet per second.

If x is small compared to d, as is the case in *vertical shooting*,

$$d = \tfrac{1}{2} v_1 t \tag{7}$$

These equations assume that the average velocity in the upper layer is the same as the velocity at the very surface of the earth. If a well penetrates one or more layers, it is possible to get the velocity by detonating dynamite at an appropriate depth in the well. It is possible, however, to calculate the *effective average velocity* by the equation

$$V_a = \sqrt{\frac{x_2^2 - x_1^2}{t_2^2 - t_1^2}}$$
(8)

Here V_a is the average effective velocity, x_2 and x_1 are the distances from the shot point to two separate receiving stations, and t_2 and t_1 are the travel times for the same receiving stations. More commonly, the squares of the distances are plotted against the squares of the travel-times; the cotangent of the angle that the average curve makes with the abscissa is the square of the average velocity.

Figure 23-9 is a structure contour map of the Loudon, Illinois, oil field, based on a reflection survey.[1]

In the last decade instruments and techniques have been developed for preparing *continuous seismic profiles*. These are essentially structure sections of the strata and extend to depths of many thousands of feet.

At sea an energy source is towed behind the ship. One device is an eletric spark in the water. This signal is sent out at regular intervals, such as once every 9 seconds. After a few seconds the receiver on the ship receives a reflection from the sea floor and a whole series of reflections from strata beneath the sea floor down to depths of thousands of feet. The velocity of

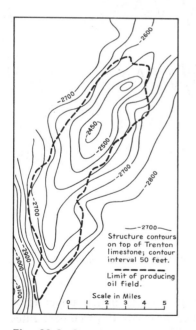

Fig. 23-9. Structure contour map made by seismic reflection method. Loudon Field, Illinois. (After P. L. Lyons.)

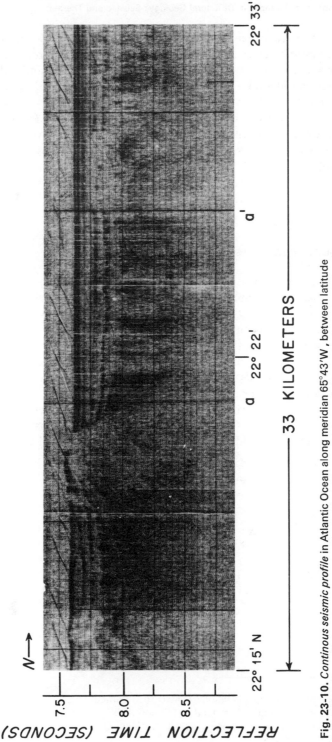

Fig. 23-10. *Continous seismic profile* in Atlantic Ocean along meridian 65° 43′ W, between latitude 22°15′ and 22°33′N, north of Puerto Rico. Vertical exaggeration about 30. Depth of water 5800 meters. Bottom of ocean is at 7.6 seconds (time for sound to travel from ship to bottom and back to ship). Much of area is underlain by flat strata a few hundred meters thick. Somewhat to left of center a hill of "basement," several hundred meters high, reaches the sea floor. The diagonal lines above 7.6 seconds are instrumental. (Courtesy Elizabeth T. Bunce. Bunce, Elizabeth T., and J. B. Hersey, 1966; *Geol. Soc. Amer. Bull.* 77 : 803–11.)

sound in sea water is a function of temperature and depth, ranging from 1500 meters per second near the surface in temperate regions to 1800 meters per second at great depths. The velocity in sediments and sedimentary rocks differ considerably, depending on the rock and depth, ranging from 1.7 to 4.0 km/sec. If the velocity in the sedimentary rocks is known, the time for each reflection can be converted into depths (equation (6), p. 483) and thickness determined. But because of the uncertainty in the velocities, the ordinate is generally expressed in time. Figure 23-10 illustrates a profile made from shipboard.

A somewhat similar method is used on land.[2] A series of profiles are run, with successive shot points and a large number of receivers for each shot.

FOCAL MECHANISM OF EARTHQUAKES

It is generally accepted that earthquakes result from the rapid movement along faults. From a study of seismograms it is possible to locate the focus of an earthquake, that is, the point within the crust where the displacement began. The epicenter is the point on the surface of the earth directly above the focus. The depth of focus generally ranges from 0 to 30 km, but may be as great as 700 km. In recent decades the seismologists have developed ingenious methods for determining the orientation of the fault causing the earthquake and also the slip.

Only a very brief and elementary outline of the method can be presented here. The records from as many seismograph stations as possible from all parts of the world are utilized.[3,4] Obviously, small earthquakes will be

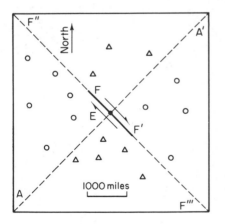

Fig. 23-11. Compressions and dilatations related to a vertical strike-slip fault. E = Epicenter. FF′ = fault; FF″ and F′F‴ = projection of trace of fault. AA′ = trace of auxilliary plane. O = compressions, Δ = dilatations.

recorded only at nearby stations. The energy from the earthquake is transmitted in many ways, but here we shall confine ourselves to a discussion of the so-called *body waves*. In the transmission of primary or *P* waves, the particles vibrate in the direction of transmission. In the transmission of secondary or shear waves, that is, the *S* waves, the particles vibrate in a plane at right angles to the direction of transmission. They are polarized in this plane, that is, vibrate parallel to a line in this plane. The *P* waves travel at a rate of 4 to 6 km/sec, whereas the *S* waves travel at half this velocity.

Figure 23-11 is a map. *E* is the epicenter of a shallow focus earthquake along a vertical strike-slip fault (*FF'*) striking northwest. The fault is a few hundred miles long. The northeast side moved southeast relative to the southwest side, that is, it is a right-slip fault. *FF"* and *F'F'''* are the projections of the fault. *AA'* is the so-called *auxilliary plane*, which passes through the epicenter and is perpendicular to the fault. In this particular instance it strikes northeast and is vertical. The intersection of the fault and the auxilliary plane is the *null vector;* in this particular case it is vertical.

The map may be divided into four quadrants (Fig. 23-11). In any seismograph station in the eastern quadrant the first arrival of *P* would be a compression, as implied by the arrows. The same is true of the western quadrant. But in the northern and southern quadrants the first movement will be a stretching or dilatation.

Figure 23-12 is a map prepared after an earthquake in an effort to determine the focal mechanism. Good records were obtained at nineteen stations. The epicenter was located and is placed at the center of the map. The nineteen stations are plotted on the map on the basis of their azimuth and distance from the epicenter. The character of the first arrivals of the *P* waves is also

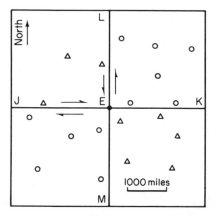

Fig. 23-12. Study of focal mechanism. Nineteen seismograph stations are located relative to epicenter. O = compression; Δ = dilatation. LEM and JEK divide the area into four quadrants. Either of these lines could be the trace of the fault. See text for methods of resolving ambiguities.

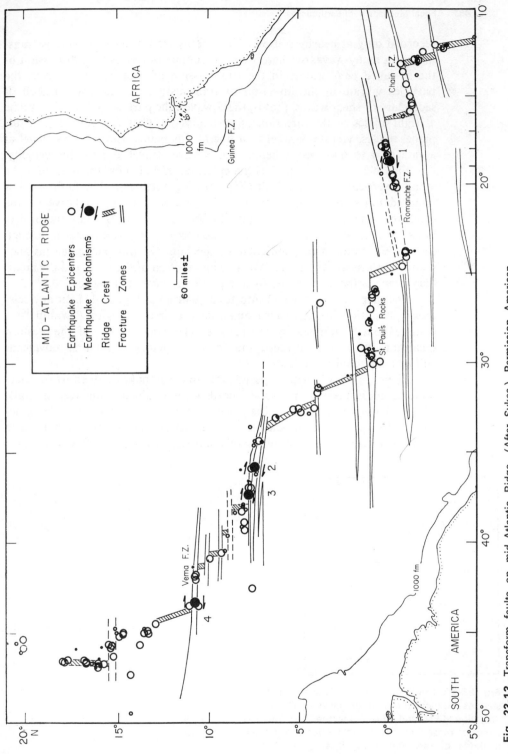

Fig. 23-13. Transform faults on mid-Atlantic Ridge. (After Sykes.) Permission American Geophysical Union.

shown, whether compression or dilatiation. It is now clear that the area may be divided into four quadrants by lines *JK* and *LM*. The arrows are added to be consistent with the compressions and dilatations. But we do not know which of the two lines is the trace of the fault. Regional geology may be useful in attaining an answer, or a solution based on *S* waves may be significant.

The presentation given above is greatly simplified. Many other factors must be considered. (1) The *extended distance* from the epicenter should be plotted rather than the true distance, because rays radiating from the earthquake are bent as they pass through the earth. (2) The problem is three-dimensional and normally neither of the two planes nor the null vector is vertical. (3) Many other *P* phases may be used; that is, *P* waves that are reflected off the base of the crust, off the core of the earth or refracted through the core. (4) The data are now commonly plotted on an equal-area projection based on the *focal sphere*, that is, the imaginary sphere representing the expanding front reached by the vibrating particles (in this case, the front reached by the *P* waves). (5) Analysis of the *S* waves theoretically gives an unambiguous solution, but the method is laborious.[5]

The focal mechanism of thousands of earthquakes has now been studied[6,7]. Thus much new and valuable information is being obtained on the present tectonic activity of the crust. Such studies show that the transverse faults on the mid-Atlantic ridge are transform faults (Fig. 23-13). Conversely, the longitudinal fractures between the transform faults are normal faults.[4]

It has long been known that the depth of foci of earthquakes increases progressively from the outer edge of island arcs toward the continent. These foci define a plane dipping about 45° under the island arc. This plane has been called the Benioff Zone, after its discoverer (Fig. 23-14). This zone has

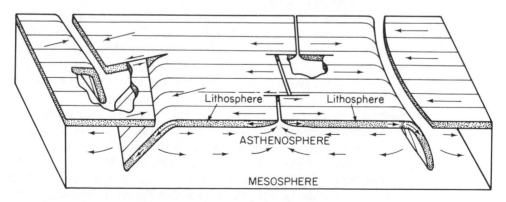

Fig. 23-14. Plate tectonics and sea-floor spreading. The lithosphere plunges downward along the Benioff zone. (After Isacks et al[7,8]) Permission American Geophysical Union.

been interpreted as a major fault zone. Moreover, first-motion studies indicate that the displacement is a relative underthrusting of the oceanic segment beneath the continental segment. But where this zone approaches the surface of the earth, the displacement is normal rather than thrust; that is, the hanging wall moves down relative to the footwall. It has been suggested that this normal faulting is due to stretching where the flat plate must bend in order to slide under the Benioff Zone.[8]

Electrical Methods

The electrical methods employed in geophysical prospecting are many and varied, both in technique and in the properties measured. Some methods utilize natural electrical currents, whereas other methods are based on artificial currents that are introduced into the rocks by direct contact or induction. Electrical methods in general are effective only for shallow depths not exceeding 1500 feet. Consequently, they are more useful in the search for shallow ore deposits or water supply than in the search for petroleum. Since the application to general structural features is limited, they will not be described.

Radioactive and Thermal Methods

Aeroscintillometer maps are comparable to aeromagnetic maps. Radioactive disintegration involves emission of α, β, and γ rays. The α-rays are essentially positively charged helium atoms that lack two electrons. The β-rays are electrons (negatively charged). The γ-rays are electromagnetic radiations.

The chief sources of such radiation from rocks are uranium, thorium, and potassium. Inasmuch as the amounts of these elements are unlike in different rocks, a means of mapping formations is suggested. Geiger counters and scintillometers are instruments that measure mostly γ-radiation. For airborne surveys a scintillometer is mounted in a plane that is flown a few hundred feet above the ground. Such airborne surveys show that the intensity of the radiation is related to the distribution of the rocks.

Measurements of terrestrial heat flow are of interest in many aspects of geology, such as comparing the oceanic crust with the continental crust.[9,10] But they may also be used to calculate the depth to which rock bodies extend, or, more correctly, the depth to which a contrast extends. Much of the terrestrial heat flowing out into space at the surface of the earth is from the deep interior. But some of the heat is generated within the crust. Although numerous variables are involved, differences in these values may give a clue to the shape and size of rock bodies.

The principles are relatively simple. A vertical bore hole penetrates the rocks in the area being investigated. On land this bore hole is ideally at least 1000 feet deep. On the ocean floor the hole is only a few tens of meters deep. A core an inch or so in diameter should be recovered in order to study the thermal conductivity and other properties of the rock. After sufficient time has elapsed to establish equilibrium—this again depends on conditions—the temperature is measured to determine the temperature gradient. If the temperature at the top of the hole is 0°C and at a depth of 300 m is 10°C, the gradient is 10°C/300 m or 33.3°C per kilometer.

To determine the heat flow through the crust it is necessary to measure the thermal conductivity of the rocks. This can be done on the core. Small sections of the core are measured; in stratified rock there may be a range in conductivities, and analysis may be complex. The conductivity of a few rocks in calories per centimeter per second per degree centigrade are: granite, 7.8×10^{-3}; syenite 7.7×10^{-3}; dolerite, 4.8×10^{-3}; quartzite, 16×10^{-3}; limestone, 6×10^{-3}.

The heat flow is given by the following equation:

$$Q = k\frac{\delta T}{\delta Z} \tag{9}$$

where Q is amount of heat in cal/cm² sec that flows across a unit area of surface normal to the flow, k is the thermal conductivity, and $\delta T/\delta Z$ is the thermal gradient.

In a bore hole in the Conway Granite in Conway, New Hampshire,[11] $k = 7.8 \times 10^{-3}$ cal/cm sec deg, and $\delta T/\delta Z = 26.10$ deg/km.

Thus, Q is 2.04×10^{-6} cal/cm² sec or 2.04 microcalories cm² sec. However, with corrections for topography and other factors this becomes reduced to 1.95 μcal/cm² sec (μcal = microcalories); country rock is 1.3 μcal/cm².

Because of its unusually high content of radiocative elements the Conway Granite produces 17.5×10^{-13} cal/cm³ sec. The value for the country rock is about 7×10^{-13} cal/cm³ sec. The excess heat production by the Conway Granite over the country rock is 10.5×10^{-13} cal/cm³ sec. We can calculate the depth to which the contrast extends by determining the thickness of the layer to give this excess heat.

$$t = \frac{h}{p} \tag{10}$$

where t = thickness of Conway Granite, h is excess heat flow above the Conway Granite, and p is excess heat production by Conway Granite, the excesses referring to excesses over country rock.

$$t = \frac{0.6 \times 10^{-6}}{0.105 \times 10^{-11}} = 5.7 \times 10^5 \text{ cm} = 5.7 \text{ km} \tag{11}$$

This is the depth to which the Conway Granite, extends.

References

1   Lyons, P. L., 1948, Geophysical case history of the Loudon field, Illinois, pp. 461–70, *in* Nettleton, L. L., 1948, *Geophysical case histories*, vol. 1, 671 pp., Houston: Society Exploration Geophysicists.

2   Froelick, A. J., and Krieg, E. A., 1969, Geophysical-geologic study of northern Amadeus Trough, Australia, *Amer. Assoc. Petrol. Geol. Bull.* 53: 1978–2004.

3   Hodgson, John H., 1957, Nature of faulting in large earthquakes, *Geol. Soc. Amer. Bull.* 68: 611–44.

4   Sykes, L. R., 1967, Mechanism of earthquakes and nature of faulting on the mid-oceanic ridges, *J. Geophys. Res.* 72: 2131–53.

5   Studer, W., S. J., and Bollinger, G. A., The S-wave project for focal mechanism studies, earthquakes of 1964, *Seismol. Soc. Amer. Bull.* 56: 1363–71.

6   Chandra, Umesh, 1970, Comparison of focal mechanism solutions obtained from *P*- and *S*-wave data, *J. Geophys. Res.* 75: 3411–30.

7   Kasahara, Keichi, and Stevens, Anne E., eds., 1969, *A symposium on processes in the focal region*, Publication of the Dominion Observatory, Canada, vol. 37, no. 7, pp. 183–235.

8   Isacks, Bryan, Oliver, Jack, and Sykes, Lynn R., 1968, Seismology and the new global tectonics, *J. Geophys. Res.* 73: 5855–98.

9   Lee, William H. H., ed., 1965, *Terrestrial heat flow*, Amer. Geophys. Union, Geophys. Monogr. 8.

10   Simmons, Gene, and Roy, Robert F., Heat flow in North America, *in* Hart, Pembroke J., ed., 1969, *The earth's crust and upper mantle*, Amer. Geophys. Union, Geophys. Monogr. 13.

11   Birch, Francis, Roy, Robert F., and Decker, Edward R., *Heat flow and thermal history in New England*, *in* E-an Zen et al., eds. 1968, *Studies of Appalachian geology, northern and maritime*, pp. 437–51.

LABORATORY
EXERCISES

OUTCROP PATTERN OF HORIZONTAL AND VERTICAL STRATA

Patterns

If strata are horizontal, each stratigraphic horizon,* such as the top or bottom of a bed, is at the same altitude everywhere. Horizontal beds are shown in Fig. 4-7A. The outcrop of the top or bottom of a horizontal bed thus follows topographic contours. The width of outcrop of a horizontal bed depends upon the thickness of the bed and upon the topography. The width of outcrop of a bed of uniform thickness is greatest where the slopes are gentle—that is, where the contours are far apart.

The relation between the attitude of a bed, the outcrop pattern displayed by the bed, and the topography may be remembered by the "rule of V's." This rule states in part that the outcrop of a horizontal bed forms a V as it crosses a valley and that the apex of the V points upstream (formation

---

* Horizon is used in this book to refer to a surface having no thickness.

shown by circles in Fig. 4-7A). The top and bottom contacts of the bed are parallel to topographic contours.

If an horizon is exposed at one place in a region of horizontal beds, it is obviously possible to predict the location of the horizon everywhere else on the map. If the exposure of the horizon coincides with a contour, the location and the pattern of the horizon will be identical with that of the contour. If the altitude of the exposure of the horizon falls between two contours, interpolation is necessary in order to locate the position of the horizon on the map. In regions of low relief the error of the predicted location may be considerable.

The top and bottom of a vertical bed will appear on the map as straight lines parallel to the strike of the bed. Topography has no control on the outcrop pattern of vertical beds. The outcrop pattern of a vertical bed is shown in Fig. 4-7B.

Structure Sections

Structure sections show the structure as it would appear on the sides of an imaginary vertical trench. Sometimes they portray the structure as observed in cliffs, highway cuts, quarries, or mine workings, either as observed on a vertical face or projected from an inclined face onto a vertical section. More often, structure sections are based on exposures at the surface of the earth or on drill records. The accuracy of the section depends on the complexity of the geology, the amount of data available, and the skill of the geologist.

In regions of low relief the surface of the earth may be represented by a straight line, but ordinarily the top of the structure section shows the topography. The first step, therefore, in preparing a structure section is to make a topographic profile. Such a profile is prepared from the topographic map in the following way. In Fig. E1-1A the structure section is to be made along

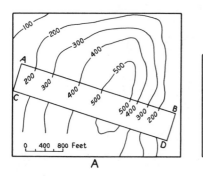

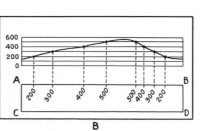

Fig. E1-1. Preparation of topographic profile. (A) *AB* is line along which profile is to be made. *ABCD* is strip of blank paper laid over topographic map. (B) Plotting profile; scale along left side of profile is same as horizontal scale of the map.

the line *AB*. A strip of blank paper is laid across the topographic map, the top of the strip coinciding with the line of the section. A mark is made on this strip of paper at each place that a contour crosses the line of the section. The altitude represented by each of these marks may be written on the strip.

A base on which to plot the profile is then prepared, as shown in the

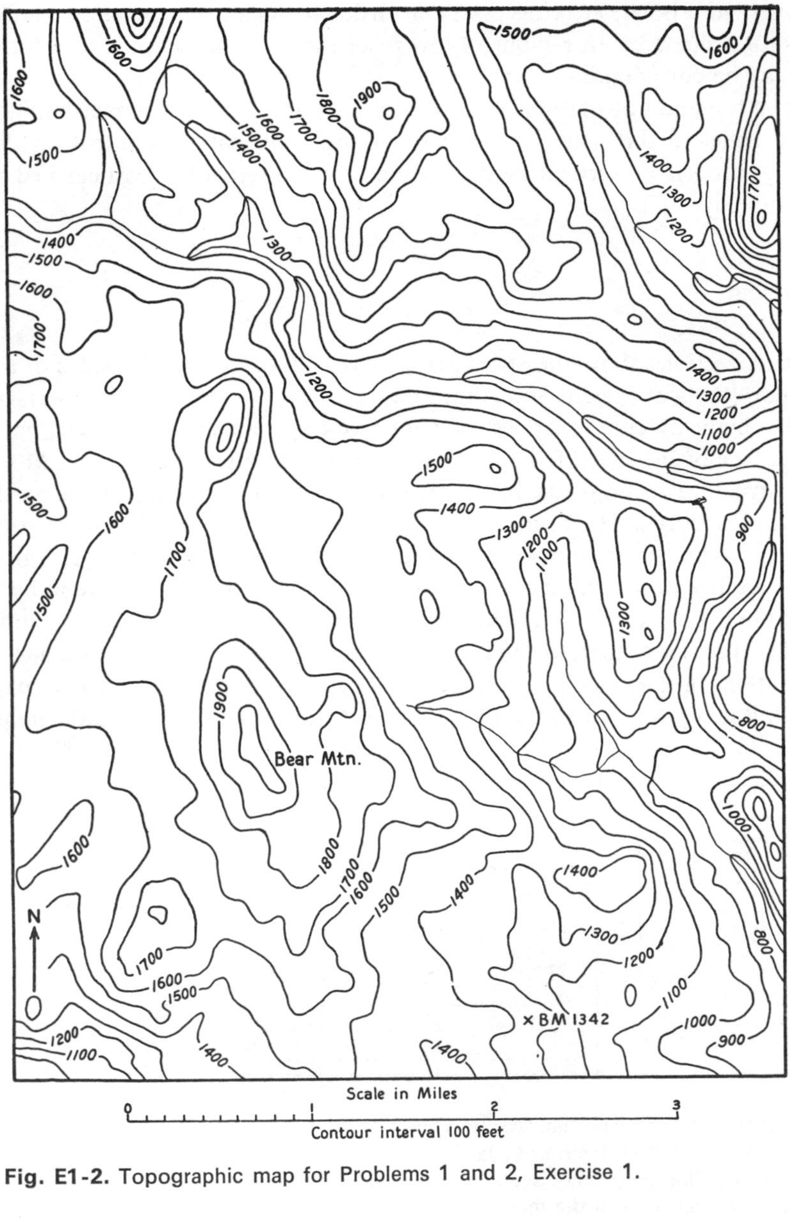
Bear Mtn.
× BM 1342
N
Scale in Miles
0
1
2
3
Contour interval 100 feet

Fig. E1-2. Topographic map for Problems 1 and 2, Exercise 1.

upper part of Fig. E1-1B. The horizontal scale is the same as that of the map. The altitude above sea level of each horizontal line is written on this diagram; each line thus corresponds to a contour. Generally the vertical scale should be the same as the horizontal scale; otherwise the geological structure becomes distorted. If the strata are horizontal, however, it is sometimes necessary to exaggerate the vertical scale.

The strip of paper prepared from the topographic map is then placed under the diagram (Fig. E1-1B). Each pencil mark on the strip is then projected upward to the appropriate altitude line and a dot is made. These dots are then connected by a smooth line to make the topographic profile.

To add the geology to the profile, place a new strip of paper along the line of the section on the map and make a pencil mark wherever a geologic contact crosses the line of section. This strip is then placed beneath the topographic profile, and the contacts are projected vertically upward to the profile. If the beds are horizontal, horizontal lines drawn through these points are the geological conatcts in the profile.

Problems

1. In Fig. E1-2 assume that the base of a horizontal bed of sandstone 150 feet thick is exposed at an altitude of 1550 feet about 1.2 miles northwest of the top of Bear Mountain. On a piece of tracing paper placed over the figure draw the upper and lower contacts of the sandstone and color those parts of the area in which the sandstone is the surface formation.

2. In Fig. E1-2 assume that the base of a series of vertical limestone beds 1500 feet thick is exposed at B. M. 1342 near the southern end of the map. The beds strike N.25°E. and are younger toward the northwest. On a piece of tracing paper placed over the figure draw the upper and lower contacts of the limestones and color those parts of the area in which the limestone is the surface formation.

3. In Fig. E1-3 different patterns are used to show various geological formations. (a) What is the attitude of the strata? (b) Describe briefly the relationship of the outcrop pattern to the topography. (c) Draw a topographic profile and geologic section along the line *AB*.

The instructor may also want to use some of the many maps published by the U. S. Geological Survey, especially some of those in the Geological Quadrangle Maps series, showing flat or gently dipping strata.

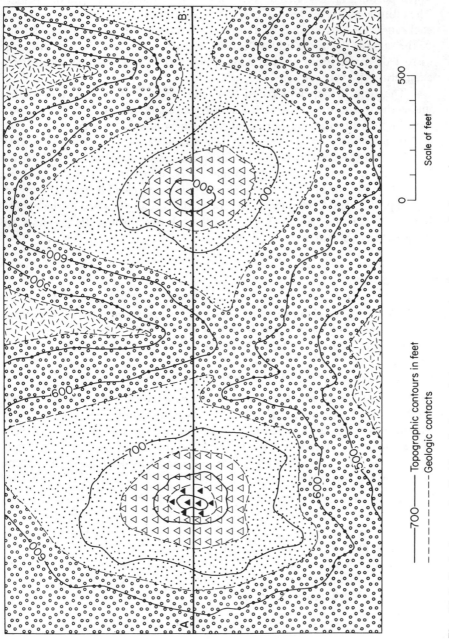

700 —— Topographic contours in feet

– – – – – Geologic contacts

Scale of feet

0   500

Fig. E1-3. Geologic map for Problem 3. Open circles represent conglomerate; stipple, sandstone; open triangles, rhyolite; filled triangles, basalt; diverse dashes, granite.

498

PATTERNS OF
DIPPING STRATA;
THREE-POINT PROBLEMS

Rule of V's

It has been shown in Exercise 1 that the contacts of horizontal beds follow topographic contour lines, and that wherever such strata cross a valley their outcrop pattern forms a *V* that points upstream. It was also shown that topography has no influence on the pattern of vertical beds, the outcrop pattern of which forms straight lines parallel to the strike of the beds.

The outcrop pattern of beds that dip upstream forms a *V* that points upstream, as shown in Fig. 4-8. The contacts of the bed are not parallel to topographic contours. Figure E2-1A shows the same thing. The horizon, known to outcrop at *a* and *b*, strikes N.90°E. and dips 11.3°N.; that is, it drops 20 feet vertically every 100 feet horizontally. The heavy line shows the expected outcrop pattern.

The outcrop pattern of beds that dip downstream at an angle greater than the stream gradient forms a *V*, the apex of which points downstream. In Fig. E2-1B let us assume that an horizon strikes N.90°E. and dips down-

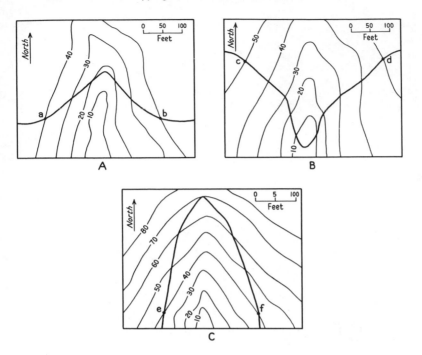

Fig. E2-1. Rule of *V*'s (see text).

stream at an angle of 11.3 degrees; that is, it drops 10 feet vertically every 50 feet horizontally. Moreover, the stream gradient is 10 feet vertically every 100 feet horizontally. If the horizon is known to outcrop at *c* and *d*, the heavy line shows the expected outcrop pattern.

The outcrop pattern of beds that dip downstream at an angle less than the gradient of the stream forms a *V* that points upstream. The contacts of the beds are not parallel to topographic contours. In Fig. E2-1C let us assume that the horizon strikes N.90°E. and dips south (downstream) at an angle of 5.7°; that is, it drops 10 feet vertically every 100 feet horizontally. Moreover, the stream gradient is 10 feet vertically every 50 feet horizontally. If the horizon outcrops at *e* and *f*, the heavy line shows the predicted outcrop pattern.

Application of the rule of *V*'s to geological maps enables one to determine the approximate dip of beds from the outcrop patterns. By use of the three-point method, which will be described later, it is possible to determine the value of the dip with considerable precision.

Outcrop Pattern of Dipping Bed

The outcrop pattern of an horizon can be predicted if a contour map showing the topography is available, if the dip and strike of the horizon are known, and if the location of one exposure of the horizon is given. This is possible, however, only if the horizon is truly a plane surface—that is, if its dip and strike are constant.

Figure E2-2 illustrates the procedure that may be followed. The horizon outcrops at X. The ground surface is represented by 100-foot contours. Inasmuch as the horizon is known to strike N.90°E. and to dip 20°S., it is possible to predict its position at any place in the area. The position of the horizon may be represented by structure contours, which are more fully described in Exercise 6.

Draw the line SS' through the outcrop X parallel to the strike of the horizon (N.90°E.). Inasmuch as the outcrop is at an altitude of 800 feet, at every place on this line the horizon has an altitude of 800 feet. Now make a vertical section at right angles to the strike by drawing AB perpendicular to the strike of the bed at any convenient distance from the map. The intersection of AB and SS' may be designated by C. At C lay off the angle BCE equal to the dip of the horizon, in this instance 20 degrees. CE is the trace of the horizon on the vertical section. Along SS' from point C lay off 100-foot units (equal to the topographic contour interval), using the same scale as that of the map.

Through each 100-foot point above or below C draw a line parallel to AB to an intersection with line CE. The intersections are points on the bedding plane; they are 100 feet apart vertically. From each of these intersections draw lines parallel to SS'. These lines are 100-foot structure contours on the horizon. At each point where a structure contour intersects a topographic contour of the same altitude, the horizon will outcrop. The locations of these intersections have been marked by small circles. Connecting these circles, as shown in Fig. E2-2, shows the predicted outcrop pattern.

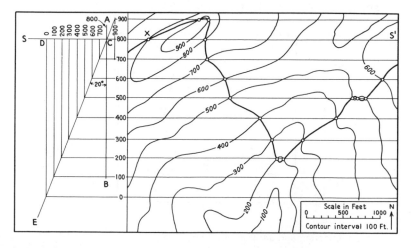

Fig. E2-2. Prediction of outcrop pattern. If a stratigraphic horizon at X strikes N90°E., and dips 20°S., the expected outcrop pattern is shown by the heavy line passing through the open circles.

Three-Point Problems

The method of working a three-point problem is the opposite of constructing an outcrop pattern. It is possible to calculate the dip and strike of an horizon if the location and altitude of three points on that horizon are known and if the horizon is truly a plane and not a warped surface.

A simple illustration of a three-point problem will be given first. Figure E2-3A is a map giving the location and altitude of three points on an horizon; these points are *A*, *B*, and *C*. Inasmuch as the strike of any plane is a line connecting points of equal altitude on that plane, line *AB* is the strike of the horizon under consideration because *A* and *B* are at the same altitude. The dip is measured at right angles to the strike, and in this case it is toward the southeast. A perpendicular is dropped from *C* to *AB*, the intersection being labeled *D*. To find the value of the dip a vertical triangle is rotated to the surface around *DC* as an axis. *CF* is erected perpendicular to *DC*. The difference in altitude between points *C* and *D*, 600 feet, is set off, on the same scale as the map, along the line *CF*. The angle *CDE* is the dip of the horizon.

A more general problem is illustrated by Fig. E2-3B. The location and altitude of three points on the horizon are shown. Some point, to be determined, between points *B* and *C*, will have the same altitude as *A* (1050 feet); a line connecting that point with *A* will be the strike of the horizon. The unknown point can be located by proportion:

$$\frac{\text{Altitude of } A \text{ minus altitude of } B}{\text{Altitude of } C \text{ minus altitude of } B} = \frac{\text{Distance } BD}{\text{Distance } BC}$$

where *D* is the point we wish to find. Solving the equation, we obtain *BD* = 1100 feet. This distance is set off from point *B* using the same scale as the

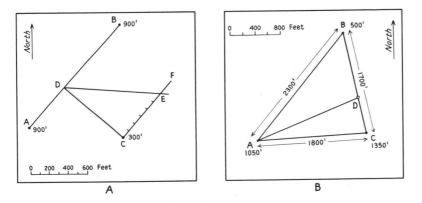

Fig. E2-3. Three-point method. Location and altitude of a plane are given at *A*, *B*, and *C*. Dip and strike of the plane can be determined.

map. AD is the strike of the horizon. The dip may be found in the same way as in Fig. E2-3A.

Problems

1. Figure E2-4 is a topographic map in which two geologic horizons are shown, one by a broken line and the other by a dotted line. What is the attitude of the horizons at a, b, c, and d?

2. In another area a north-south drainage tunnel 10 feet in diameter is to be driven in bedrock at an altitude of 500 feet above sea level. The tunnel will go directly under a point to be designated A. Three vertical drill

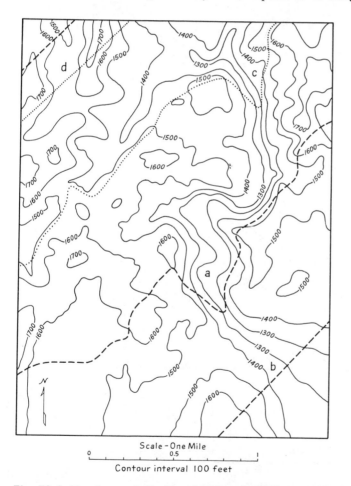

Scale - One Mile

Contour interval 100 feet

Fig. E2-4. Map for use in Problem 1 in Exercise 2. Topographic contour interval is 100 feet.

Scale of feet

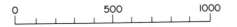

Fig. E2-5. Topographic map to be used in Problem 4.

holes were driven to locate a probable fault. The location of the drill holes and altitude of the fault is as follows.

Drill Hole	Location	Altitude of Fault above Sea Level
B	1000 feet east of A	900 feet
C	1000 feet north of A	100 feet
D	1200 feet N.60°W. of A	700 feet

a.   What is the attitude of the fault?
b.   Where will the tunnel cross the fault?

3.   It is proposed to construct an east-west water supply tunnel in bedrock at an altitude of 1800 feet above sea level. The surface is level and at an altitude of 2000 feet. The proposed tunnel is to be 1000 feet north of a point that may be designated A, which is located on a very weak shale. But a few hundred feet south of it there is a well-consolidated conglomerate. Three drill holes were made; the data are as follows. In each case conglomerate is found beneath the shale.

Location	Altitude of Contact of Shale and Conglomerate Relative to Sea Level
A Reference point	1800 feet
B 700 feet N.50°W. of A	1400 feet
C 2000 feet N.10°E. of A	0 feet

a.   What is the attitude of the contact of the conglomerate and shale?
b.   In what rock would the proposed tunnel be located?
c.   Where would you suggest locating the tunnel?

4.   On Fig. E2-5 a thin bed of limestone, striking N.90°E. and dipping 20° north, crops out at the X. Show the trace of the limestone on the map on an overlay of tracing paper; altitude of X is 2050 feet.

5.   Figure E2-6 is an area of limited outcrop. The actual outcrops are surrounded by dotted lines. Four formations are exposed: conglomerate, marble, quartzite, and amphibolite.
a.   The base of the conglomerate is well exposed at A, B, and C. Assuming that the base of the conglomerate can be treated as a plane, calculate the attitude of this contact.
b.   The trace of the contact of the marble and quartzite is shown in two of the southerly outcrops. Assuming a planar contact, calculate the attitude of this contact.

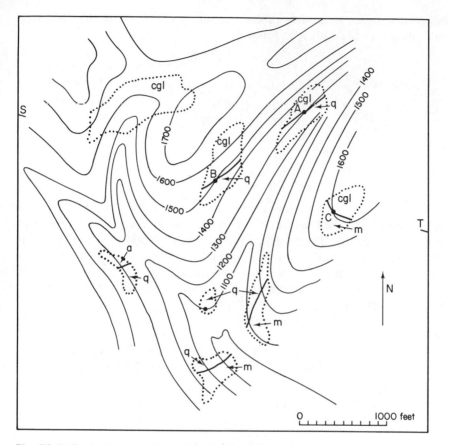

Fig. E2-6. Geologic map to be used in Problem 5.

c. On an overlay, draw a geologic map for the entire area. Assume that the quartzite-amphibolite contact is parallel to the marble-quartzite contact.
d. What is the nature of the basal contact of the conglomerate?

The instructor may also want to use some of the many maps published by the U. S. Geological Survey, especially some of those in the Geological Quadrangle Maps series.

THICKNESS AND
DEPTH OF STRATA

Introduction

It is essential in structural, stratigraphic, and economic problems to be able to calculate the thickness and depth of strata. A stratum is tabular in shape, like a sheet of paper. Strata may be horizontal, inclined, or vertical. The thickness is measured perpendicular to the plane of the bedding (t of Fig. E3-1). The depth is the vertical distance measured from any defined point on the surface of the earth to the top of the desired stratum. In Fig. E3-1, d is the depth at a, and d' is the depth at b.

If the whole stratum is favorably exposed in a cliff, the thickness may be measured directly by tape. Ordinarily, however, a direct measurement is impossible, and the thickness must be calculated by means of data obtained from a map, such as the top of the block in Fig. E3-1.

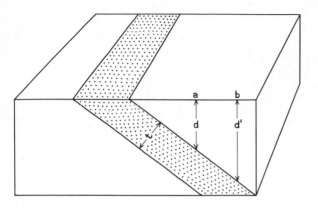

Fig. E3-1. Thickness and depth. Bed of sandstone shown by dots; shale above and below is left blank. *t* represents thickness of sandstone: *d*, depth of top of sandstone at point *a*; *d'*, depth of top of sandstone at point *b*.

Equations

The fundamental equations for calculating the thickness of beds and the depth of beds are given below. The derivation of some of the simpler equations is given. For a more complete discussion the reader is referred to some older papers.[1,2,3] Numerous papers have also been written on graphical solutions of these equations.[4,5] Where a large amount of data are involved, especially from drilling, a computer program may be justified.

Case a

Thickness, if the ground surface is horizontal, and if the breadth of outcrop of the bed is measured at right angles to its strike. As shown in cross section in Fig. E3-2A,

$$\sin \delta = \frac{t}{s}$$

or
$$t = s \sin \delta \tag{1}$$

where t = thickness of the bed, δ = angle of dip of the bed, and s = breadth of the outcrop at right angles to the strike, measured along a horizontal surface.

Case b

Thickness, if the ground surface slopes in the same direction that the bed dips, and if the breadth of outcrop is measured at right angles to the strike of the bed. As shown in cross section in Fig. E3-2B,

$$\sin (\delta - \sigma) = \frac{t}{s}$$

or
$$t = s \sin (\delta - \sigma) \tag{2}$$

where t = thickness of bed, δ = angle of dip of the bed, σ = angle of slope of the surface of the ground, and s = breadth of outcrop measured at right angles to the strike and along the surface of the ground, *not* the map distance.

Case c

Thickness, if the ground surface slopes in the opposite direction to that in which the bed dips, and if the breadth of outcrop is measured at right angles to the strike of the bed. As shown in a cross section in Fig. E3-2C,

$$\sin (\delta + \sigma) = \frac{t}{s}$$

or
$$t = s \sin (\delta + \sigma) \tag{3}$$

where symbols are the same as for equation (2).

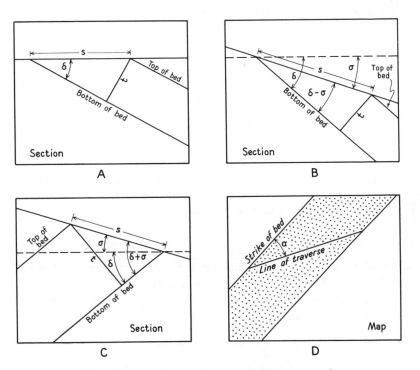

Fig. E3-2. Measurement of thickness. t represents thickness; δ, angle of dip of bedding; s, slope distance between outcrop of top of bed and outcrop of bottom of bed; σ is the angle of slope of ground; α, angle between strike of bedding and trend of traverse. (A) Ground horizontal. (B) Bedding dips in same direction that ground slopes. (C) Bedding dips in opposite direction to that in which ground slopes. (D) Map for case in which traverse is not at right angles to strike of bedding.

Case d

Thickness, if the ground surface is sloping, and if the breadth of outcrop is not measured at right angles to the strike of the bed. A plan is shown in Fig. E3-2D.

$$t = s \,(\sin \delta \cos \sigma \sin \alpha + \sin \sigma \cos \delta) \tag{4}$$

and
$$t = s \,(\sin \delta \cos \sigma \sin \alpha - \sin \sigma \cos \delta) \tag{5}$$

where t = thickness, s = slope distance (*not* map distance), α = azimuth of traverse—that is, the horizontal angle between the strike of the stratum and the direction in which the slope distance is measured, δ = dip of the bed, and σ = angle of slope of the surface of the ground in the direction of the traverse.

Equation (4) is used if the dip of the bed and the slope of the ground are in opposite directions. Equation (5) is used if the dip of the bed and the slope of the ground are in the same direction.

Case e

Depth, if the ground surface is horizontal, and if the distance is measured at right angles to the strike of the bed. As shown in a cross section in Fig. E3-3A, in which p is the point at which the depth is to be determined,

$$\tan \delta = \frac{d}{s}$$

or
$$d = s \tan \delta \tag{6}$$

where d = depth to the bed, δ = angle of dip of the bed, and s = distance along surface of ground between the outcrop of the bed and the point at which the depth of the bed is to be calculated.

Case f

Depth, if the ground surface slopes in the same direction that the bed dips, and if the distance is measured at right angles to the strike of the bed. As shown by a cross section in Fig. E3-3B, in which p is the point at which the depth is to be determined,

$$\cos \sigma = \frac{x}{s} \quad \text{and} \quad x = s \cos \sigma$$

$$\sin \sigma = \frac{y}{s} \quad \text{and} \quad y = s \sin \sigma$$

$$\tan \delta = \frac{d+y}{x} = \frac{d+y}{s \cos \sigma}$$

$$d + y = s \cos \sigma \tan \delta$$

or
$$d + s \sin \sigma = s \cos \sigma \tan \delta$$

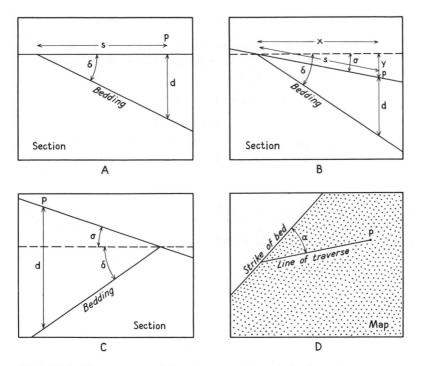

Fig. E3-3. Measurement of depth to a straigraphic horizon. d is the depth at point p; δ is the angle of dip of bedding; σ is the angle of slope of ground; s is the slope distance between outcrop of bed and point p; α is the angle between strike of bedding and trend of traverse. (A) Ground horizontal. (B). Bedding dips in same direction that ground slopes. (C) Bedding dips in opposite direction to that in which ground slopes. (D) Map for case in which traverse is not at right angles to strike of bedding.

or$$d = s \cos \sigma \tan \delta - s \sin \sigma$$

or$$d = s \left(\cos \sigma \tan \delta - \sin \sigma \right) \tag{7}$$

where d = depth, s = slope distance, σ = angle of slope, and δ = angle of dip of bed.

Case g

Depth, if the ground surface slopes in the opposite direction from that in which the bed dips, and if the distance is measured at right angles to the strike of the bed. From Fig. E3-3C, in which p is the point at which the depth is to be determined, the following equation may be derived:

$$d = s(\cos \sigma \tan \delta + \sin \sigma) \tag{8}$$

where the symbols have the same meaning as in equation (7).

Case h

Depth, if the ground surface is sloping and the distance is not measured at right angles to the strike of the bed. A plan is shown in Fig. E3-3D, *p* is the point at which the depth is to be determined.

$$d = s(\tan \delta \cos \sigma \sin \alpha + \sin \sigma) \tag{9}$$

and
$$d = s(\tan \delta \cos \sigma \sin \alpha - \sin \sigma) \tag{10}$$

where d = depth, s = slope distance, α = azimuth of traverse (that is, the horizontal angle between the strike of the bed and the direction of the traverse), δ = dip of the bed, and σ = slope of the surface of the ground in the direction of the traverse.

Equation (9) is used if the dip of the beds and the slope of the ground are in opposite directions. Equation (10) is used if the dip of the bed and the slope of the ground are in the same direction.

Alignment Diagrams

Figures E3-4, E3-5, E3-6, and E3-7 are alignment diagrams. Figures E3-4 and E3-5 may be used if the breadth of outcrop or distance perpendicular to the strike of the bed is known and if the ground is horizontal; Figure E3-4 is for thickness, and Fig. E3-5 is for depth. In Fig. E3-4 a line drawn from the "width of outcrop" on the left-hand scale to the "dip" on the right-hand scale gives the thickness on the central scale. Figure E3-5 is used in the same way to determine depth. These two diagrams may be used also where the ground is sloping if the breadth of outcrop perpendicular to the strike of the bedding is known. But "width of outcrop" on the diagram is the slope distance, and "dip" is the dip plus (or minus) the slope angle. If the bedding dips in the opposite direction to that in which the ground slopes, the slope angle is added to the dip; if the bedding dips in the same direction as that in which the ground slopes, the slope angle is subtracted from the dip.

Figures E3-6 and E3-7 are more complicated diagrams and may be used for sloping ground where the breadth of outcrop is not measured perpendicular to the strike.

Using Fig. E3-6, the thickness diagram, we first locate a point which we may call *a* on the left-hand scale, on which "azimuth of traverse" means the angle between the strike of the beds and the line along which the breadth of outcrop is measured. Thus if the breadth of outcrop is measured perpendicular to the strike of the beds, the "azimuth of traverse" is 90 degrees. If the strike of the beds is north, whereas the breadth of outcrop is measured in a northeasterly direction, the "azimuth of traverse" is 45 degrees. The upper half of the scale is used if the bed and the ground slope in opposite directions;

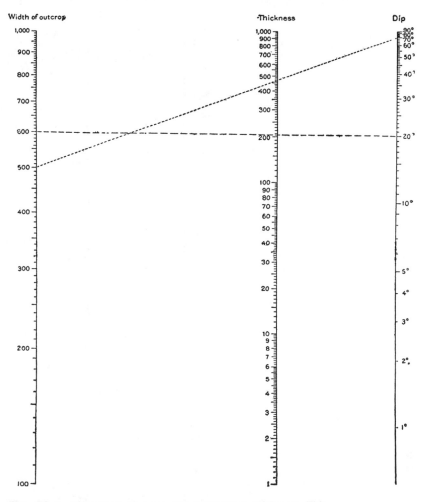

Fig. E3-4. Alignment diagram for computing thickness. This form is to be used only where breadth (width) of outcrop is measured at right angles to strike of bed. If ground surface is horizontal, if width of outcrop is 500 feet, and if dip is 70 degrees, the thickness is 470 feet. If the ground surface is horizontal, if width of outcrop is 600 feet, and if dip is 20 degrees, the thickness is 205 feet. (After H. S. Palmer.[1])

the lower half of the scale is used if the bed and the ground slope in the same direction.

A second point, which for convenience we may call *b*, is located on the triangle near the center of the diagram. The intersection of the angle of dip, given on the horizontal scale, and the angle of slope, given on the vertical scale, is located; this is point *b*.

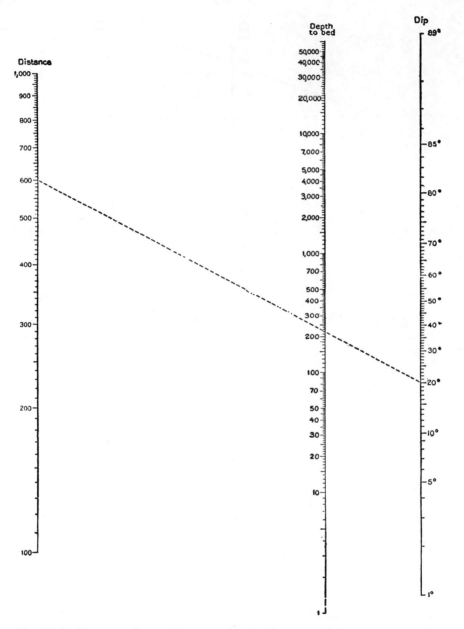

Fig. E3-5. Alignment diagram for computing depth to a stratigraphic horizon. This diagram is to be used only where distance to the outcrop of the horizon is measured at right angles to the strike of the horizon. If the ground surface is horizontal, if the distance to the outcrop is 600 feet, and if dip of bed is 20 degrees, the depth is 220 feet. (After H. S. Palmer.[1])

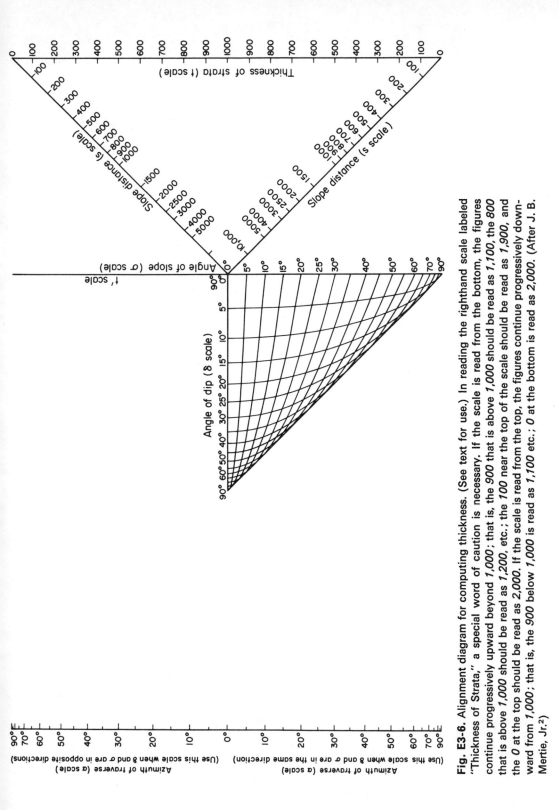

Fig. E3-6. Alignment diagram for computing thickness. (See text for use.) In reading the righthand scale labeled "Thickness of Strata," a special word of caution is necessary. If the scale is read from the bottom, the figures continue progressively upward beyond 1,000; that is, the 900 that is above 1,000 should be read as 1,100, the 800 that is above 1,000 should be read as 1,200, etc.; the 100 near the top of the scale should be read as 1,900, and the 0 at the top should be read as 2,000. If the scale is read from the top, the figures continue progressively downward from 1,000; that is, the 900 below 1,000 is read as 1,100 etc.; 0 at the bottom is read as 2,000. (After J. B. Mertie, Jr.[2])

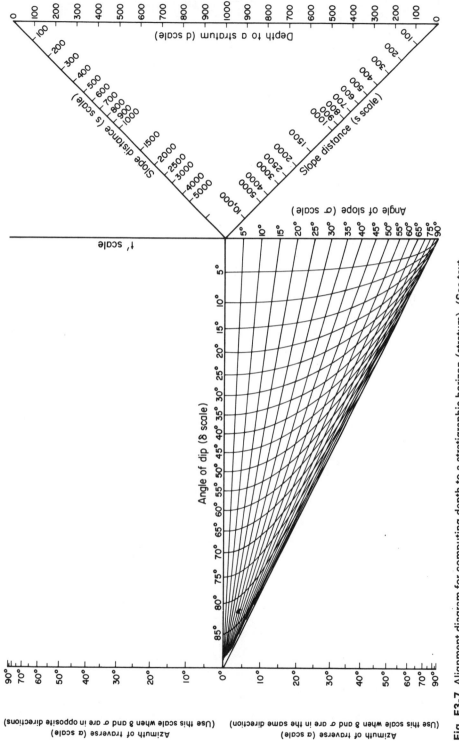

Fig. E3-7. Alignment diagram for computing depth to a stratigraphic horizon (stratum). (See text for use.) For reading the right-hand scale labeled "Depth to a Stratum," see legend under Fig. E3-6. (After J. B. Mertie, Jr.[2])

A line drawn from *a* to *b* is extended to its intersection with the scale labeled *t'*. The point of intersection may be called *c*.

A new point, which we may call *d*, is located on the slope-distance scale; this point is the breadth of outcrop of the bed measured along the sloping surface of the ground. It is *not* the map distance, unless the surface of the ground is horizontal. If *c* is on the upper half of the diagram, *d* is likewise located on the upper half. If *c* is on the lower half, so is *d*.

A line connecting *c* and *d* is extended to intersect the right-hand vertical scale (*t* scale); this intersection is point *e*, which is the thickness of the bed. If points *c* and *d* are on the upper half of the diagram, read the *t* scale as though it were numbered from 0 at the top to 2000 at the bottom. If points *c* and *d* are on the lower half of the diagram, read the *t* scale as though it were numbered from 0 at the bottom to 2000 at the top.

The alignment chart for depth, Fig. E3-7, is used in a similar way.

Problems

A neat diagram should be submitted with each problem to show the relations graphically.

1. Determine the thickness of a shale that strikes N.90°E. and dips 42°S.; the breadth of outcrop, measured in a due north direction, is 425 feet. The region is one of no relief. Solve this problem in the following order: (a) by construction of a diagram to scale; (b) by the alignment diagram in Fig. E3-4; (c) by the alignment diagram in Fig. E3-6; and (d) by the equation.

2. Determine the thickness of a sandstone that is exposed on the east side of a mountain; the sandstone strikes N. and dips 32°W. The top of the sandstone is exposed at an altitude of 2450 feet and the bottom is exposed at an altitude of 2100 feet. The distance between the top and bottom of the sandstone, measured along the slope and perpendicular to the strike is 2900 feet. Solve in the following order: (a) by the alignment diagram in Fig. E3-4; (b) by the alignment diagram in Fig. E3-6; and (c) by the equation.

3. A stream flows in a southerly direction across a limestone that strikes N.30°W. and dips 50°SW. Determine the thickness of the limestone if the base of the limestone is exposed at an altitude of 2900 feet, and the top is exposed at an altitude of 2000 feet. The breadth of the limestone along the stream, as shown on a map, is 2100 feet. Solve in the following order: (a) by the alignment diagram in E3-4; (b) by the alignment diagram in Fig. E3-6; and (c) by the equation.

4. In a region of no relief a conglomerate strikes N.50°E. and dips 55°SE. Calculate the depth to the conglomerate 500 feet S.40°E. of its outcrop in the following order: by Fig. E3-5, E3-7, and the equation. Calculate the depth by Fig. E3-7 and the equation for a point 900 feet due east of the outcrop.

5. A sandstone is exposed on the west slope of a mountain range. The sandstone strikes north and dips 46°E. The top of the bed is exposed at an altitude of 1710 feet, the bottom at an altitude of 1115 feet. The *map distance* between the top and bottom of the bed, measured in a direction N.70°W., is 1300 feet. Calculate the vertical distance between the top and bottom of the bed. Calculate the depth to the bottom of the bed at a point 600 feet east of where it outcrops; this point is at an altitude of 1300 feet. Use Fig. E3-5, E3-7, and the equation, in this order. Do the same for a point 1900 feet S.60°E. of the outcrop; this point is at an altitude of 1500 feet. Use Fig. E3-7 and the equation, in this order.

6. Bore hole *B* in an oil field is 5000 feet due north of bore hole *A* and bore hole *C* is 10,000 feet due east of bore hole *A*. The tops and bottoms of a key sandstone bed are reached at the following altitudes relative to sea level in the three holes: *A*, −2500 and −2700 feet; *B*, −2800 and −3000 feet; and *C*, −3000 and −3200 feet. What is the attitude of the sandstone and how thick is it?

References

[1] Palmer, H. S., 1918, *New graphic method for determining the depth and thickness of strata and the projection of dip*, U. S. Geol. Surv. Prof. Paper 120, pp. 122–28.

[2] Mertie, J. B., Jr., 1922, *Graphic and mechanical computation of thickness of strata and distance to a stratum*, U. S. Geol. Surv. Prof. Paper 129, pp. 39–52.

[3] Mertie, J. B., Jr., 1940, Stratigraphic measurements in parallel folds, *Geol. Soc. Amer. Bull.* 51: pp. 1107–34.

[4] Miller, F. S., 1944, *Graphs for obtaining true thickness of a vein or bed*, American Institute of Mining and Metallurgical Engineers, Contrib. No. 136, 6 pp.

[5] Mandelbaum, H., and Sanford, J. T., 1952, Table for computing thickness of strata measured in a traverse or encountered in a bore hole, *Geol. Soc. Amer. Bull.* 63: 765–76.

APPARENT DIPS
AND STRUCTURE SECTIONS
OF FOLDED STRATA

Method

This exercise describes the construction of structure sections that are made along vertical planes. Under certain conditions it may be desirable to prepare structure sections along inclined planes, especially planes that are normal to the fold axes in those areas where the plunge of the axes is relatively uniform; this subject is discussed in Exercise 5.

The topographic profile is prepared as indicated in Exercise 1. The contacts between geologic formations are also transferred in the same way.

If the structure section that is to be constructed is approximately at right angles to the general trend of the structure, dips can be used as they appear on the map without any correction. The strike-dip symbols on the map are seldom located right on the line of the structure section. How close a symbol must be in order to be used depends upon the complexity of the geology. In regions in which the structure is simple, symbols a mile or so

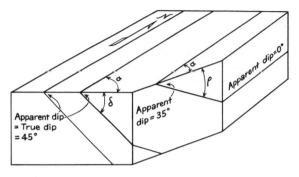

Fig. E4-1. Apparent dip and true dip. (See text.)

away may be used. The student should use a protractor in laying off the dips on the structure section.

The depth of the section depends on many variables. In this exercise it is suggested that they go to about half a mile below sea level. The beds should be connected above the ground surface by broken lines in order to emphasize the structure. Care must be taken to connect the horizons properly. The thickness of a formation should be approximately constant throughout the section.

A title, a legend, the scale, and the orientation should appear on the structure sections.

If the strike of a bed is not at right angles to the strike of the structure section, the apparent dip in the structure section is not the same as the true dip. The difference between true dip and apparent dip is illustrated by Fig. E4-1. If a bed strikes N. and dips 45°E, the apparent dip on a vertical east-west face is the same as the true dip. But on a vertical face striking northeast the apparent dip is 35°. On a vertical north-south face the apparent dip is zero. If the strike of the bed is within 10 or 20 degrees of perpendicular to the strike of the structure section, the discrepancy ordinarily may be neglected.

The apparent dip of a bed in any desired direction may be calculated from the true dip by the equation

$$\tan p = \tan \delta \sin \alpha$$

where p is the apparent dip on a vertical plane, δ is the true dip, and α is the angle between the strike of the bed and the direction of the apparent dip.

Ordinarily the apparent dip can be determined with sufficient precision by using the alignment diagram of Fig. E4-2. The true dip is given on the left-hand scale; the angle between the strike of the bed and the strike of the

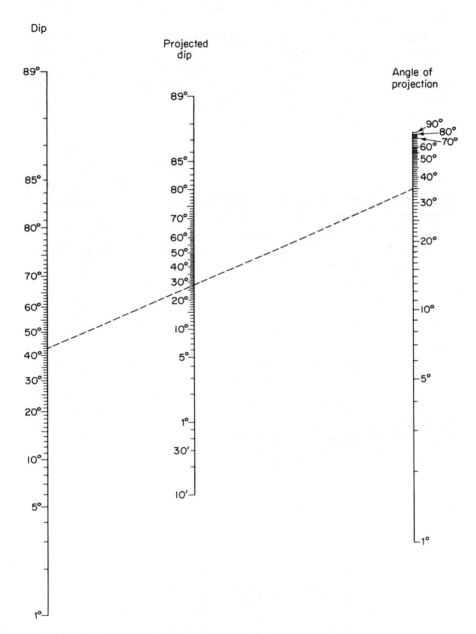

Fig. E4-2. Alignment diagram for computing apparent dip (projected dip). If true dip is 43 degrees, the apparent dip on a vertical section making a 35-degree angle with the strike of the bedding would be 28 degrees. (After H. S. Palmer.)

vertical section upon which the apparent dip is plotted is given in the right-hand scale. A line connecting the points on these two scales gives the apparent dip on the central scale.

Examples of the use of the alignment diagram may be cited. If the strike of an horizon is N.45°E. and the dip is 30°NW., what is the apparent dip on a vertical section that trends N.5°W? The point on the left-hand scale is 30 degrees. The point on the right-hand scale is 50 degrees. A line connecting these two points crosses the central scale at 24 degrees. Therefore 24 degrees is the apparent dip in a N.5°W. direction.

Tables and charts for apparent dip are given in Lahee.[1]

The reverse problem is to calculate the true strike and dip if two apparent dips are known. The apparent dips must be measured on vertical faces. Moreover, the two vertical faces cannot have the same strike. The general principles are illustrated in Fig. E4-3A. Suppose that the apparent dip is known on two vertical faces, one striking in the direction OB and the other in the direction OA. Draw two rays from a central point O. These two rays, OA and OB, are parallel to the vertical faces on which the apparent dips are measured. Moreover, they extend from O in the direction of the apparent dip. Lay off along each ray, using any convenient scale, the value of the *cotangent of the apparent dip*, OA' on OA, and OB' on OB. Connect A' and B'; the direction of this line is the direction of strike of the bed. Drop a perpendicular, OC, from $A'B'$ to O. The magnitude of this line CO equals the cotangent of the true dip.

Figure E4-3B is a specific example. On a vertical face trending N.80°E. the apparent dip is 36° in the direction N.80°E. On a vertical face trending N.60°W. the apparent dip is 47° in a direction N.60°W. On OB plot OB', which is 1.38 units long (the cotangent of 36°). On OA plot OA', which is 0.93 unit long (the cotangent of 47°). The strike (the trend of $A'B'$) is N.85°W. The true dip is the angle whose cotangent is the length of CO, which is 0.40 unit; that is, the true dip is 68°N.

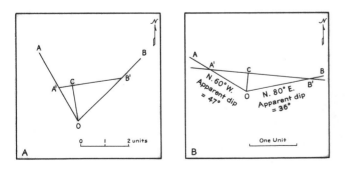

Fig. E4-3. Calculating strike and true dip from two apparent dips. (See text.)

Problems

1. From Fig. E4-4 draw a series of structure sections along the lines *AB*, *AC*, *AD*, *AE*, and *AF*. Connect the beds above the ground surface by broken lines in order to show the structure more clearly.

2. The apparent dips on different pairs of vertical faces are given below. Determine the strike and dip of the bedding for each of the three cases.

	Value of Apparent Dip	*Direction of Apparent Dip*
(a)	30°	N.90°W.
	45°	N.
(b)	55°	N.60°E.
	65°	S.45°E.
(c)	10°	N.70°E.
	35°	N.30°E.

3. The strike of a bed can be measured on the flat top of an outcrop, but the true dip cannot be measured. The apparent dip may be observed on a

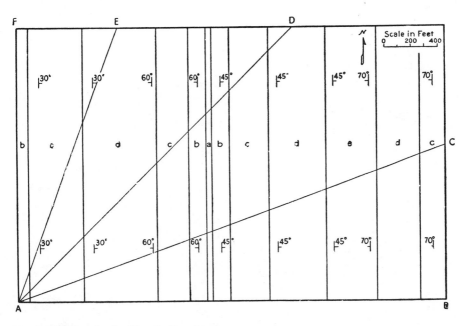

Fig. E4-4. Map for Problem in Exercise 4.

vertical face *not* perpendicular to the strike of the bedding. For each of the two cases given below determine the true dip.

	Value of Apparent Dip	*Direction of Apparent Dip*	*Strike of Bed*
(a)	60°	N.55°E.	N.90°E.
(b)	55°	N.40°W.	N.10°W.

4. Make structure sections on an assigned geologic map. The instructor can choose suitable maps that are available.

Reference

[1] Lahee, F. H., 1961, *Field geology*, 6th ed., 926 pp., appendixes 13 and 14, New York: McGraw-Hill Book Company, Inc.

5

GEOMETRICAL CONSTRUCTION OF FOLDS

Folds With Little or No Plunge

APPLICATION

Folds with little or no plunge can be reconstructed graphically by the so-called Busk method.[1] The basic assumption is that the folds are parallel, that is, that the thickness of each bed remains relatively constant during the folding. A somewhat similar method that may be applied to folds in which the thickness changes[2] will not be considered here. The method described here may be used to reconstruct parallel folds if the attitude of the beds is known in a number of places. Ideally the exposures should be along the line of the cross section, but data may be projected from either side. The method can be applied to areas where overturning of the beds occurs, but this often involves thinning of the beds. Obviously the direction in which the beds become younger must be known. The distance between exposures must be less than the wavelength of the folds.

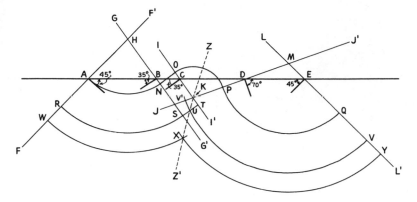

Fig. E5-1. Geometrical construction of folds. Dip of beds is known at *A*, *B*, *C*, *D*, and *E*. (After Busk.[1])

PRINCIPLE

Figure E5-1 shows the method of reconstructing parallel folds. The attitude of strata at points *A*, *B*, *C*, *D*, and *E* is shown. These points are on a line perpendicular to the trend of the folds. If the strike of the section is not at right angles to the strike of the bedding, the apparent dip in the direction of the section must be calculated for each outcrop. Perpendiculars to the bedding at each outcrop are constructed and extended to intersect the perpendiculars to the bedding of adjacent outcrops. *FF'*, the perpendicular to *A*, intersects *GG'*, the perpendicular to *B*, at *H*. The dip of the bedding is the same at any point along *FF'* as it is at *A*. Similarly, the dip at any point on *GG'* is the same as that at *B*. With *H* as a center, and *HA* as the radius, an arc is swung from *A* to intersect *GG'* at *N*. Inasmuch as the dip at *B* and *C* is the same in this problem, the perpendiculars to the bedding at these points will not intersect. The horizon is extended from *N*, parallel to the dip of the beds, to intersect *II'* at *O*. With *K*, which is the intersection of *II'* and *JJ'* as a center, and *KO* as the radius, an arc is constructed to intersect *JJ'* at *P*. With *M*, which is the intersection of *JJ'* and *LL'*, as a center, and *MP* as a radius, an arc is swung from *P* to intersect *LL'* at *Q*. *ANOPQ* is the position of the horizon outcropping at *A*.

In reconstructing an horizon at *R*, we find that the horizon falls at *T* below the intersection *K*. In such a case the distance *QV* is set off on line *LL'* equal to the distance *AR*. From *V*, with *M* as a center and *MV* as the radius, an arc is swung to intersect *JJ'* at *V'*. The position of the horizon that occurs at *R* is *RSUV*.

At the depth at which horizon *W* occurs, the dip at *C* no longer has any influence. *WXY* shows the form assumed by the horizon at *W*.

ZZ' is the trace of the axial plane of the central anticline on the vertical section. *ZK* bisects angle *IKJ'*; *KZ'* passes through points *U* and *X*.

INTERPOLATION

The accuracy of the reconstructed fold depends upon the number of dip and strike readings in the section. If outcrops are missing in a critical locality, the reconstruction may not be accurate. A key bed, for example, may be recognizable in several different localities. Its position, as predicted by the method outlined above, may not correspond to its actual position. Dips may then be interpolated to make the fold pattern fit the field facts. The method is outlined below.

In Fig. E5-2, points *A*, *B*, *C*, and *D* show the location and attitude of folded beds. When the fold is reconstructed according to the method given above, the horizon at *A* should reappear at *J*. The theoretical position of the horizon is *AHIJ*. If field mapping has shown that the horizon at *A* is the same as that at *D*, then a dip must be interpolated at some point in the section to erase the apparent discrepancy. A dip should be interpolated for a point between the two outcrops that are farthest apart—that is, between *B* and *C*.

With *G* as a center and *GD* as a radius, an arc is constructed to intersect *EE'* at *K*. For arc *HI* two arcs that are tangent to each other and tangent to *H* and *K* will be substituted. From *K*, *KL* is drawn perpendicular to *EE'*. From *H*, *HM* is drawn perpendicular to *VV'*. These two lines intersect at *N*. *H* and *K* are connected by a straight line, *HK*. A perpendicular to *HK*, drawn through point *N*, is extended to intersect *EE'* at *R* and *VV'* at *S*. With *R* as a center and *RK* as a radius, arc *KT* is constructed. Arc *HT* is drawn with *S* as a center and *SH* as a radius. *AHTKD* is the pattern of

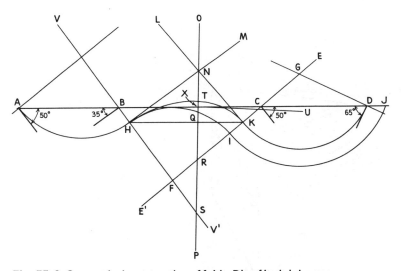

Fig. E5-2. Geometrical construction of folds. Dip of beds is known at *A*, *B*, *C*, and *D*. Same stratigraphic horizon crops out at *A* and *D*. (After Busk.)

the fold. A perpendicular to *OP* at *X* gives the interpolated dip, which is the plunging fold *UXC*.

PLUNGING FOLDS

On page 79 it was shown, in conjunction with Fig. 4-12, that in areas of plunging folds it is possible to utilize the value of the plunge in reconstructing folds. In order to use the method successfully, the bearing and value of the plunge must be relatively uniform. The basic principle is illustrated in Fig. E5-3. The top surface of the block diagram is a map of a series of folds plunging east about 10°. Any point on the map can be projected upward at an angle of 10° to the vertical section, NS. In the diagram the fold hinges are shown by broken lines. Similarly, the folds can be projected downward to the right to the vertical section N′S′. Because Fig. E5-3 is a perspective view, the structure is somewhat distorted.

Figures E5-4 and E5-5 illustrate the simple trigonometry involved. A fold is plunging at a 25° angle, *p*, in a direction N.60°E. A vertical cross section is to be made along the line *AB*. The hinge of the fold at *a* projects upward at an angle of 25° to a′.

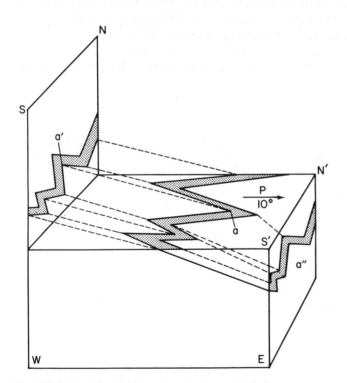

Fig. E5-3. Plunging folds projected onto vertical planes.

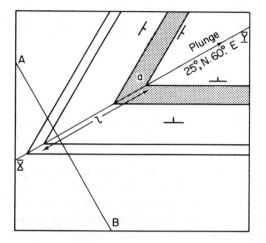

Fig. E5-4. Geologic map of symmetrical fold plunging 25°N. 60°E.

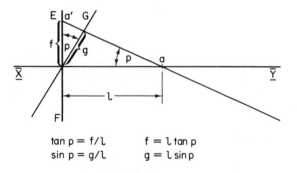

$$\tan p = f/l \qquad f = l \tan p$$
$$\sin p = g/l \qquad g = l \sin p$$

Fig. E5-5. Relation between breadth of exposures in map, vertical cross section, and cross section perpendicular to plunge of fold.

$$\tan p = \frac{f}{l} \qquad\qquad (1)$$

$$f = l \tan p \qquad\qquad (2)$$

That is, in the cross section a′ will be a vertical distance f higher than a. Thus all points can be projected upward to the plane of the cross section.

To transfer every point on the map would be a laborious process. A simpler procedure is to project key points, and sketch in the rest of the structure. But a still simpler method is available. The cross section is a reproduction of the map, except that any distance parallel to the direction of the plunge is changed in the ratio f/l (Fig. E5-5). In a sense a new map is made, leaving unchanged the scale at right angles to the bearing of the plunge, but changing the other scale in the ratio f/l. This can be done by using a pantograph or certain types of drafting equipment.

For some purposes it may be desirable to draw an inclined section that is perpendicular to the plunge, that is, a section on the plane *GH*. In this case

$$\sin p = \frac{g}{l} \qquad\qquad (3)$$

$$g = l \sin p \qquad\qquad (4)$$

In this case, the ratio of the scales on the map to the vertical scale on the section is g/l.

The accuracy of sections constructed in the manner indicated above is limited by several factors: (1) the map; (2) the uniformity of the plunge; and (3) the uniformity of the structure up and down the plunge. One other word of caution is necessary. The plunge that produced the map pattern is the one that must be used. For example, if the axes of a series of large recumbent folds were plunging northwest, it would be quite erroneous to use minor folds plunging north to predict the larger structure.

Problems

1.   A traverse is made due east across a series of rocks that strike north-south. The following data are obtained on the dip of the beds. The left-hand column gives the distance in feet from the west end of the traverse; the right-hand column gives the dip.

Distance in Feet	Dip
0	25°E.
1000	15°W.
2100	30°E.
2400	10°W.
3100	15°E
4200	30°E.
5000	30°E.
5900	30°W.
7000	30°E.
7500	25°W.

(a)   Reconstruct the folds, assuming parallel folding. (b) If a distinctive bed outcrops at the west end of the traverse, at what depth (or height, in which case it would be eroded) would the bed be located at the east end of the section (distance 7500)?

2.   A traverse is made due south across a series of rocks that strike N.90°E. The following data are obtained on the dip of the beds. The left-hand column gives the distance in feet from the north end of the traverse; the right-hand column gives the dip.

Distance in Feet	Dip
0	40°S.
2300	25°N.
3500	30°S.
5300	45°N.
7000	30°S.

(a) Reconstruct the folds, assuming parallel folding. (b). If a distinctive bed outcrops at the north end of the traverse, at what depth (or height, in which case it would be eroded) would the bed be located at the south end of the section? (c) Actually the distinctive bed outcrops at the south end of the traverse (distance 7000). Adjust the section to allow for the known position of the distinctive bed. Do this by interpolating a dip between the 5300-foot and 7000-foot points. What is the location and value of the interpolated dip?

3. A traverse is made in a due east direction across a series of rocks that strike N. The following data are obtained on the dip of the beds. The left-hand column gives the distance in feet from the west end of the traverse. The right-hand column gives the dip.

Distance in Feet	Dip
0	50°E.
1600	20°W.
2400	20°W.
2700	15°W.
3600	15°E.
4000	40°E.
5700	50°E.
6800	0°

(a) Reconstruct the folds, assuming parallel folding. (b) On the surface of the earth what is the horizontal distance of the crest of the anticline from the west end of the section? (c) At a depth of 1000 feet what is the horizontal distance of the crest of the anticline from the west end of the section? (d) At a depth of 2000 feet?

4. Using Fig. 4-12, draw a series of cross sections along the line *MN* assuming the fold axes plunge NNE. at the following angles: (a) 10°; (b) 30°; (c) 60°; and (d) 90°. Assume the bedding generally dips steeply NW, as shown in Fig. 4-12B. Assign an arbitrary scale.

References

[1] Busk, H. G., 1929, *Earth flexures*, Cambridge: Cambridge University Press; reprinted.

[2] Gill, W. D., 1953, Construction of geological sections of folds with steep-limb attenuation, *Amer. Assoc. Petrol. Geol. Bull.* 37: 2389–2406.

6

STRUCTURE CONTOURS AND ISOPACHS

Structure Contours

A structure contour is an imaginary line connecting points of equal altitude on a single horizon, usually the top or bottom of a sedimentary bed. A structure contour map thus shows the form of the horizon. Structure contours are analogous to topographic contours, which show the form of the surface of the earth.

The dip of an horizon represented by structure contours can be calculated with great precision. For a given contour interval, the closer the structure contours are to each other on a map, the steeper is the dip, just as the closer topographic contours are to each other, the steeper is the slope. The method of determining the dip quantitatively is discussed on page 113.

The method of preparing a structure contour map is illustrated by Fig. E6-1. Here we are concerned with the basic principles involved. In practice, especially in petroleum geology, these basic principles are adapted to a

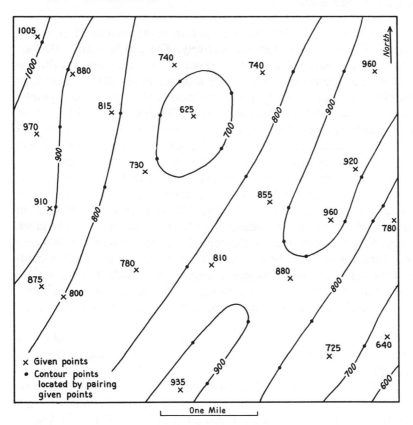

Fig. E6-1. Preparation of structure contour map.

computer program and the map can be made by computer, as discussed on page 114.

Some horizon must be chosen to be represented by the structure contours; it may be the top or bottom of a coal bed, the top of an oil-bearing stratum, or the top or bottom of some bed that is readily recognizable. Wherever this horizon is exposed at the surface of the earth, the altitude may be plotted on the map. The data may also come from drill holes or mines. If the thicknesses of the various stratigraphic units have been precisely determined, it is possible to predict at what depth the key horizon occurs, even though it is not exposed or penetrated by drill holes or mines.

The altitudes of numerous points on the top of a bed of limestone are given in Fig. E6-1. It is decided to use a structure contour interval of 100 feet. In the lower left-hand corner of Fig. E6-1 the altitude of the top of the limestone at one point is shown to be 800 feet. The 800-foot structure contour will pass through this point. Otherwise it is necessary to interpolate proportionally between each pair of points for which data are given. In the northeast corner of the map, two points on the horizon being contoured have the altitudes of 740 feet and 960 feet, respectively. The 800-foot and 900-foot contours will pass between these two points. The difference in altitude between these two points is 220 feet. The distance of the 800-foot contour from the 740-foot point will be $\frac{6}{22}$ of the total distance between the two points. The distance of the 900-foot contour from the 740-foot point will be $\frac{16}{22}$ of the total distance between the two points. The location of contours over all the map may be found in the same way. The contours obtained in this way may be modified somewhat in order to smooth out sharp curves.

If the altitudes are given on some bed other than the one being contoured, the thickness of the beds must be taken into consideration. That is, if altitudes are given on a bed 1000 feet stratigraphically above the horizon being contoured, the structure contour is 1000 feet below the altitude given on the map. Strictly, we should use the depth factor rather than the thickness (Figs. E3-2 and E3-3). Since, however, the dip is unknown prior to the completion of the structure contour map, it is impossible to make this correction. Moreover, if the dip does not exceed 10 degrees the correction is negligible. The error is usually much less than the probable error in the thicknesses of the formations.

If an area is entirely enclosed by one or more contours, it is known as a *closed structure;* this usage is not to be confused with the use of the term *closed fold* (p. 54). The closure of a fold is the vertical distance between the highest and the lowest contours that completely enclose the fold. The precision of the measurement of the closure of a fold, as determined from a structure contour map, depends upon the contour interval. If a fold has a closure less than the contour interval that is used, the closure may not appear on the map. A minus (−) before a contour indicates depth below some level plane, usually sea level.

Isopachs

A bed or formation is not constant in thickness. If the thickness is known at many localities, it is possible to draw *isopachs*, which are lines connecting points at which the bed is of equal thickness. Suppose, for example, that a formation is known to be 100 feet thick at three localities that are designated *a*, *b*, and *c*, but 300 feet thick at three other localities designated *e*, *f*, and *g*. The 100-foot isopach would pass through *a*, *b*, and *c*, and the 300-foot isopach would pass through *e*, *f*, and *g*. The 200-foot isopach would lie between the 100-foot and the 300-foot isopachs; if no other data were available, it would be placed halfway between them. In the preparation of isopach maps from data at numerous localities, interpolation is performed in the same way as in the preparation of structure contour maps. Isopach maps, like structure contour maps, can be made by computer when the magnitude of the data justifies the time invested in preparing a computer program.

Problems

1.   Figure E6-2 shows a series of structure contour maps. Assume that they are all on the same horizon, such as the top of a bed of sandstone, and that

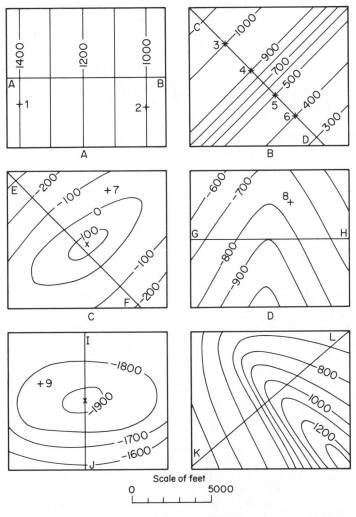

Scale of feet

0   5000

Contour Interval = 100 feet

Fig. E6-2. Structure contour maps for Problem 1 in Exercise 6.

the surface of the earth is 1500 feet above sea level. The structure contours are relative to sea level. Note that negative structure contours are given in some cases.

a. What is the dip in feet per mile and in degrees at the following places: between *1* and *2*, between *3* and *4*, between *4* and *5*, and between *5* and *6*?

b. What is the depth to the sandstone bed at *7*, *8*, and *9*?

c. Describe the structure represented by each of the six maps.

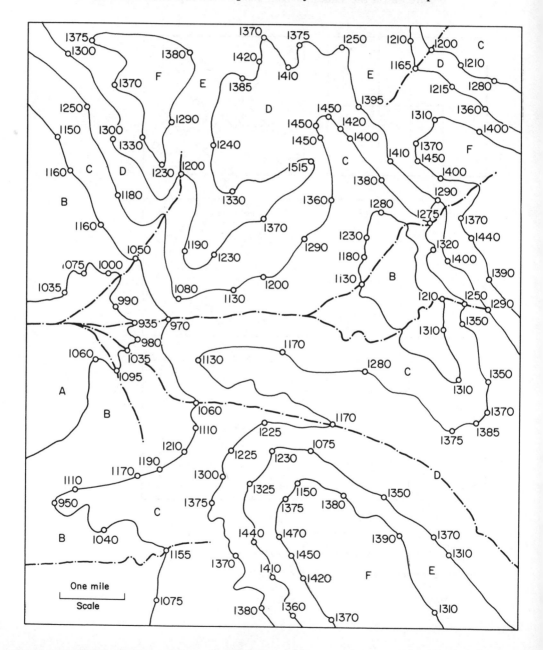

d. Draw a cross section along the lettered lines in each of the six maps. A vertical exaggeration of 10 will be necessary.

2. Figure E6-3 is a geologic map of an area with several hundred feet of relief. Formation *A* is the oldest, *F* is the youngest. The altitudes above sea level of many points on the contacts are given. Well records show that the formations have the following thicknesses. *E* is 50 feet, *D* is 160 feet, *C* is 80 feet, *B* is 250 feet. Draw a structure contour map on the top of formation A, using a contour interval of 100 feet. It will be necessary to assume that vertical distance between contacts is the same as thickness. How great an error is introduced by this assumption?

3. In the northwest corner of a square area 10 miles on a side a formation is 100 feet thick. In the other corners the thicknesses are: southwest, 400 feet; northeast, 300 feet; and southeast, 1000 feet. In the middle of each of the sides the thicknesses are: north, 150 feet; east, 500 feet; south, 600 feet; and west, 250 feet. Prepare an isopach map.

Fig. E6-3. Geologic map for Problem 2 in Exercise 6. *A, B, C, D, E,* and *F,* are geologic formations; boundaries between formations are solid lines. Dash-dot lines are streams. Figures are altitude of points on contacts.

exercise

7

TRIGONOMETRIC SOLUTION OF FAULT PROBLEMS

Method

The types of faults and the definition of terms relating to the movement along faults are given on pages 177-198.

The solution of the problems should be accompanied by a plan, as well as by one or more structure sections that show the fault and the disrupted beds.

The law of sines, which can be used to solve any triangle, is useful in this exercise. This law states that the sines of angles are proportional to the sides opposite the respective angles. As shown in Fig. E7-1, if *abc* is the triangle, and x is a perpendicular to c,

$$\sin \alpha = \frac{x}{b}$$

$$\sin \beta = \frac{x}{a}$$

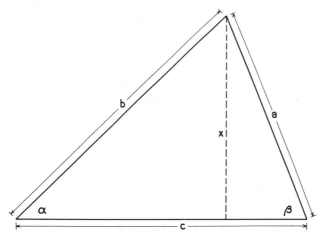

Fig. E7-1. Triangle to illustrate the *law of sines*.

$$\frac{\sin \alpha}{\sin \beta} = \frac{\dfrac{x}{b}}{\dfrac{x}{a}}$$

$$\frac{\sin \alpha}{\sin \beta} = \frac{xa}{xb}$$

and
$$\frac{\sin \alpha}{\sin \beta} = \frac{a}{b}$$

The following problem, illustrated by a structure section given in Fig. E7-2A, will be solved to illustrate the method used.

A horizontal tunnel extends east-west and intersects a fault that strikes north and dips 40 degrees to the west. At a distance of 500 feet east of this intersection the tunnel cuts a bed of sandstone that strikes north and dips 30 degrees east. At a distance of 700 feet west of the fault the tunnel cuts the same sandstone, also striking north and dipping 30 degrees east. The slickensides on the footwall rise from north to south (Fig. E7-2B) and make an angle of 30 degrees with a horizontal line in the fault plane; that is, the rake of the slickensides is 30 degrees north.

Calculate the net slip, dip slip, strike slip, heave, throw, stratigraphic separation, vertical separation in a plane perpendicular to the fault, and horizontal separation in the same plane.

From Fig. E7-2A, which is a vertical section at right angles to the strike of the fault, we obtain

$$Dip \ slip = DB + BE$$

In triangle *ABD*

$$\frac{\sin 110^\circ}{\sin 30^\circ} = \frac{AB}{DB}$$

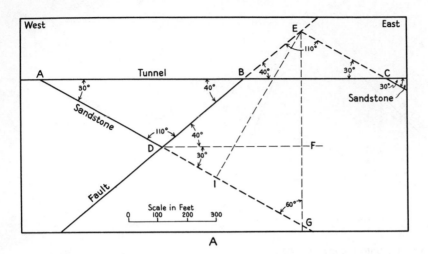

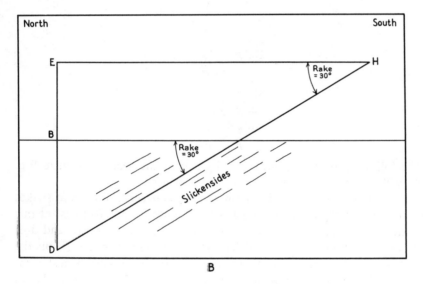

Fig. E7-2. Trigonometric solution of fault problems. (A) Vertical cross section of a thin bed of sandstone that is displaced along a fault. (B) Slickensides on fault rake 30°N.

and

$$DB = \frac{AB \sin 30°}{\sin 110°}$$

In triangle EBC

$$\frac{\sin 110°}{\sin 30°} = \frac{BC}{BE}$$

and

$$BE = \frac{BC \sin 30°}{\sin 110°}$$

$$\text{Dip slip} = DB + BE = \frac{700 \sin 30°}{\sin 110°} + \frac{500 \sin 30°}{\sin 110°}$$

$$\text{Dip slip} = 638 \text{ feet}$$

From Fig. E7-2B

$$\textit{Strike slip} = EH$$

$$\tan 30° = \frac{DB + BE}{EH} = \frac{638}{EH}$$

$$EH = 1105 \text{ feet}$$

$$\textit{Net slip} = DH$$

$$\sin 30° = \frac{DB + BE}{DH} = \frac{638}{DH}$$

$$DH = 1276 \text{ feet}$$

From Fig. E7-2A, triangle *DEF*

$$\textit{Throw} = EF$$

$$\sin 40° = \frac{EF}{DB + BE}$$

$$EF = (DB + BE) \sin 40°$$

$$EF = 638 \sin 40° = 410 \text{ feet}$$

$$\textit{Heave} = DF$$

$$\cos 40° = \frac{DF}{DB + BE}$$

$$DF = (DB + BE) \cos 40°$$

$$DF = 638 \cos 40° = 488 \text{ feet}$$

From Fig. E7-2A, triangle *DEI*

$$\textit{Stratigraphic separation} = EI$$

$$\sin (30° + 40°) = \frac{EI}{DB + BE}$$

$$EI = 638 \sin 70° = 599 \text{ feet}$$

From Fig. E7-2A, triangle *DEG*

$$\textit{Vertical separation} = EF + FG$$

$$\frac{\sin (30° + 40°)}{\sin 60°} = \frac{EF + FG}{DB + BE}$$

$$\frac{\sin 70°}{\sin 60°} = \frac{EF + FG}{638}$$

$$EF + FG = 692 \text{ feet}$$

From Fig. E7-2A

$$Horizontal\ separation = AB + BC$$
$$AB + BC = 1200\ feet$$

Problems

1. A horizontal tunnel that trends N.20°E. intersects a thin commercial talc bed that strikes N.70°W. and dips 40°NE. Seven hundred fifty feet further N.20°E. the tunnel intersects a vertical fault striking N.70°W. and with vertical slickensides. Twelve hundred feet still further N.20°E. the tunnel intersects the same talc bed as before, striking N.70°W. and dipping 40°NE.
a.   What are the following horizontal separations: (1) parallel to the tunnel; (2) east-west; and (3) north-south?
b.   What is the vertical separation?
c.   What is the net slip, dip slip, and strike slip?

2. A hill slopes 20° in a direction S.45°W. At the foot of the hill a limestone strikes N.45°W. dips 60°SW. Twelve hundred feet directly up the slope, a fault strikes N.45°W., dips 40°NE. Fifteen hundred feet still further uphill the same limestone crops out, striking N.45°W. and dipping 60°SW. Note that the distances are slope distances.
a.   What are the following horizontal separations: (1) north; (2) N.45°E.; and (3) N.90°E.?
b.   What is the vertical separation?
c.   What are the heave, throw, and stratigraphic throw?
d.   Assuming dip-slip displacement, what is the net slip, dip slip, and strike slip?

3. A traverse was made in a S.60°E. direction across a level surface. The following data were obtained; the left-hand column gives the distance in feet from the starting point at the northwest end of the traverse. Wherever data are available, the attitude of the bedding averages N.30°E., 35°SE.

Distance from Northwest End of Traverse	Lithologic and Structural Data
0	Contact of sandstone overlain by shale.
300	Shale.
400	Contact of shale overlain by limestone.
600	Limestone
900	Fault, N.90°E., 90°. Limestone north side, sandstone south side. Slickensides plunge 25°W.
1300	Sandstone
1500	Contact of sandstone overlain by shale.
1900	Contact of shale overlain by limestone.

The contact at 0 is the same as that at 1500; similarly, the contact at 400 is the same as that at 1900.

Draw a map showing these relations.

Draw two vertical sections, one parallel to the traverse, the other parallel to the fault.

a. What are the horizontal separations: (1) parallel to the traverse; (2) parallel to the fault; and (3) north-south?

b. What is the vertical separation?

c. Using the cross section parallel to the fault, calculate the following: net slip, dip slip, strike slip, throw, heave.

d. Calculate the thickness of the sandstone, shale, and limestone.

8

PROJECTIONS

Method

An object having three dimensions may be shown on a single plane by means of projection. The projection of a point on a plane is a point. The projection of a line on a plane is generally a line. The plane upon which points or lines are projected is the *plane of projection*. The *direction of projection* is the direction in which a point is projected into the plane of projection. *Normal projection*, in which the direction of projection is perpendicular to the plane of projection, is used in this exercise.

Normal projections are illustrated in Fig. E8-1. Points and lines have been projected into the horizontal plane represented by the upper surface of the blocks. In Fig. E8-1A, point *B* is the projection of point *A*; line *EF* is the projection of line *CD*. In Fig. E8-1B, line *GI* is the projection of line *GH* into this plane.

The line of intersection of two planes is the trace of one plane upon the other. The trace of one plane of projection upon a second plane of projec-

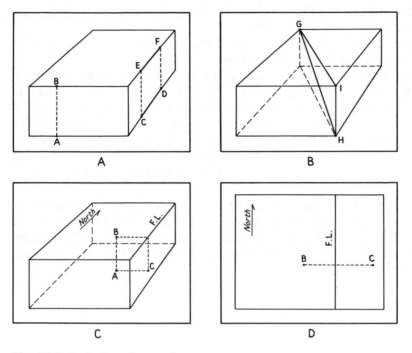

Fig. E8-1. Projections (see text).

tion is called a *folding line*. In normal projection the angle between two planes having a common folding line is always 90 degrees.

In most cases one plane of projection is horizontal, whereas the others are vertical. In geology the horizontal plane is a plan, corresponding more or less precisely with the map. A map portrays the relations on the surface of the earth, regardless of its topographic irregularities. A plan is made on a plane surface. In areas of low relief the geological map may be treated as a plan, and, whenever the term *map* is used in the ensuing descriptions, it is to be considered synonymous with the word *plan*. In geology the other planes of projection are structure sections. However, a plane of projection does not have to be either horizontal or vertical.

To represent a plan (map) and a section on one plane (the plane of the paper), it is necessary to rotate the section into the plane of the map around the folding line as an axis. In the following problems it is best to consider the section as lying below the folding line; the section will then be rotated upward into the horizontal plane. In Fig. E8-1C, point *B* is the projection of point *A* into a horizontal plane; point *C* is the projection of point *A* into the vertical plane represented by the side of the block. Figure E8-1D shows the projection after the vertical plane has been rotated into the plane of the paper about the folding line (*F. L.*) as an axis.

The attitude of a line is defined by the bearing of its horizontal projection and by its plunge (see p. 58).

The following problem, similar to those at the end of the exercise, illustrates the procedure to be followed.

Problem: As shown in Fig. E8-2, a vein that strikes N.40°E. and dips 40°NW. intersects a vein that strikes N.30°W. and dips 55°NE. Draw the projections of the intersection of the two veins (a) on a horizontal plane, (b) on a vertical plane striking north-south, and (c) on a vertical plane striking parallel to the direction of plunge of the intersection. (d) Show the trace of the N.40°E. vein on the N.30°W. vein. (It will be necessary to rotate the N.30°W. vein into the horizontal to show the projection.) (e) What is the attitude of the intersection?

Solution: Draw *AB* and *CD* parallel to the respective strikes of the veins, as shown in Fig. E8-2. The veins intersect at point *O*. Construct a folding line *FF'* perpendicular to *AB*; construct a folding line *GG'* perpendicular to *CD*. These folding lines may be constructed at any convenient point along *AB* and *CD*.

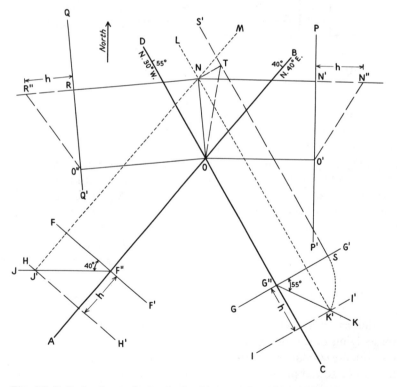

Fig. E8-2. Projection to find attitude of intersection of two veins: one striking N.40°E., dipping 40°NW.; the second striking N.30°W., dipping 55°NE.

Using these folding lines as the horizontal, now make cross sections. Draw *HH'* parallel to and at an arbitrary distance *h* from *FF'*; construct *II'* at the same distance from *GG'*. *HH'* and *II'* represent a level which will be called the *lower reference plane;* it is *h* distance below the level of the plan (map). Throughout any one problem the elevation of the lower reference plane remains constant. *F"* is the intersection of *FF'* with *AB*; *G"* is the intersection of *GG'* with *CD*. Draw angles *FF"J* and *G'G"K* equal to the respective dips of the two veins. *F"J* intersect *HH'* at *J'*; *G"K* intersects *II'* at *K'*. Draw *J'M* and *K'L* parallel to *AB* and *CD*, respectively. These two lines represent the horizontal projections of contours on the two veins at *h* distance below the plan; that is, these two lines are the horizontal projection of the intersection of the two veins with the lower reference plane. Point *N* is the horizontal projection of the intersection of the two veins on the lower reference plane, and point *O* is the intersection of the two veins on the plan. Therefore line *ON* is the horizontal projection of the intersection of the two veins. The bearing of the intersection in this case is N.6°W.

Draw line *PP'* in a north-south direction through some arbitrary point. This line is the trace of a north-south vertical plane on the plan. The vertical plane must be rotated into the upper reference plane (the plan) around *PP'* as the axis of rotation to show the projection of the intersection *ON* on the vertical plane.

Project point *O* to *O'* in the north-south plane; similarly project point *N* to *N'* in the north-south plane. Draw a line through *NN'* and extend it to the right of *N'*. Lay off point *N"* at a distance *h* from point *N'*. Point *N"* is the intersection of the two veins in the lower reference plane projected into a north-south vertical plane; line *O'N"* is the intersection of the two veins projected into the north-south vertical plane.

Through some arbitrary point construct line *QQ'* parallel to line *ON*. Line *QQ'* is the trace on the plan (map) of a vertical plane that strikes parallel to the direction of the plunge of the intersection of the two veins. This vertical plane must be rotated into the horizontal around the folding line *QQ'* in order to show the projection of the intersection of the two veins upon it. Project point *O* into this plane to *O"*; similarly project point *N* to *R* in this same plane and extend line *NR* to the left of *R*. From point *R*, lay off *RR"* equal to *h*. Point *R"* is the intersection of the two veins with the lower reference plane projected into a vertical plane that strikes parallel to the intersection of the two veins. Line *O"R"* is the intersection of the two veins projected to this same plane. Angle *RO"R"*, 31°, is the plunge of the intersection of the two veins.

To draw the trace of the N.40°E. vein on the N.30°W. vein, it is necessary to rotate the N.30°W. vein into the plan (map). *CD*, the trace of the vein on the plan, is used as the axis of rotation. Using the line *G"K'* as the radius and the point *G"* as the center, draw an arc to intersect *GG'* at *S*. Through point *S* draw a line *SS'* parallel to the strike of the N.30°W. vein.

This line, *SS'*, is the trace of the N.30°W. vein on the lower reference plane, now rotated into a horizontal position. From point *N*, construct *NT* perpendicular to *SS'*. *N* is the horizontal projection of a point on the lower reference plane; this point on the lower reference plane falls at *T* when it is rotated to the surface. Line *OT* is the trace of the N.40°E. vein on the N.30°W. vein. Angle *DOT*, 39°, is the rake of the trace of the N.40°E. vein on the N.30°W. vein.

Problems

1.   A vertical vein that strikes N.45°W. intersects a vein that strikes N.30°E. and dips 50°SE. An ore shoot is located at the intersection of the two veins. Determine the bearing and plunge of the ore shoot.

2.   The two limbs of a fold strike N.30°W. and N.40°E. and dip 40°NE and 70°NW, respectively.

a.   What is the bearing and plunge of the hinge of the fold?

b.   Let *x* be the point on the ground where a hinge emerges. Imagine an east-west vertical cross section 1000 feet south of *x*. How far west of *x* and how high above *x* would this same hinge appear in the cross section?

3.   A diabase dike that strikes N.40°E. and dips 50°SE. intersects a sandstone bed that strikes N.60°W. and dips 50°SW. Draw the projections of the intersection (a) on a horizontal plane; (b) on a vertical plane striking north-south; (c) on the plane of the diabase dike; and (d) on a vertical plane striking parallel to bearing of the intersection. (e) In which of these projections can the plunge be measured directly? (f) What is the bearing and amount of the plunge of the intersection?

9

MEASUREMENTS BY DESCRIPTIVE GEOMETRY

Introduction

As will be shown in some later exercises, angular relations of planes and lines may be measured rather rapidly by the use of the stereographic and equal-area projections. But lengths, including distances, thicknesses, and areas, cannot be measured by such projections. It is necessary to used descriptive geometry.

Lengths of Lines

To measure the length of a line we project the line onto a vertical plane parallel to its direction of plunge. The method for constructing such a projection was described in Exercise 8. An example is given here in Fig. E9-1.

A thin, cylindrical body of copper ore crops out at point A at an altitude of 3050 feet. The same ore-body was located underground in a tunnel at

2.

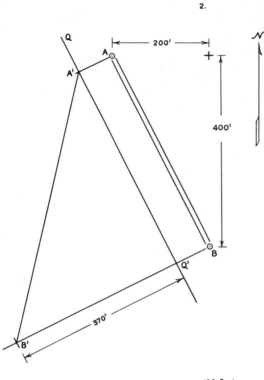

Fig. E9-1. Length of a line.

point B, 400 feet south and 200 feet east of its outcrop at point A, and at an altitude of 2680 feet. To find the length of the ore-body above the tunnel, draw a folding line QQ' parallel to AB, the map projection of the ore-body. From A, draw a line perpendicular to QQ', locating A'. From B, draw a similar line and extend it beyond Q'. Measure the distance $Q'B'$ equal to the difference in altitude, 370 feet, using the same scale as the map. The line $A'B'$ is the length of the ore-body, 580 feet, and the angle $B'A'Q'$, 40 degrees, is its plunge.

Areas of Planes

To measure the area of a plane figure, we must make a projection in the plane in which the figure lies. Such a projection can be made by one of two methods.

The first method was described in Exercise 8 (see Fig E8-2, p. 548). An arc is swung from the edge view of the N.30°W. vein (line $G''K'$) to determine a segment of $G''G'$ equal in length to $G''K'$. This segment is $G''S$. Through S the line SS' is passed parallel to $G''D$. This line is the projection in the plane of the N.30°W. vein of the line $K'N$; and the line OT is the projection in the plane of the N.30°W. vein of the line whose projection in the horizontal plane is ON. This method of obtaining projections in sloping planes by drawing arcs is very useful where only a few points have to be projected. However, if a large number of points have to be projected, the finished drawing, which is essentially two different views superimposed on each other, is hard to read.

A method of solving the problem in complicated cases is to make a completely new projection from the edge view of the plane. In Fig. E9-2, a

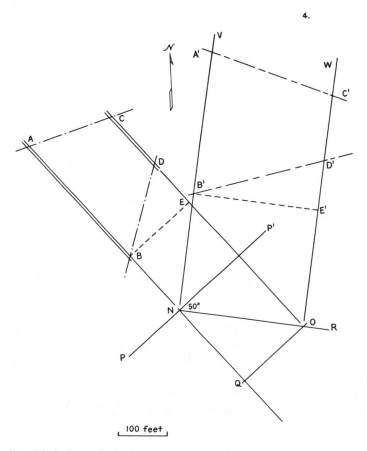

Fig. E9-2. Area of a bed.

coal bed striking N.40°W. and dipping 50°NE. is explored by two horizontal tunnels, lying in the bed, along the lines AB and CD. The tunnel along the line CD is 100 feet lower than the tunnel along the line AB. In the upper tunnel the coal bed is cut off by faults at A and B; in the lower tunnel it is cut off by the same faults at C and D. The intersections of the faults and the coal bed are shown in plan by the dot-dash lines. The coal bed is 5 feet thick. We wish to find the volume of coal in the block between the two tunnels and between the two faults, by drawing a projection of the coal bed in the plane of the bed.

First we draw an edge view of the plane of the coal bed. To do this, draw the folding line PP' through the point N perpendicular to AB extended. Draw NR making an angle of 50 degrees with PP' (the dip of the coal bed). Measure NQ on ABN extended, equal to 100 feet (the difference in altitude of the two beds), and through Q draw a line parallel to PP'. In the vertical plane perpendicular to the coal bed, represented by PP' and NO, this last line we have drawn is 100 feet below PP'. The point O, where this line intersects the line NR, is a point on the bed 100 feet below the tunnel AB. As is to be expected, this point lies on the line CD extended.

Now, to obtain a projection in a plane parallel to the plane of the coal bed, we must project the points A, B, C, and D onto a plane perpendicular to the vertical plane and passing through NO. To do this, draw lines NV and OW perpendicular to NO. At any convenient distance on NV, locate the point B'. Returning to the plan view, erect a line BE perpendicular to AB at B, and intersecting CD at E. Similarly at B' erect a line perpendicular to NV and intersecting OW at E'. Measure $B'A'$ equal to BA; $D'E'$ equal to DE; and $C'D'$ equal to CD. The lines $A'B'$ and $C'D'$ are the projections of the tunnels in the plane of the coal bed, and the lines $A'C'$ and $B'D'$ are the traces of the faults on the coal bed. If the area of $A'B'C'D'$ is now measured, at the scale of the map, and multiplied by the thickness of the coal bed, the volume of coal can be determined.

Stratigraphic Thicknesses

To measure stratigraphic thicknesses in regions of bedded rocks by the methods of descriptive geometry, we make a projection of the contacts between different rocks on a vertical plane perpendicular to the strike of the beds. The thickness of a bed is then measured along a line perpendicular to the trace of the bedding in this vertical section. The construction of such vertical projections has already been described in Exercise 2, p. 499. In Fig. E2-2, the line AE is the projection of the bed passing through X, striking N.90°E. and dipping 20° south, on the due north vertical plane SCB. An example of the use of such constructions to obtain stratigraphic thicknesses

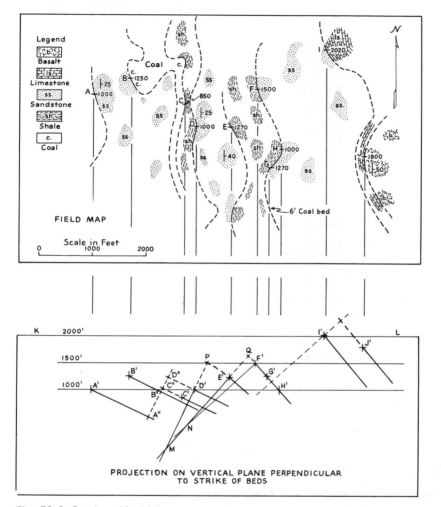

Fig. E9-3. Stratigraphic thickness.

is illustrated in Fig. E9-3, which is a plane-table map of a series of bedded rocks striking due north and dipping east. The points *A* through *J* were carefully located by surveying and their altitudes determined as indicated on the map. Dips and strikes on the beds were carefully measured; all the reliable ones are plotted on the map. The points *F*, *G*, and *H* are on the top of the same 6-foot coal bed. The scale of the map is shown graphically.

To measure graphically the thickness of the beds, draw a folding line *KL* perpendicular to the strike of the beds (Fig. E9-3). Consider this line to have the altitude 2000 feet. Draw lines parallel to *KL* at altitudes of 1500 feet and 1000 feet using the same scale as the map. Through the points *A*, *B*,

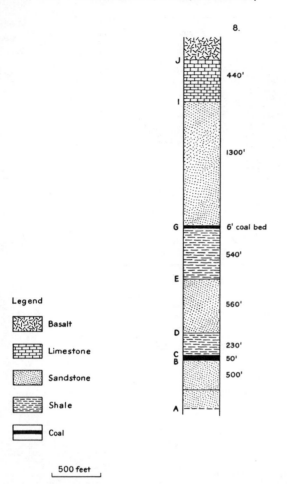

8.

J   440'

I   1300'

G   6' coal bed

540'

E

560'

D

C   230'
B   50'

500'

A

Legend

Basalt

Limestone

Sandstone

Shale

Coal

500 feet

Fig. E9-4. Stratigraphic column determined from Fig. E9-3.

C, etc., draw lines parallel to the strike and extend them to cut line KL. On these lines locate the points A', B', C', etc., at their proper altitudes in the vertical projection. Through the points A', D', E', and J' draw lines making angles with the line KL equal to the dip of the beds at those points. The points F', G', and H' fall on the same straight line; the angle between this line and the 1500 foot line parallel to KL is the dip of the coal bed at F, G, and H. Where the dip changes between two adjacent points, reconstruct the bedding by the geometric method described in Exercise 5.

The thicknesses of the various beds are measured along lines perpendicular to the trace of the bedding in the vertical projection plane. Thus the thickness of the lowest sandstone bed is the distance $A''B''$; of the lower coal bed, the distance $B''C''$; and of the lower shale bed $C''D''$. The thickness of the middle sandstone bed is $D'P$, where P is the point of intersection of the arc $E'P$ (swung about the center M) with $D'M$ extended. Similarly, the thickness of the upper shale bed is the distance $E'Q$.

When the stratigraphic thicknesses have been measured, a stratigraphic column can then be drawn up, using any convenient scale. It should show lithology by appropriate symbols and thicknesses by figures along the side. Such a column is illustrated in Fig. E9-4.

Problems

1. On a featureless plain a vertical dike trending N.45°W. cuts across an older vertical fault that trends N.45°E. A thin vein striking N. and dipping 40°W. is exposed in the quadrant east of the intersection of the fault and dike. The vein terminates against the fault 300 feet N.45°E. of the intersection of the fault and dike.

a. Determine the area of the vein beneath the surface of the plain in this easterly quadrant.

b. An inclined shaft is driven from the intersection of the dike and fault to the vein; it is as short as possible. What are the bearing and plunge of the shaft? What is its length?

2. On the 300-foot level of a mine veins A (striking N.90°E., dipping 25°S.) and B (striking N.45°W., dipping 60°SW) are truncated on the east by a vertical fault striking N.20°E. The two veins intersect 250 feet west of the point where vein A is truncated by the fault.

a. Determine the bearing and plunge of the intersection of the two veins.

b. Determine the length of this intersection below the 300 foot level.

c. Determine the area of both veins below the 300 foot level, between their mutual intersection and the fault.

3. On the 1000 foot level of a gold mine, a vein striking N.90°E. and dipping 40°S. is cut on the west by a younger dike striking N.20°W. and dipping 50°NE. Fifteen hundred feet east of this intersection the vein is truncated by a fault striking N.45°E. and dipping 50°SE.

a. What is the area of the vein below the 1000 foot level?

b. If the vein averages 4 feet thick, what is the volume below the 1000 foot level?

c. If the ore is $12\frac{1}{2}$ cubic feet per ton and runs 10 ounces of gold per ton, what is the total value of the gold?

4. Figure E9-5 is a plane table map of a badland area in a coal field. Points *1* through *7* were located by triangulation from the two plane table stations on the west side of the creek. Point *1* is at the base of the coal measures, where they rest on granite. Points *2*, *3*, and *4* are on the base of a coal bed 20 feet thick. The section between the coal bed and the base of the formation consists of sandstone. Point *5* is the top of a coal bed 40 feet thick. The section between the coal bed at *2* and the coal bed at *5* is shale. Point *6* is at the base of a coal bed 10 feet thick. The section between *5* and *6* is sandstone. From *6* to *7* is interbedded sandstone and shale. At point *7* is the base of a very thick conglomerate formation. Assuming the dip of all the beds to be the same, prepare a stratigraphic column of the coal measures, on a scale of 1 inch equals 100 feet, showing lithology by appropriate symbols and thickness by figures at the right of the respective lithologic units.

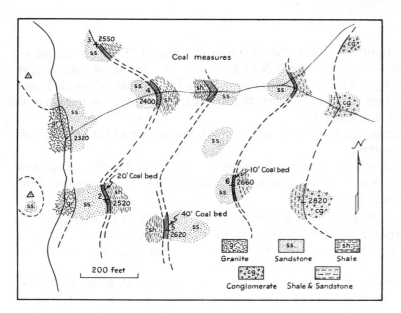

Fig. E9-5. Map for Problem 4 in Exercise 9.

SOLUTION OF THREE-POINT PROBLEMS AND VERTICAL FAULT PROBLEMS BY DESCRIPTIVE GEOMETRY

Three-Point Problem

The method used in this exercise for the solution of three-point problems is based on the graphic solution of similar triangles. The following problem illustrates the procedure to be followed:

Given: Points *A*, *B*, and *C*, all on top of a sandstone bed. Point *B* lies N.45°W. of *A* at a *map distance* of 300 feet. Point *C* lies N.60°E. of *A* at a *map distance* of 400 feet. The elevations of points *A*, *B*, and *C* are 950, 1100, and 1350 feet, respectively.

To Find: The strike and dip of the top of the bed of sandstone. In all such cases the top of the bed must be a plane surface.

Construction (Fig. E10-1): Locate the three points according to the data given, using some convenient scale. Draw line *AC*, connecting the *highest* and the *lowest* of the three points. Some point along this line has the same altitude as the intermediate point *B*.

Along line *CD*, drawn in some convenient direction from point *C*, lay

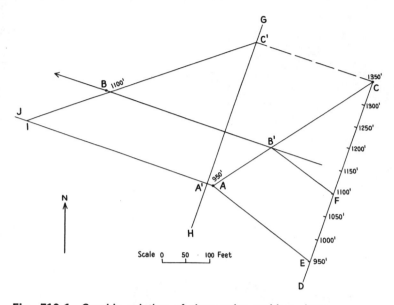

Fig. E10-1. Graphic solution of three-point problem. Location and altitude of *A, B,* and *C* are known.

off *CE* equal to the difference in elevation between points *A* and *C* using any convenient scale. On the same line, using the same scale, lay off *CF* equal to the difference in elevation between points *B* and *C.*

Connect points *A* and *E* by a line. Through point *F* draw a line parallel to line *AE* to intersect line *AC*. This intersection, point *B'*, is the point having the same altitude as point *B*. The line connecting points *B* and *B'* is the strike of the top of the bed.

At any convenient place draw line *GH* perpendicular to the strike of the bed. Project points *A* and *C* into this line to points *A'* and *C'*. Line *CC'* is the strike line on the top of the bed at an altitude of 1350 feet, and line *AA'* is the strike line on the top of the bed at an altitude of 950 feet. The dip is therefore from point *C'* toward point *A'*—that is, toward the southwest.

Now make a vertical section along line *GH*. Erect the perpendicular *A'J* to *GH* at point *A'*, and on this line, using the same scale as the horizontal map scale, lay off *A'I* equal to the difference in altitude between points *A* and *C*. Connect points *I* and *C'*. Angle *A'C'I* is equal to the dip of the top of the sandstone.

Vertical Fault

The following example illustrates the graphic method of determining the displacement on a fault if the attitude and location of two displaced horizons on both sides of the fault are known.

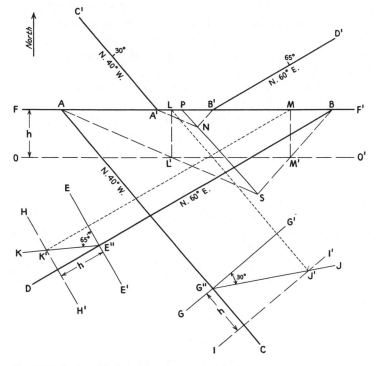

Fig. E10-2. Graphic solution of a vertical fault problem.

Given (Fig. E10-2): On a surface of no relief, fault *FF′* strikes N.90°E. and dips vertically; a vein striking N.40°W. and dipping 30°NE. is exposed at *A* and *A′* on the south and north sides of the fault, respectively; another vein, striking N.60°E. and dipping 65°NW., is exposed at *B* and *B′* on the south and north sides of the fault, respectively.

Find: (a) the net slip, (b) the plunge of the net slip, (c) the strike of the horizontal projection of the net slip, and (d) the relative movement along the fault plane.

Construction: Draw *AC* and *A′C′* parallel to the strike of the N.40°W. vein, and draw *BD* and *B′D′* parallel to the strike of the N.60°E. vein (Fig. E10-2). Draw *EE′* and *GG′* perpendicular to the strikes of the veins at convenient places, such as points *E″* and *G″*. Using these lines as the horizontal, we shall now make structure sections.

Draw *HH′* parallel to and at some arbitrary distance *h* from *EE′*; draw *II′* parallel to and at a distance *j* from *GG′*. These lines represent a level which we will call the *lower reference plane;* it is *h* distance below the surface plan. Draw angles *G′G″J* and *EE″K* equal to the respective dips of the N.40°W. and N.60°E. veins. Draw *J′L* and *K′M* parallel to *AC* and *BD*, respectively. These lines represent the horizontal projection of contours on the two veins; at all points on these contours the two veins are exactly distance *h* below the horizontal surface represented by the plan. Point L is the horizontal projection of the intersection of the fault, the lower reference plane, and the N.40°W. vein on the south side of the fault. Point M is the horizontal projection of the intersection of the fault, the lower reference plane, and the N.60°E. vein on the south side of the fault.

We will now make a vertical section along line FF'—that is, in the plane of the fault. Draw OO', h distance away from and parallel to FF'. This construction, in reality, amounts to rotating the vertical fault into the plane of the map (plan). Draw LL' and MM' perpendicular to FF'. Inasmuch as L' and M' are points on the section, h distance below L and M on the surface plan, they represent points on the veins in the cross section.

Draw AL' and BM'; extend them to an intersection at S. AS and BS represent the traces of the two veins on the fault plane (in the cross section), and S represents their intersection on the south wall. Draw $A'N$ and $B'N$ parallel to AS and BS, respectively. $A'N$ and $B'N$ represent the traces of the two veins on the fault plane on the north wall, and N is their intersection.

Draw NS. This line is the total displacement in the fault plane between the two points N and S, which were together before the faulting; NS is therefore the net slip.

Extend NS to intersect FF' at P. The angle SPF' is the rake of the net slip; in a vertical fault the rake and plunge are equal.

The net slip must lie in the fault plane. Inasmuch as the fault plane is vertical and strikes N.90°E., the bearing of the horizontal projection of the net slip is also N.90°E. Point S is at a lower elevation in the vertical section along the fault than point N and is to the east of a vertical line through point N; therefore the south side of the fault has moved down toward the east relative to the north side of the fault.

Problems

1.   A key bed in an oil field is exposed in well A at an altitude of -1500 feet. In well B, which lies 1000 feet north of A, the bed is reached at -800 feet. In well C, which lies 500 feet west of A, the bed is reached at -1100 feet. The wells are vertical. Determine the strike and dip of the key bed: (a) by descriptive geometry; (b) the method described on page 559.

2.   A vertical fault strikes N.90°E. across a featureless plain. Veins A and B are disrupted by the fault. The basic data are as follows. The distance where each segment abuts against the fault is measured easterly.

Vein	Strike	Dip	Distance on South Wall	Distance on North Wall
A	N.25°W.	35°NE.	0 feet	200 feet
B	N.50°E.	55°NW.	550 feet	450 feet

Determine: (a) the bearing, plunge, and value of the net slip; (b) the relative movement, that is, which block has gone up.

3. A horizontal east-west water supply tunnel is to be constructed 100 feet north of an east-west vertical fault and at a depth of 300 feet. At the surface, two dikes, each 50 feet thick, are exposed along the fault. Dike A, on the south wall of the fault, strikes N.20°W., dips 40°NE. Three hundred feet east of where the west contact of dike A abuts against the fault, the west contact of dike B is exposed on the north wall; it strikes N.40°E., dips 55°NW. Evidence elsewhere shows that the rake of the net slip along the fault is 60°W., that the value of the net slip is 50 feet, and that the north wall moved up relative to the south wall.

The country rock is granite and the rock will stand in the tunnel without artificial support. But the dikes are badly jointed and structural steel support will be necessary. Where will these dikes be crossed by the tunnel? The tunnel is 10 feet in diameter. For what distance in crossing each of the two dikes will the tunnel have dike rock somewhere on its walls, roof, or floor?

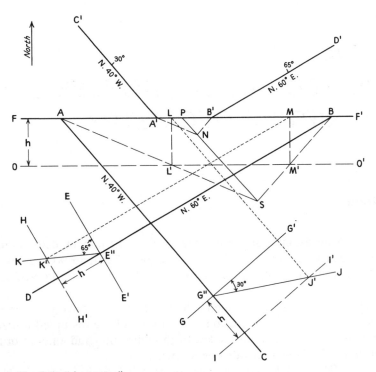

Fig. E10-2 (repeated)

SOLUTION
OF INCLINED FAULT PROBLEMS
BY DESCRIPTIVE GEOMETRY

Method

The following example illustrates the graphic method of determining the displacement on a fault that dips at an angle other than 90 degrees, if the attitude and location of each of two displaced horizons are known on both sides of the fault.

The example also shows the method of locating a disrupted horizon on one side of a fault, if the location and attitude of the disrupted horizon on the opposite side of the fault are known; the direction and amount of movement along the fault must also be known.

Given (Fig. E11-1): On a surface of no relief, fault *FF'* strikes N.90°E. and dips 40°S.; a vein striking N.30°W. and dipping 35°NE. is exposed at *A* and *A'* on the south and north sides of the fault, respectively; another vein, striking N.30°E. and dipping 60°NW., is exposed at *B* and *B'* on the south and north sides of the fault, respectively; a third vein, striking N.40°E.

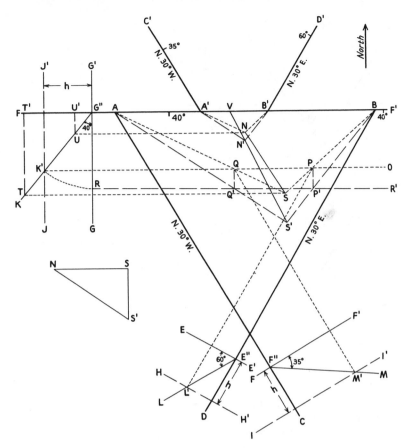

Fig. E11-1. Graphic solution of an inclined fault problem. Two veins, one striking N.30°W. and dipping 35°NE., the other striking N.30°E. and dipping 60°NW., are displaced along the fault *FF'*.

and dipping 70°SE., is exposed at *X* (Fig. E11-2) on the south side of the fault.

To Find: (a) the net slip of the fault, (b) the bearing of the horizontal projection of the net slip, (c) the plunge of the net slip, (d) the relative movement along the fault—that is, which block moved up—and (e) the location of the third vein on the north side of the fault.

Construction: As shown in Fig. E11-1, draw *AC* and *A'C'* parallel to the strike of the N.30°W. vein, and draw *BD* and *B'D'* parallel to the strike of the N.30°E. vein. Draw *EE'*, *FF'*, and *GG'* perpendicular to *BD*, *AC*, and *FF'*, respectively, at any convenient points such as *E''*, *F''*, and *G''*. Using these lines as horizontals, we shall now make cross sections.

Draw *HH'* parallel to and at an arbitrary distance *h* from *EE'*; construct

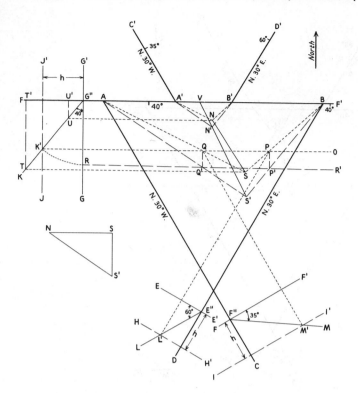

Fig. E11-1. Graphic solution of an inclined fault problem. Two veins, one striking N.30°W. and dipping 35°NE., the other striking N.30°E. and dipping 60°NW., are displaced along the fault *FF'*.

II' and *JJ'* in a similar way. These lines represent a level that we shall call the *lower reference plane;* it is *h* distance below the surface plan. Lay off angle *GG'K* equal to the dip of the fault; *G"K* intersects *JJ'* at *K'*. Lay off angle *EE"L* equal to the dip of the N.30°E. vein; *E"L* intersects *HH'* at *L'*. Lay off angle *F'F"M* equal to the dip of the N.30°W. vein; *F"M* intersects *II'* at *M'*. From *K'* draw *K'O* parallel to *FF'*. From *L'*, draw a line parallel to *BD* to intersect *K'O* at *P*. From *M'*, draw a line parallel to *AC* to intersect *K'O* at *Q*. *K'O*, *L'P*, and *M'Q* represent contours on the fault and the veins; at all points on these lines the respective fault or vein is exactly *h* distance below the surface of the plan. The intersections of the two veins with the fault on this lower reference plane are at *P* and *Q*.

We shall now make a section in the plane of the fault. It is necessary to rotate the fault plane into the plane of the map in order to show this section. Line *FF'* is used as the axis of rotation. Using *G"K'* as the radius and *G"* as the center, draw an arc to intersect *GG'* at *R*. *G"R* is the true slope distance in the fault plane between the surface plan and the lower reference plane. From R, draw RR' parallel to FF'. $BG''RR'$ is a section in the plane of the fault.

From Q, construct a line perpendicular to $K'O$ to intersect RR' at Q'. From P, construct a line perpendicular to $K'O$ to intersect RR' at P'. Q' and P' represent the intersection of the veins with the lower reference plane, viewed in the fault plane.

Draw AQ' and BP' to an intersection at S'. AS' and BS' represent the trace of the two veins on the south wall of the fault plane; S' represents their intersection on the south wall. Draw $A'N'$ and $B'N'$ parallel to AS' and BS', respectively. N' represents the intersection of the trace of the two veins on the north wall. Draw $N'S'$, which is the net slope.

Draw AQ and BP to an intersection at S. AS is the horizontal projection of the intersection of the N.30°W. vein and the fault. BS is the horizontal projection of the intersection of the N.30°E. vein and the fault. S is the horizontal projection of the intersection of the two veins on the south wall of the fault plane. Draw $A'N$ and $B'N$ parallel to AS and BS, respectively. Draw NS, which is the horizontal projection of the net slip.

To determine the plunge of the net slip, it is necessary to find the altitude of the points for which N and S are the horizontal projections. Draw a line from S, parallel to FF', to intersect $G''K$ at T; and draw a line from N, parallel to FF', to intersect $G''K$ at U. Lines perpendicular to FF' are then erected, one from T and the other from U. These perpendiculars intersect FF' at T' and U'. $T'U'$, on the same scale as the map, whatever it may be, is the difference in altitude between the points of which N and S are the horizontal projections.

Lay off NS on a separate part of the paper. Drop a perpendicular from S to S' such that $S'S = T'U'$. Draw NS'. SNS' gives the vertical angle that the net slip makes with its horizontal projection; it is the plunge of the net slip. As a check on your work, NS' in this diagram should equal $N'S'$ in the main construction.

The point of intersection of the two veins on the south side of the fault with the fault plane lies at a lower altitude and to the east of the corresponding intersection on the north side of the fault. The relative movement along the fault is such that the south side has moved down and to the east in relation to the north side. Extend $N'S'$ to intersect FF' at V. Angle $S'VB$ is the rake of the net slip.

Figure E11-2, which illustrates the method used to locate the N.40°E. vein on the north side of the fault, shows data concerning the fault identical with that in Fig. E11-1. FF' is the trace of the fault on the map. $K'O$ is the horizontal projection of the intersection of the fault and the lower reference plane, and RR' is this intersection after it has been rotated into the plane of the map.

The location of the N.40°E. vein on the south side of the fault is shown.

Draw *XY* parallel to the strike of the vein. At some convenient point along *XY*, such as *A''*, draw *AA'* perpendicular to *XY*. At *h* distance (the same distance as used in Fig. E11-1) from *AA'*, draw *BB'* perpendicular to *XY*. Lay off angle *A'A''C* equal to 70°, the dip of the N.40°E. vein. From *C'*, the intersection of *A''C* with *BB'*, draw a line parallel to the strike of the N.40°E. vein to intersect *K'O* at *D*. *C'D* represents a contour on the vein; at all points on this line the vein is exactly *h* distance below the surface of the plan. The intersection of the vein with the fault on this lower reference plane is at *D*. From *D* draw a line perpendicular to *K'O* to intersect *RR'* at *D'*. *D'* represents the intersection of the vein with the lower reference plane, viewed in the fault plane.

Draw *EE'*, which passes through points *X* and *D'*. *EE'* is the trace of the N.40°E. vein on the south wall of the fault plane. Figure E11-1 shows that the north block has moved up and to the west; therefore the trace of the vein on the north wall of the fault plane will be above and to the west of *EE'*. At any convenient point along *EE'*, as at *X*, lay off angle *G''XH* equal to the rake of the net slip, which is the angle *S'VB* of Fig. E11-1; and on *XH*, lay off *XH'* equal to the net slip, which is *N'S'* of Fig. E11-1. Through *H'*, draw *II'* parallel to *EE'*. *II'* is the trace of the vein on the north side of the fault. From *X'*, draw *X'Y'* parallel to *XY*. Point *X'* represents the intersection of the vein with the fault on the north side of the fault, and *X'Y'* is the trace of the N.40°E. vein on the plan, north of the fault.

Problems

1. A vertical fault strikes due east across a featureless plain. Dikes *A*, *B* and *C* are disrupted by the fault. The basic data are as follows, distance being measured toward the east from zero.

Dike	Strike	Dip	Distance on North Wall	Distance on South Wall
A	N.20°W.	20°E.	0 feet	150 feet
B	N.	50°W.	400 feet	250 feet
C	N.30°E	45°SE.	450 feet	?

(a)   Determine the bearing of the net slip, the rake of the net slip, the value of the net slip, and the relative movement—that is, which block moved relatively upward?

(b)   Locate vein *C* on the south side of the fault.

2. A fault plane exposed on a featureless plain strikes N.90°E. and dips 50°S. Veins A and B are disrupted by the fault. The basic data are as follows.

Vein	Strike	Dip	Distance on North Wall	Distance on South Wall
A	N.30°E.	35°SE.	150 feet	0 feet
B	N.30°W.	65°SW.	350 feet	300 feet

(a)   Determine the bearing, plunge, rake, and value of the net slip and the relative movement on the fault.

3.   A fault plane, exposed on a featureless plain, strikes due east and dips 55°N. The bedding strikes N.10°E. and dips 45°E. A dike strikes N.45°W. and dips 65° SW. A vein strikes N.35°W., and dips 70°SW. A thin distinctive fossiliferous limestone bed occurs in the sedimentary rocks. The basic data are as follows.

	Distance on North Wall	Distance on South Wall
Limestone	0 feet	150 feet
Dike	250 feet	200 feet
Vein	350 feet	?

(a)   Determine the bearing, plunge, rake, and value of the net slip, and the relative movement on the fault.

(b)   Locate the vein on the south side of the fault.

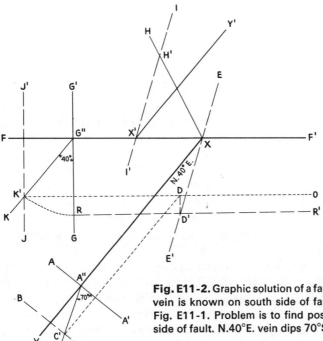

Fig. E11-2. Graphic solution of a fault problem. Position of N.40′E. vein is known on south side of fault, *FF′*. Net slip is same as in Fig. E11-1. Problem is to find position of N.40′E. vein on north side of fault. N.40°E. vein dips 70°SE.

EQUAL-AREA NET,
PART I

Introduction

This and the following two exercises are concerned with the use of the equal-area net in solving structural problems. Some of the problems that have been attacked in preceding exercises could have been solved more rapidly by the use of the equal-area net. But although it is possible to determine directions and angles by the use of this projection, it is not possible to determine distances and area. On the other hand, use of the equal-area projection and descriptive geometry may be effectively combined.

A small equal-area projection is shown in Fig. 5-3 and a larger one to use is Fig. 5-4 on page 589.

The general principles involved in the use of the equal-area net are presented in Chap. 5. Moreover, the method of projecting lines, planes, and the perpendiculars to planes are discussed there in some detail. The method of determining the attitude of the intersection of planes is also shown. The preparation and significance of *pi diagrams* and *beta diagrams* are discussed there. We are now prepared to show how additional problems may be solved. The pertinent section of Chap. 5 should be carefully reviewed.

In the ensuing figures, such as figure E12-1, a piece of tracing paper (the overlay) has been placed over an equal-area net. For clarity, the overlay is represented by a square, the edges of which trend east-west and north-south. North, east, south, and west are marked by N, E, S, and W.

Plotting a Line That Lies in a Plane

In these projections all planes and lines are treated as if they pass through the center of the sphere from which the projection is made.

Figure E12-1A illustrates how to plot a line with a rake of 50° to the north in a plane that strikes N.30°E. and dips 30°NW. A line AB bearing N.30°E. is drawn on the overlay. The overlay is then rotated 30° in a counterclockwise direction and the appropriate arc for a plane dipping 30°NW is drawn (ACB). The line we are seeking may be constructed in one of two ways. The plane may be figuratively rotated 30° to the horizontal position ADB. The line OF is then drawn, making an angle of 50° with AB; this is most readily done by plotting F 50° from A as measured on the circumference. The plane is then figuratively rotated 30° back into its original position —that is, D moves back to C. F migrates along a small circle to G. A much quicker way is to plot G directly at 50° from A along the arc ACB. The bearing (OH) of the horizontal projection of OG is N.46°W. The plunge of OG is measured by rotating OG 46° clockwise to coincide with the YY' axis of the net (Fig. 5-3). The plunge 24°N. is measured from A.

Determining the Angle Between Two Lines

The angle between two lines is always measured in the plane that contains the two lines. Visualize two lines, one plunging 23° N.76°W., the other 13° S.4°E. As in Figure E12-1B, line OM is drawn with a bearing N.76°W. The overlay is rotated clockwise so that OM coincides with the YY' axis of the net. The point a is established by measuring 23° from the circumference on the YY' axis. The line OP is drawn with a bearing S.4°E. The overlay is rotated clockwise so that OP overlays the YY' axis. The point b is established by measuring 13° from the circumference. The overlay is now rotated so that a and b fall on the same great circle (Fig. 12-1C). The angle between the two lines can then be measured along the same great circle; it is 100°. The angle between the two lines plus the rake of each line (50° and 30° as measured along the great circle) is 180°.

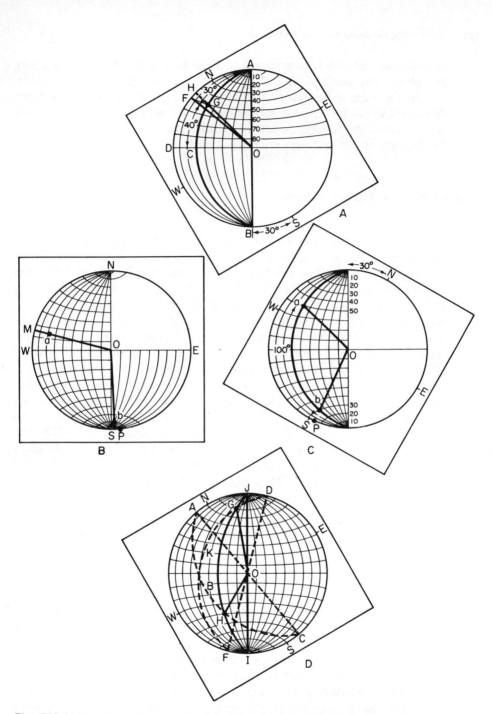

Fig. E12-1. Use of equal-area projection. (A) A line in a plane.
(B) Angle between to lines before rotation. (C) Angle between
two lines after rotation. (D) Apparent dip on inclined surfaces.
(E) Apparent dip on inclined surfaces after final rotation.

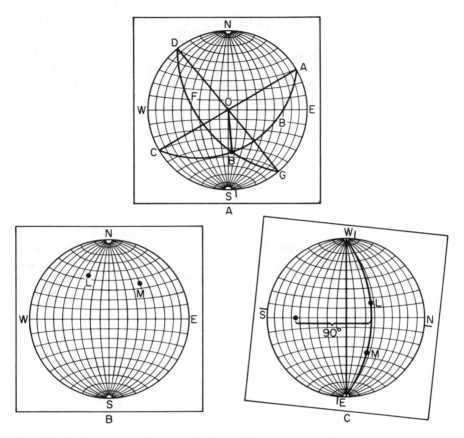

Fig. E12-2. Use of equal-area projection. (A) Intersection of two planes by plotting planes. (B) Intersection of two planes by plotting perpendiculars to two planes. (C) One step in determining attitude of intersection of two planes.

True Strike and Dip from Apparent Dips

The apparent dip is usually measured on a vertical face (Fig. E4-1). The true strike and dip of a plane can be determined from two apparent dips as long as they are measured on two nonparallel vertical planes. For brevity, the same data will be used as in the preceeding paragraph but will be stated somewhat differently. On a vertical plane striking N.76°W. the apparent dip is 23°W; on a vertical plane striking S.4°E. the apparent dip is 13°S. The data are plotted as before in Figures 12-1B and 12-1C. The strike of the plane, as shown in Figure 12-1C is N.30°W. and true dip, as measured along the *XX'* axis from the circumference of the net, is 30°SW.

Sometimes the rake of apparent dips are measured on inclined surfaces such as joints. A general case is shown in Figure 12-1D. Joint *ABC* strikes N.10°W. dips 37°W; the rake of the apparent dip on that plane is 80°S. Joint *DKF* strikes N.44°E., dips 30°NW; the rake of the apparent dip on that plane is 30°N. The joints are plotted first as *ABC* and *DKF*. Also the apparent dips, *OG* and *OH* are plotted as above. The overlay is rotated until *G* and *H* lie on the same great circle; *GHIJ* is the plane. It strikes N.30°E. (the bearing of *JI*) and dips 50°NW (measured along the *XX'* axis from the circumference).

Attitude of Intersection of Two Planes

One method of determining the attitude of the intersection of two planes has been described in Chapter 5. An alternative method requires that only the poles of the perpendiculars to the two planes be plotted. One plane (*ABC*) strikes N.60°E., dips 40°SE. The other plane (*DFG*) strikes N.45°W. and dips 50°SW. A plot of the planes is shown in Fig. E12-2A; this intersection plunges 38° S.6°E., as shown in Fig. E12-2A. The plot of the perpendiculars to the two planes is shown in Fig. E12-2B. *M* represents the perpendicular to *DFG*, whereas *L* represents the perpendicular to *ABC*. The overlay is rotated until the two points (*L* and *M*) fall on the same great circle (Fig. E12-2C). The perpendicular to that plane is the intersection of the planes *ABC* and *DEF*. It plunges 38° S.6°E.

Problems

1. One limb of a fold strikes N.45°E. and dips 50°SE. The other limb strikes N.80°W. and dips 30°S. Determine: (a) the bearing and value of the plunge of the fold and (b) the apparent dips of each limb in a vertical structure section that strikes N.63°W.

2. The attitudes of four planes are as follows: *A*, N., 50°W.; *B*, N.38°E., 10°NW.; *C*, N.78°E., 60°S.; and *D*, N.56°W., 41°SW.

(a) How many intersections are there? (b) How do the attitudes of the intersections compare? (c) Is it possible that a fifth plane is perpendicular to all four planes? What is its attitude? (d) What is the angle between the perpendiculars to planes *C* and *D*? (e) What is the angle between the perpendiculars to planes *B* and *C*?

3. The apparent dips on a sandstone bed were measured in two vertical joints. One joint strikes N.50°E; the rake of the apparent dip is 40°NE. The other joint strikes due north; the rake of the apparent dip is 30°S. Determine:

(a) the strike and dip of the sandstone; (b) the apparent dip of the sandstone as seen in a vertical joint striking N.38°W.; and (c) the bearing and plunge of the apparent dip as seen in a joint striking N.90°E. and dipping 20°N.

4. An amphibolite bed that crops out on a horizontal surface strikes N.20°W. On a nearby outcrop surface that trends N.52°E. and is inclined 35°SE., the rake of the trace of the bed is 35°SW. Determine; (a) the bearing and plunge of the apparent dip on this inclined surface; (b) the strike and dip of the bed; and (c) the apparent dip of the bed on a plane that strikes N.90°E. and dips 30°N.

5. On an outcrop surface that trends N.70°E. and is inclined 35°SE., the rake of the trace of the axial plane of a fold is 34°SE. The hinge of the fold plunges 34° N.43°E. Determine: (a) the attitude of the axial plane; (b) the bearing of the trace of the axial plane on a horizontal surface; and (c) the plunge of the trace of the axial plane on a vertical cliff trending east-west.

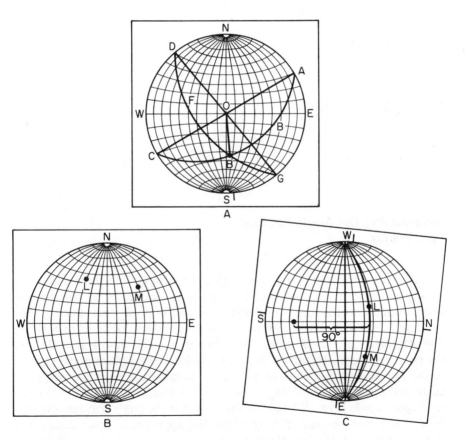

Fig. E12-2.

exercise

13

EQUAL-AREA NET, PART II

Determining the Angle between Two Planes

The true angle between two planes is always measured in a third plane that is perpendicular to the first two. That angle is unique, being the smallest possible one between the two planes. The obtuse angle also describes the true angle. Figure. E13-1A illustrates the problem. Plane *ABC* strikes N.50°E., dips 45°NW. Plane *DEF* strikes N.80°W., dips 10°S. Points *H* and *I* are the poles to planes *ABC* and *DBF*, respectively. A third plane perpendicular to the two planes must contain their poles, because the poles are perpendicular to the planes. The overlay is rotated clockwise until *H* and *I* fall on the same great circle (Fig. E13-1B). In that position the angle between the perpendiculars may be read directly along the appropriate great circle. The true angle between the two planes is 59°.

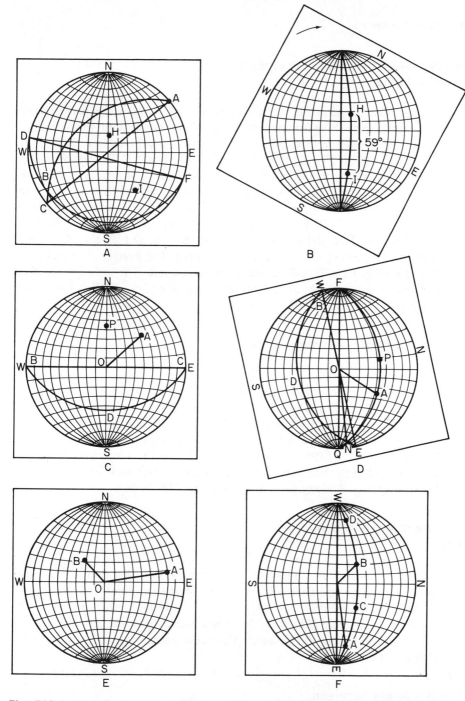

Fig. E13-1. Use of equal-area projections. (A) and (B) Determining angle between two planes. (C) and (D) Orthographic projection of a line on a plane. (E) and (F) Bisecting the angle between two lines.

Determining the Orthographic Projection
of a Line on a Plane

A plane that is perpendicular to another must contain the perpendicular to the other plane. An infinite number of planes could fulfill this condition. But one plane is specified if it must also contain a given line. Such a line (*OA*) is shown in Fig. E13-1C. It plunges 40° N.47°.E The problem is to project *OA* orthographically onto the plane *BCD* that strikes N.90°E and dips 45°S. *OA* must be projected in such a way that the distance from any point on *OA* to the projection plane is minimal. This is equivalent to stating that *OA* (Fig. E13-1D) must lie in a plane (*FPH*) perpendicular to the plane of projection. Thus the orthographic projection of *OA* on the plane *BCD* is the intersection of the planes *BCD* and *FPH*. The overlay is rotated until A and *P* (Fig. E13-1D) fall on the same great circle. The intersection of the the plane represented by that great circle and the plane *BCD* is the desired projection of *OA*; it is *OQ*.

Determining the Angle between a Line
and a Plane

From the preceding it is clear that the smallest angle between a line and a plane is measured between the line and its orthographic projection on the plane. Figures E13-1C and E13-1D may be used again to illustrate the problem. The true angle between OA and the plane *BDE* is found by first finding the plane (*FPH*) that is perpendicular to the projection plane and contains *OA* (Fig. E13-1D). The angle between *OA* and *OQ* is measured between *Q* and *A* along the great circle *HQAPF* as 56°.

Bisecting the Angle between
Two Lines

The angle between two lines may be determined by first finding the plane that contains the two lines. Figure E13-1E shows two lines. *OA* plunges 20° N.80°E. *OB* plunges 60° N.45°W. Rotate the overlay until *A* and *B* fall on the same great circle (Fig. E13-1F). Then bisect the angle between *AO* and *OB* at points *C* and *D*. *C* bisects the acute angle, whereas *D* bisects the obtuse angle.

Bisecting the Angle between
Two Planes

The plane that bisects the angle between two other planes will include the bisection of the true angle between the planes. This is shown in Fig. E13-2A. Plane *ABC* strikes N.15°W and dips 65°W. Plane *DBE* strikes N.90°E. and

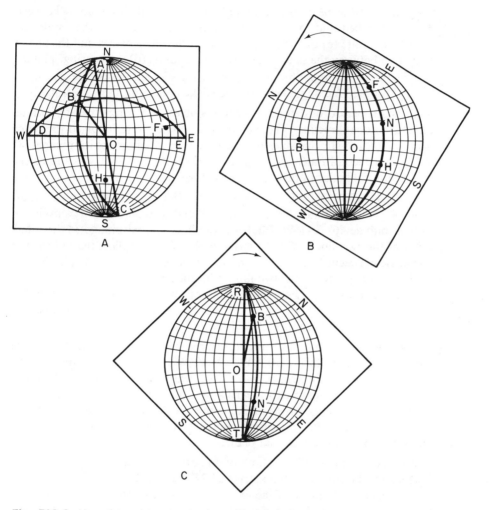

Fig. E13-2. Use of equal-area projections. Bisecting the angle between two planes.

dips 40°N. The projections of the poles of the perpendiculars to these planes are at F and H, respectively. The overlay is rotated so that F and H fall on the same great circle (Fig. E13-2B). ON is the bisector of the angle between OF and OH. The bisecting plane must also contain OB, the line of intersection of planes ABC and DBE. The overlay must be rotated again, until B and N fall on the same great circle (Fig. E13-2C). That great circle is the projected trace of the bisecting plane ($TORBN$).

Problems

1. A vertical fault that strikes N.35°E truncates a sandstone bed that strikes N.15°W and dips 40°E. Slickensides in the fault plane rake 55°S.

Determine: (a) the bearing and plunge of the line of intersection between the sandstone bed and the fault; (b) the true angle between the sandstone bed and the fault; (c) the true angle between the slickensides and the line of intersection between the sandstone bed and the fault; (d) the bearing, plunge, and rake of the orthographic projection of the slickensides on the sandstone bed; and (e) the true angle between the slickensides and the sandstone bed.

2.   During a period of deformation the pebbles in a conglomerate became elongated in one direction. These elongated pebbles were involved in a later period of folding. In one outcrop the long axes of the pebbles plunge 38° S.70°E. In a second outcrop they plunge 22° S.10°W. What was the orientation of the fold axes in this period of deformation?

3.   Fault A strikes N.90°E and dips 75°N. The slickensides in this fault plunge 50° N.72°E. Fault B strikes N.20°W and dips 50°SW. Slickensides in this fault plunge 46° S.41°W. Find: (a) the bearing and plunge of the intersection of the two faults; (b) the strike and dip of the plane that bisects the acute angle between the faults; (c) the strike and dip of the plane that bisects the obtuse angle between the two faults; (d) the bearing and plunge of the bisector of the obtuse angle between the line of intersection of faults A and B and the slickensides in fault A; (e) the true acute angle between the bisector mentioned in A and the plane of fault B.

4.   The following attitudes were measured on a folded limestone bed: N.16°W., 61°E.; N.48°E., 70°NW.; N.49°W., 51°NE.; N.78°E., 55°N.; and N.75°W., 49°N. Determine: (a) the bearing and plunge of the fold axis by preparing a pi diagram; (b) bearing and plunge of the fold axis by preparing a beta diagram; (c) the strike and dip of the axial plane of the fold, given that the axial trace on a horizontal surface trends due N., and (d) the rake of the fold axis in the axial plane.

5.   The beds on the northwest limb of an anticline strike N.60°E., dip 45°N. The beds on the southeast limb strike N.25°W. and dip 55°NE. Determine: (a) the bearing and plunge of the axis of the anticline; (b) the angle between the two limbs of the fold in a vertical section trending east-west; (c) the rake of the axial plane of the fold in a vertical section trending east-west; (d) the angle between the two limbs of the fold as viewed in a "down-plunge" projection; and (e) the rake of the axial plane in a "down plunge" projection.

USE OF EQUAL-AREA NET INVOLVING ROTATION

Introduction

Many of the problems in the preceding exercises include steps that require rotation of lines and planes. Geologic processes such as some types of folding and faulting commonly give rise to bedding orientations that cannot be restored to their original state by translational movement alone. Stereographic or equal-area projections provide a means by which rotation of structural data may be easily and accurately accomplished.

Projection of a Cone

A cone is the locus of lines at a constant angle from another line. The latter line is the axis of rotation about which the surface of the cone may be generated. Figure E14-1A illustrates the projection of a cone whose axis (*ON*) trends due north and is horizontal. The surface of the cone is at an angle of

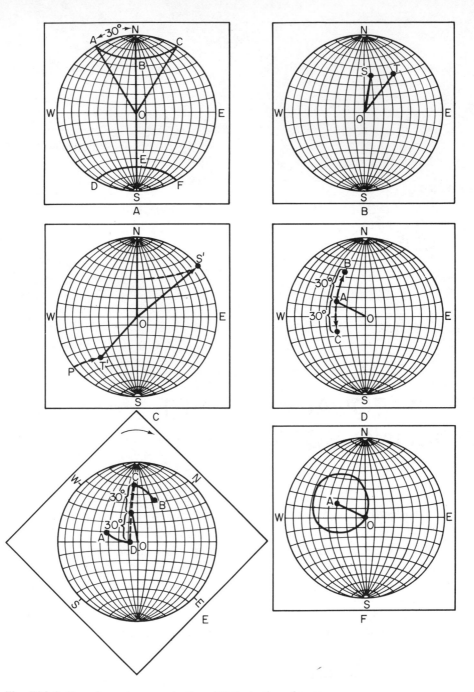

Fig. E14-1. Use of equal area projection. (A) Projection of cone, axis plunging 0° N., apical half-angle = 30°. (B) Projection of two lines on a cone. (C) Rotation of lines in figure (B). (D) *AO* is projection of axis of cone plunging 60° N.60°W. Apical half-angle is 30°. *B* and *C* are projections of two lines in the cone. (E) Step in rotation of cone. (F) Projection of entire cone.

30° to the axis. This angle may be measured from point *N* in either direction around the circumference or down along any one of the great circles. Each point on the small circle inclined 30° to *ON* represents the projected pole of each line in the surface of the cone. Thus the trace of the cone on the lower hemisphere is the small circle *ABC*. The upper part of the cone, that is, the part lying above the projection plane, will project through the origin (*O*) as the small circle *DEF*.

Any line in the surface of a cone may be rotated on the small circle representing the projected trace of the cone, and the angle between the line and the axis of the cone will be preserved. Figure E14-1B shows two lines on the same small circle. Line *OS* plunges 49° N.12°E. Line *OT* plunges 30° N.42°E. The problem is to rotate both lines to the east until *OS* is horizontal. The line *ON* is the axis of rotation in this case since the poles of the lines to be rotated are on the same small circle. Figure E14-1C shows the two lines in their rotated positions. *OS* is horizontal after 80° rotation. *OT* however, reached a horizontal position after only 40° rotation on the small circle. The additional rotation of 40° takes *OT* into the upper hemisphere. However, there is also a pole in the lower hemisphere, but in the southern part. Thus it is rotated 40° from *P* to *T'*.

A cone whose axis is inclined to the horizontal may also be projected. Suppose the axis of a cone plunges 60° N.60°W. The surface of the cone makes an angle of 30° with the axis of the cone. In Fig. E14-1D *A* is the projection of the pole of the axis and *AO* is the projection of the axis. *A* is on a great circle. A distance of 30° is measured in both directions along the great circle to establish points *B* and *C*. The overlay is then rotated some arbitrary amount such as 20°. *A* now falls on another great circle. As before, 30° is measured in both directions to establish points *D* and *E*. The arcs *BD* and *CE* represent the partial locus of lines 30° to the axis of the cone. Rotation of the overlay to successive great circles will establish a series of points and thus the projected trace of the whole cone, as shown in Fig. E14-1F.

Small Circle Rotation of Planes

Planes as well as lines may be rotated along small circles provided the axis of rotation is the *YY'* axis of the net. An example is illustrated in Fig. E14-2A Plane *ABC* strikes N.20°E., dips 40°SE. The problem is to rotate the plane to the horizontal. As the strike-line of the plane is the only line about which the plane may be rotated to the horizontal, the overlay is rotated 20° counterclockwise to the position shown in Fig. E14-2B. In that position the trace of the plane on the lower hemisphere may be rotated through the dip angle to the horizontal. Each line in the plane, such as *OA* or *OB*, will rotate up along small circles through the same angle.

The problem is more complicated in cases where the axis of rotation is

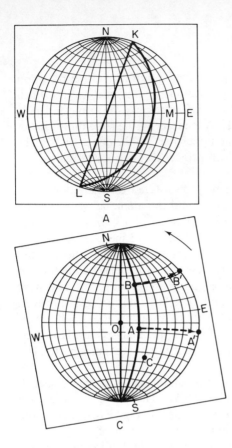

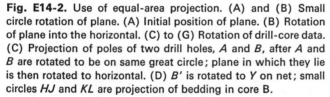

Fig. E14-2. Use of equal-area projection. (A) and (B) Small circle rotation of plane. (A) Initial position of plane. (B) Rotation of plane into the horizontal. (C) to (G) Rotation of drill-core data. (C) Projection of poles of two drill holes, *A* and *B*, after *A* and *B* are rotated to be on same great circle; plane in which they lie is then rotated to horizontal. (D) *B'* is rotated to *Y* on net; small circles *HJ* and *KL* are projection of bedding in core B.

plunging. The two rotations will in general be needed to bring the axis of rotation into registry with the *YY'* axis of the net. In other cases, it may prove easier to rotate only the projection of the pole of the perpendicular to the plane rather than the trace of the plane itself. For example, by rotating the projection of the perpendicular to a plane about the *YY'* axis of the net, the plane is rotated to the horizontal.

Rotation of Drill-Core Data

The attitude of bedding or any other planar feature may be determined from subsurface data obtained from drill cores, provided the attitude of the planar feature is uniform within the area penetrated by the borings. The orientation

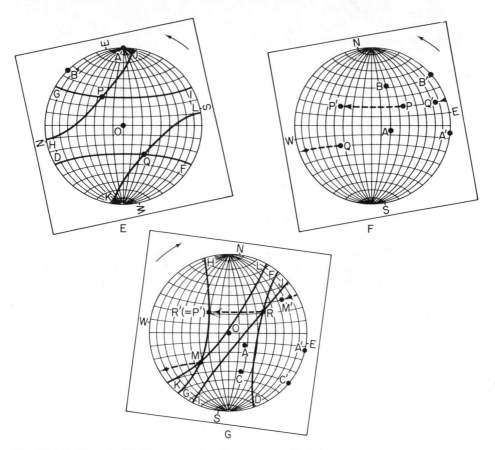

Fig. E14-2. (Cont.). (E) *A'* is rotated to *Y* on net; small circles *GI* and *DF* are projection of bedding in core *A*. *P* and *Q* are the perpendiculars to the two possible positions of the bedding. (F) Steps in rotating drill cores back toward original position. (G) Projection of bedding based on drill holes *A* and *C*.

of the bore holes must be known. But the orientation of the core within a hole is unknown, because it turns while being brought up. However, the angle between the axis of the core and the perpendicular to the bedding may be measured. This is the same as the maximum angle between the normal to the axis of the core and the trace of the bedding.

Normally the data from three differently oriented drill cores are required for a unique solution.

However, if the angle between the perpendicular to the bedding and the axis of the core is zero, one bore hole is sufficient. The bedding is perpendicular to the bore hole. If this angle is zero in a vertical bore hole, the bedding is horizontal.

If this angle is zero in an inclined bore hole, the strike is perpendicular to the bearing of the bore hole and the dip is the complement of the plunge.

The normal procedure is illustrated in the following example.

Drill Hole	Bearing	Plunge	Angle Between Perpendiculars to Bedding and Axis of Core
A	S.65°E.	70°	60
B	N.30°E.	45°	52
C	S.28°E.	48°	82

The orientation of each of the drill holes is given as well as the angle between the axis of the core and the perpendicular to the bedding. This latter angle is unique, being the minimum angle between a line and a plane (Exercise 13). Since each of the drill holes penetrates the bedding whose attitude is desired, each core contains a line common to the bedding. But because the cores have been unavoidably rotated about their own axes, only the locus of possible lines at a constant angle about the axis of the core is known. The locus in each case is a cone defined by the angle between the bedding and the axis of the core. The axis of each core is oriented according to the bearing and plunge of each drill hole.

Figure E14-2C shows on an overlay the projection of the pole of each of the three drill holes, but rotated 10° counterclockwise so that A and B fall on the same great circle. This is done so that A and B may be rotated simultaneously to the horizontal on small circles. Thus A and B are rotated 70° about the YY' axis of the net so that they assume positions A' and B'. Then the overlay is rotated counterclockwise 48° so that B' occupies point Y on the YY' axis of the net. The two small circles 52° from B' are then traced; they are HJ and KL. These small circles are the loci of possible poles of the perpendiculars to the bedding; these perpendiculars form a cone about OB'.

The overlay is rotated again counterclockwise 45° (Fig. E14-2E) so that A' occupies point Y on the YY' axis of the net. The two small circles 60° from A' are then traced; they are GI and DF. These small circles are the loci of possible poles of the perpendiculars to the bedding; these perpendiculars form a cone about OA'.

The cone in each case may be thought of as the loci of perpendiculars to the bedding if it is rotated through 360° while the axis of the cone occupies the position of the axis of the net. Projections DQF-GPI and HPJ-KQL interesect at points P and Q. These points are the projections of the poles of the perpendiculars to the only two possible orientations of the bedding.

In order to find the true orientation of P and Q, the original orientation of the drill holes must be restored. This must be in a series of steps. (a) B' is restored to its position in Fig. E14-2C by rotating the overlay 94° clockwise. The line $B'PO$ will thus also rotate 94° about O; and P, remaining on this line, but the same distance from O and B', assumes the position shown in Fig. E14-2F. (b) B' is rotated back to B by a 70° rotation about the axis of the net; P thus moves to P' (Fig. E14-2F). P' represents the perpendicular to a plane striking N.33°E and dipping 57°S.E.

Q must go through a similar procedure. Q' in Fig. E14-2F represents the perpendicular to bedding that strikes N.16°W., dips 58° SW.

To determine whether P or Q is correct, a new solution must be made using the data for drill hole C. This may be paired with either drill hole A or C. In Fig. E14-2G A has been combined with C. *DNF-GMH* and *INJ-KML* are the projections of the cones drawn about OA' and OC'' as axes. Two interesections result, M and R. As before, the drill cores are rotated back to their initial position. M moves to M', R moves to R'. R' corresponds to P'. That is, the final answer is that the bedding strikes N.33°E., dips 57°SE.

As a further check on the results, the data from drill holes B and C may be combined. Of course, if the attitude of the planar feature—bedding, schistosity, flow structure, etc.—is not uniform, a consistent answer will not be obtained.

Problems

1. A fold axis plunges 35°S. 53°W. What is the minimum angle through which it must be rotated about a north-south horizontal axis to bring it to a horizontal plunge?

2. The contact between a limestone and a shale strikes N.55°E. and dips 35°SE. What are the attitudes of the contacts after the following rotations? (a) A rotation of 50° about a horizontal axis that trends due north (two answers); (b) clockwise rotation of 80° about a vertical axis; (c) rotation of 21° about a line whose rake in the plane of the contact is 30° (two answers).

3. A fault that strikes N.40°W. and dips 75°NE has cut across horizontal beds. The movement on the fault was rotational, one block having rotated 15° about an axis perpendicular to the plane of the fault. Assume that in looking at the fault from the southwest, the nearer block has rotated clockwise. What is the strike and dip of the rotated beds?

4. Two drill holes penetrate an ash bed. In one drill hole, which plunges 50°N.32°W., the angle between the bedding and the axis of the core is 47°. The location of and other data for the second drill hole has been lost, but it is known that between the two holes the ash bed plunges 15°S.72°E. What is the attitude of the ash bed? Is there more than one solution?

5. One limb of a concentric fold strikes N.33°E. and is overturned toward the northwest. The horizontal projection of a mineral lineation in this limb bears N.13°E. If this lineation formed before the fold, what is the bearing and plunge of the lineation in the other limb of the fold, which strikes N.12°W. and dips 15°SW?

6. Three diamond-drill holes in a mine have the following attitudes. The angle between the axes of the cores and the mineral layering is also given. What is the attitude of the layering?

Drill Core	Bearing	Plunge	Angle Between Perpendiculars to Layering and Axis of Core
A	S.48°E.	31°	48
B	S.77°E.	52°	60
C	S.18°W.	59°	23

Reference

[1] Mark, S. Lyons, 1964, *Interpretation of planar structure in drill-hole core,* Geol. Soc. Amer. Special Paper 78, 65 pp.

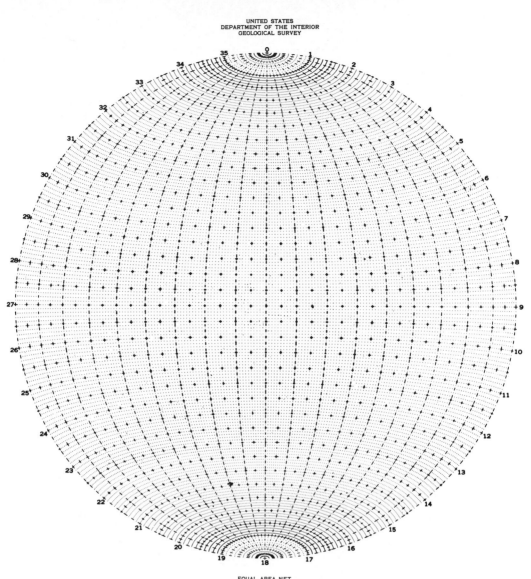

EQUAL AREA NET

Fig. 5-4. Large equal-area net. (After U. S. Geol. Surv.)

589

INDEX

Bold face type indicates principal reference. Asterisk indicates photograph.